COHOMOLOGY OF
INFINITE-DIMENSIONAL
LIE ALGEBRAS

CONTEMPORARY SOVIET MATHEMATICS

Series Editor: Revaz Gamkrelidze, *Steklov Institute, Moscow, USSR*

COHOMOLOGY OF INFINITE-DIMENSIONAL LIE ALGEBRAS
D. B. Fuks

LINEAR DIFFERENTIAL EQUATIONS OF PRINCIPAL TYPE
Yu. V. Egorov

THEORY OF SOLITONS: The Inverse Scattering Method
S. Novikov, S. V. Manakov, L. P. Pitaevskii, and V. E. Zakharov

TOPICS IN MODERN MATHEMATICS: Petrovskii Seminar No. 5
Edited by O. A. Oleinik

COHOMOLOGY OF INFINITE-DIMENSIONAL LIE ALGEBRAS

D. B. Fuks
Moscow State University
Moscow, USSR

Translated from Russian by
A. B. Sosinskii

CONSULTANTS BUREAU • NEW YORK AND LONDON

Library of Congress Cataloging in Publication Data

Fuks, D. B.
   Cohomology of infinite-dimensional Lie algebras.

   (Contemporary Soviet mathematics)
   Translation of: Kogomologii beskonechnomernykh algebr Li.
   Includes bibliographical references and index.
   1. Lie algebras. I. Title. II. Series.
QA252.3.F8513   1986   512'.55   86-25298

ISBN 978-1-4684-8767-1   ISBN 978-1-4684-8765-7 (eBook)
DOI 10.1007/978-1-4684-8765-7

I dedicate this book to Israel Gelfand
on the occasion of this seventieth
birthday

Foreword

There is no question that the cohomology of infinite-dimensional Lie algebras deserves a brief and separate monograph. This subject is not covered by any of the traditional branches of mathematics and is characterized by relatively elementary proofs and varied application. Moreover, the subject matter is widely scattered in various research papers or exists only in verbal form.

The theory of infinite-dimensional Lie algebras differs markedly from the theory of finite-dimensional Lie algebras in that the latter possesses powerful classification theorems, which usually allow one to "recognize" any finite-dimensional Lie algebra (over the field of complex or real numbers), i.e., find it in some list. There are classification theorems in the theory of infinite-dimensional Lie algebras as well, but they are encumbered by strong restrictions of a technical character. These theorems are useful mainly because they yield a considerable supply of interesting examples. We begin with a list of such examples, and further direct our main efforts to their study.

The work consists of three chapters. After the brief
Chapter 1 ("General Theory"), we begin the systematic computation of the cohomology of infinite-dimensional Lie algebras in Chapter 2. The main results of this chapter concern the algebras of formal and smooth vector fields, current algebras, and Kac—Moody algebras. (The first and last
sections of this chapter deal with another topic: the cohomology of finite-dimensional Lie algebras and the cohomology of Lie superalgebras; the latter, in their methods, results, and applications, are fairly close to the homology of
infinite-dimensional Lie algebras.) The concluding chapter
is devoted to applications. These applications comprise the
characteristic classes of foliations, combinatorial identities known as the Macdonald identities, invariant differential operators, cohomology, and, in particular, central extensions of Lie groups and cohomology operations in cobordism theory.

The chapters are divided into sections, and the sections
into subsections. Some of these subsections are further divided into smaller subsections, denoted by capital letters.
When we refer to a section from another chapter, we add the
number of the chapter to the number of the section; when re-
ferring to a subsection from another chapter, we add the
numbers of the chapter and section to the number of the subsection. Thus, Sub-subsection D of Subsection 3 of Section
2 of Chapter 1 is written 1.2.3D; in Chapter 1 this notation
is abbreviated to 2.3D; in §1.2, to 3D; and in Subsection
1.2.3, to D.

It is by no means necessary to read the book in sequence:
it is possible to reach any specific result avoiding irrele-

vant facts. For example, the reader who is interested only
in characteristic classes of foliations may limit himself
to §§1.1-1.3, Subsections 1.5.1, 1.5.2, and §§2.1, 2.2, and
3.1. The reader who would like to become acquainted with
the Macdonald identities may do so by reading §§1.1-1.3 and
Subsections 1.5.3, 2.5.1, and 3.2.1-3.2.3; as additional ma-
terial, he may leaf through Subsections 2.5.3A-B and 3.2.4.

I tried to limit lacunas in proofs to a minimum, but
did not succeed in avoiding them altogether. In the corresponding places references are supplied, but the fact that
completely proved theorems are interspersed with partially
proved theorems might be a certain inconvenience for the
reader. To avoid confusion, I shall say at once that the
main lacunas in the proofs are contained in §§2.3, 3.3, and
3.5. In addition, many sections are concluded by results
not incorporated in the main text and given without proof;
these results are usually grouped together in a separate subsection (such as Subsections 2.1.5, 2.2.7, 2.4.3B, 2.5.3,
2.6.3, 3.2.4, and 3.4.3). Finally, in a number of places I
omit proofs or parts of proofs because they are similar to
other proofs presented in this book; recovering these omis-
sions should be viewed as an exercise for the reader.

The standard graduate course in mathematics is sufficient for understanding the text. Systematically we use
facts from the classical theory of groups and Lie algebras,
homology algebra, and topology, but, as a rule, these are
the simplest definitions and theorems presented on the first
pages of textbooks in these subjects. An exception is §3.5,
the last section of the book, where some serious knowledge
of topology is required.

One can say without exaggeration that the book's subject matter is due to Israel Gelfand, to whom I dedicate
this book. At the outset, he was the initiator and first
enthusiast in the computation of the cohomology of infinite-
dimensional Lie algebras. He and his pupils are the authors
of many key results in this field. I do not think there is
any single theorem in this work which was not reported at
his seminar.

I am grateful to R. V. Gamkrelidze for suggesting that
I write this book. I am grateful to V. M. Bukhshtaber,
D. A. Leites, and especially B. L. Feigen for assistance. I
also thank my colleague A. B. Sosinskii for his competent
translation, Mrs. Ira Bychkova for her superior typing, and
my wife Ira for her patience and help in preparing the manu-
script.

D. Fuks

Contents

Chapter 1. General Theory

§1. LIE ALGEBRAS

This section is not a manual in the general theory of
Lie algebras. The few facts of this theory which will be
needed in the sequel will appear in the appropriate places
with references to readily available sources, mainly to the
works of the Sophus Lie Seminars [94]. Here we merely list
the Lie algebras which we shall use later on, and briefly
describe their properties.

The main field will always be either the field $\mathbb{C}$ of
complex numbers or the field $\mathbb{R}$ of real numbers. The symbol
$\mathbb{K}$ denotes one of these fields.

1. Finite-dimensional Lie algebras. The Lie algebra
corresponding to a finite-dimensional Lie group will be denoted by the same letters, except that capital Latin letters
denoting the group will be replaced by small Gothic charac-
ters. In particular, $\mathfrak{gl}(n, \mathbb{C})$ is the Lie algebra of all
complex $n \times n$ matrices with the operation

$$A, B \mapsto [A, B] = AB - BA; \tag{1}$$

1

$\mathfrak{sl}\,(n,\ \mathbb{C})$ is the subalgebra of this algebra consisting of matrices with zero trace; $\mathfrak{sp}\,(n,\mathbb{C})$ is the subalgebra of the algebra $\mathfrak{gl}\,(2n,\ \mathbb{C})$, consisting of matrices which annihilate* the
bilinear form

$$(x_1,\ldots,x_{2n}),\,(y_1,\ldots,y_{2n}) \mapsto \sum_{i=1}^{n} \begin{vmatrix} x_i & x_{i+n} \\ y_i & y_{i+n} \end{vmatrix}; \qquad (2)$$

$\mathfrak{u}\,(n)$ and $\mathfrak{su}\,(n)$ are real subalgebras of the algebras $\mathfrak{gl}\,(n,\mathbb{C})$
and $\mathfrak{sl}\,(n,\ \mathbb{C})$, consisting of skew Hermitian matrices; $\mathfrak{gl}\,(n,\ \mathbb{R})$
is the (real) algebra of all real $n \times n$ matrices with the
operation (1); $\mathfrak{sl}\,(n,\ \mathbb{R})$ and $\mathfrak{o}\,(n)$ are the subalgebras of this
algebra consisting of matrices with zero trace and of skew-
symmetric matrices; $\mathfrak{sp}\,(n,\mathbb{R})$ is the subalgebra of the algebra
$\mathfrak{gl}\,(2n,\ \mathbb{R})$, consisting of matrices which annihilate the form
(2).

   If $\mathfrak{h}$ is a real subalgebra of the complex Lie algebra $\mathfrak{g}$
satisfying the conditions $\dim_\mathbb{R}\mathfrak{h} = \dim_\mathbb{C}\mathfrak{g}$ and $i\mathfrak{h} \cap \mathfrak{h} = 0$, then
the algebra $\mathfrak{g}$ is canonically isomorphic to the complexifica-
tion $\mathbb{C}\mathfrak{h} = \mathfrak{h} \otimes_\mathbb{R} \mathbb{C}$ of the algebra $\mathfrak{h}$. For example,

$$\begin{aligned} \mathfrak{gl}\,(n,\mathbb{C}) &= \mathbb{C}\,\mathfrak{gl}\,(n,\mathbb{R}) = \mathbb{C}\,\mathfrak{u}\,(n); \\ \mathfrak{sl}\,(n,\ \mathbb{C}) &= \mathbb{C}\,\mathfrak{sl}\,(n,\ \mathbb{C}) = \mathbb{C}\,\mathfrak{su}\,(n). \end{aligned} \qquad (3)$$

   <u>2. Lie algebras of smooth vector fields</u>. Breaking
with traditions existing in differential topology, we shall
use the word "smooth" in the sense of "belonging to the class
$\mathscr{C}^\infty$."

   Suppose X is a smooth manifold. The space $\mathrm{Vect}\,X$ of
smooth vector fields on X, i.e., of smooth sections of the
tangent bundle of the manifold X, is a real Lie algebra with

---

*The statement "the operator P annihilates the bilinear form
φ" means that $\varphi\,(P\xi,\,\eta) + \varphi\,(\xi,\,P\eta) \equiv 0.$

respect to the commutation operation $\xi_1, \xi_2 \mapsto [\xi_1, \xi_2]$. The most natural description of this operation is obtained by inter-preting vector fields on X as first-order differential oper-ators in the ring of (real-valued) smooth functions on X: the operator $[\xi_1, \xi_2]$ is the commutator of the operators ξ_1 and ξ_2; in local coordinates,

$$\Big[\sum_i f_i \frac{\partial}{\partial x_i}, \sum_i g_i \frac{\partial}{\partial x_i}\Big] = \sum_i \Big(\sum_j f_j \frac{\partial g_i}{\partial x_j} - \sum_j g_j \frac{\partial f_i}{\partial x_j}\Big)\frac{\partial}{\partial x_i}. \qquad (4)$$

Another description of this operation in the case of a closed X may be obtained by viewing vector fields on X as infini-tesimal diffeomorphisms of X. In this case the commutator of vector fields is simply the commutator of infinitesimal dif-feomorphisms, and we can say that $\mathrm{Vect}\,X$ is the Lie algebra corresponding to the group $\mathrm{Diff}\,X$ of all diffeomorphisms of the manifold X. Actually, the relationship between the group $\mathrm{Diff}\,X$ and the algebra $\mathrm{Vect}\,X$ is not as closely knit as that between finite-dimensional Lie groups and their Lie algebras; for example, if $\dim X > 0$, then the image of the exponential map $\mathrm{Vect}\,X \to \mathrm{Diff}\,X$ does not cover any neighbor-hood of the unit of the group $\mathrm{Diff}\,X$ (see [82]).

Returning to the general case, note that the commuta-tion of vector fields is continuous with respect to the $\mathscr{C}^\infty$-topology, so that $\mathrm{Vect}\,\bar{X}$ is a topological Lie algebra.

If the Lie group G has a (left) smooth action on the manifold X, then vector fields on X correspond to tangent vectors at the unit element to the group G. This corre-spondence determines a homomorphism of the Lie algebra $\mathfrak{g}$ of the group G into $\mathrm{Vect}\,X$; if the action is locally effective, then this homomorphism is injective. In particular, the ac-tion of the group G on itself by left translations induces

the inclusion $\mathfrak{g} \dashrightarrow \text{Vect}\, G$.   The image of this inclusion coincides with the set of all right-invariant vector fields
on G.

Vector fields with compact supports constitute a closed
subalgebra $\text{Vect}_c X$ of the algebra $\text{Vect}\, X$.   Other important
subalgebras of the algebra $\text{Vect}\, X$ are obtained by fixing
some geometrical structure on X (e.g., volume form, Riemann
metric, complex structure, symplectic structure, contact
structure, foliation, submanifold, etc.) and considering only
those vector fields which preserve this structure.

Among the algebras $\text{Vect}\, X$ the one we shall use most
often is the algebra $\text{Vect}\, S^1$ (where S^1 is the circle).   Here
we shall indicate the following important property of the
algebra $\text{Vect}\, S^1$, or, to be more precise, its complexification
$\mathbb{C}\, \text{Vect}\, S^1$:   the sequence

$$e_k = \exp\,(2\pi i k t)\,\frac{d}{dt} \qquad (k \in \mathbb{Z}),$$

where t is the natural parameter on the circle (viewed as
$\mathbb{R}/\mathbb{Z}$),   constitutes a topological basis in $\mathbb{C}\, \text{Vect}\, S^1$ (i.e.,
linear combinations of elements of this sequence are pair-
wise distinct and dense in $\mathbb{C}\, \text{Vect}\, S^1$).   In this basis the
commutator is given by the formula

$$[e_k, e_l] = (l - k)e_{k+l}. \tag{5}$$

3. Lie algebras of formal vector fields. The space
of formal vector fields in $\mathbb{K}^n$ with the projective limit topology is a topological Lie algebra with respect to the commutation operation defined by formula (4) in which f and g
now denote formal power series in $x_1, \ldots, x_n$. This topological algebra is denoted by $\mathbb{K}W_n$ or simply by W_n.

The algebra W_1 will be particularly important in the sequel. It has a topological basis consisting of the fields $e_k = x^{k+1} \partial/\partial x$, in which the commutation is described by the already familiar formula (5).

The subalgebra of the algebra $\mathbb{K}W_n$, consisting of the vector fields $\sum f_i \partial/\partial x_i$, for which $f_1, \ldots, f_n$ belong to the $(k + 1)$-st power of the maximal ideal of the ring of formal power series, where $k = -1, 0, 1, 2, \ldots$, is denoted by $\mathbb{K}L_k(n)$; the notation $\mathbb{K}L_k(n)$ is sometimes abbreviated to $L_k(n)$, $\mathbb{K}L_k$, L_k. Obviously,

$$\mathbb{K}W_n = \mathbb{K}L_{-1}(n) \supset \mathbb{K}L_0(n) \supset \mathbb{K}L_1(n) \supset \cdots$$

and

$$[\mathbb{K}L_k(n), \mathbb{K}L_l(n)] \subset \mathbb{K}L_{k+l}(n);$$

in particular, $\mathbb{K}L_l(n)$ is an ideal in $\mathbb{K}L_k(n)$ when $l > k \geqslant 0$. The formula

$$\| a_{ij} \| \mapsto \sum a_{ij} x_i \frac{\partial}{\partial x_j} \tag{6}$$

determines the canonical inclusion $\mathfrak{gl}(n, \mathbb{K}) \to \mathbb{K}L_0(n) \subset \mathbb{K}W_n$; the composition of this inclusion with the projection $\mathbb{K}L_0(n) \to \mathbb{K}L_0(n)/\mathbb{K}L_1(n)$ yields the canonical isomorphism

$$\mathfrak{gl}(n, \mathbb{K}) = \mathbb{K}L_0(n)/\mathbb{K}L_1(n). \tag{7}$$

Note that the algebra $\mathbb{K}W_n$ itself has no ideals (i.e., it is simple).

The subalgebra A of the algebra $\mathbb{K}W_n$ is said to be transitive if $\dim A/(A \cap L_0) = n$.

The algebra $\mathbb{R}W_n$ may be interpreted as the algebra of ∞-jets of smooth vector fields on $\mathbb{R}^n$. In view of this, the choice of a point of the n-dimensional smooth manifold X and of a local coordinate system with origin at this point

induces a homomorphism $\dot{\text{Vect}}\, X \to \mathbb{R}W_n$ (to each vector field
we assign its ∞-jet at the fixed point, written in the fix-
ed local coordinate system). If, at the same time, the Lie
group G acts smoothly on the manifold X then the composi-
tion of this homomorphism with the homomorphism $\mathfrak{g} \to \text{Vect}\, X$
from Subsection 2 is a homomorphism $\mathfrak{g} \to \mathbb{R}\dot{W}_n$; this last map
is a monomorphism if the action of the group G is locally
effective, and its image is transitive if the action is
locally transitive. Thus, we obtain the following (in fact,
universal) method for constructing transitive finite-dimensional subalgebras of the algebra $\mathbb{R}W_n$. It suffices to
choose the group G and its subgroup H of codimension n, so
that the quotient group $(\text{Norm}\, H)/H$ is discrete, and fix an
isomorphism $\mathfrak{g}/\mathfrak{h} \to \mathbb{R}^n$; then the canonical action of the group
G in G/H, the point $H \in G/H$, and the local coordinate sys-
tem in G/H, defined by the exponential map $\mathfrak{g}/\mathfrak{h} \to G/H$, determine an inclusion of $\mathfrak{g}$ in $\mathbb{R}W_n$ as a transitive subalgebra. This construction has an obvious complex analog: the
choice of the complex Lie group G, its (complex) subgroup H
of codimension n with discrete $(\text{Norm}\, H)/H$, and of the iso-
morphism $\mathfrak{g}/\mathfrak{h} \to \mathbb{C}^n$ generate the inclusion of the algebra $\mathfrak{g}$ in
$\mathbb{C}W_n$ as a transitive subalgebra. The most important example
is $G = \text{SL}\,(n + 1,\ \mathbb{K})$, $H = \{\| a_{ij} \| \in \text{SL}\,(n + 1, \mathbb{K}) \,|\, a_{1,n+1} = \ldots =$
$a_{n,n+1} = 0\}$, the isomorphism $\mathfrak{g}/\mathfrak{h} \to \mathbb{K}^n$ being determined by
the assignment $\| a_{ij} \| \mapsto (a_{1,n+1}, \ldots, a_{n,n+1})$. The inclusion
$\mathfrak{sl}\,(n + 1, \mathbb{K}) \to W_n$ which arises may be described explicitly
by the formulas

$$\| a_{ij} \| \mapsto \sum_{i=1}^{n} \sum_{j=1}^{n} a_{ij} x_i \frac{\partial}{\partial x_j} + \sum_{i=1}^{n} a_{i,n+1} \frac{\partial}{\partial x_i} + \sum_{j=1}^{n} a_{n+1,j} x_j \sum_{k=1}^{n} x_k \frac{\partial}{\partial x_k}. \tag{8}$$

Note that under the usual identification of the algebra

$\mathfrak{gl}\,(n, \mathbb{K})$ with the appropriate subalgebra of the algebra $\mathfrak{sl}\,(n+1, \mathbb{K})$, the inclusion (6) turns out to be a restriction of the inclusion (8).

The following subalgebras of the algebra $\mathbb{K}W_n$ are classical. Vector fields $\sum f_i\, \partial/\partial x_i$ with trivial divergence $\sum \partial f_i/\partial x_i$ constitute the subalgebra $\mathbb{K}S_n$ of the algebra $\mathbb{K}W_n$; this algebra is simple, but is itself an ideal of co-dimension 1 in the algebra $\mathbb{K}\widehat{S}_n$, consisting of vector fields with constant divergence. If $n=2k$, then the vector fields which annihilate the 2-form

$$dx_1 \wedge dx_{k+1} + \ldots + dx_k \wedge dx_{2k} \qquad (9)$$

are said to be <u>Hamiltonian</u> and constitute the subalgebra $\mathbb{K}H_k$ of the algebra $\mathbb{K}W_n$;. this algebra is simple and is an ideal of codimension 1 in the algebra $\mathbb{K}\widehat{H}_k$, consisting of fields which map the form (9) into a form which differs from it by a constant factor. Finally, if $n=2k+1$, then the vector fields which send the 1-form

$$x_1 dx_{k+1} + \ldots + x_k dx_{2k} + dx_{2k+1} \qquad (10)$$

into the form obtained from it by multiplication by a formal power series are called <u>contact</u> fields and constitute the subalgebra $\mathbb{K}K_k$ of the algebra $\mathbb{K}W_n$. All these subalgebras are transitive. We have the following isomorphisms:

$$\begin{aligned}
(\mathbb{K}L_0 \cap \mathbb{K}S_n)/(\mathbb{K}L_1 \cap \mathbb{K}S_n) &= \mathfrak{sl}\,(n, \mathbb{K}), \\
(\mathbb{K}L_0 \cap \mathbb{K}\widehat{S}_n)/(\mathbb{K}L_1 \cap \mathbb{K}\widehat{S}_n) &= \mathfrak{gl}\,(n, \mathbb{K}), \\
(\mathbb{K}L_0 \cap \mathbb{K}H_k)/(\mathbb{K}L_1 \cap \mathbb{K}H_k) &= \mathfrak{sp}\,(k, \mathbb{K}), \\
(\mathbb{K}L_0 \cap \mathbb{K}\widehat{H}_k)/(\mathbb{K}L_1 \cap \mathbb{K}\widehat{H}_k) &= \mathfrak{sp}\,(k, \mathbb{K}) \oplus \mathbb{K},
\end{aligned} \qquad (11)$$

similar to (7). [The quotient algebra $(\mathbb{K}L_0 \cap \mathbb{K}K_k)/(\mathbb{K}L_1 \cap \mathbb{K}K_k)$ can also be easily described, but it does not possess a standard notation.]

The algebras $\mathbb{K}H_k$ and $\mathbb{K}K_k$ have the following equivalent description. Consider the space $F_n = \mathbb{K}\,[[x_1, \ldots, x_n]]$
of formal power series and define a Lie-algebra structure
in it by means of the following formulas:

for $n = 2k$,

$$[f, g] = \sum_{i=1}^{k} \left(\frac{\partial f}{\partial x_i} \frac{\partial g}{\partial x_{i+k}} - \frac{\partial g}{\partial x_i} \frac{\partial f}{\partial x_{i+k}} \right) ;$$

for $n = 2k + 1$,

$$[f, g] = \sum_{i=1}^{k} \left(\frac{\partial f}{\partial x_i} \frac{\partial g}{\partial x_{i+k}} - \frac{\partial g}{\partial x_i} \frac{\partial f}{\partial x_{i+k}} \right)$$

$$+ \frac{\partial f}{\partial x_{2k+1}} \left(\sum_{i=1}^{2k} x_i \frac{\partial g}{\partial x_i} - 2g \right) - \frac{\partial g}{\partial x_{2k+1}} \left(\sum_{i=1}^{2k} x_i \frac{\partial f}{\partial x_i} - 2f \right).$$

The algebra F_{2k} is known as a <u>Poisson algebra</u> and is also
denoted by $\mathbb{K}P_k$. This algebra has a one-dimensional center
consisting of constants. The formula

$$f \mapsto \sum_{i=1}^{k} \left(\frac{\partial f}{\partial x_{i+k}} \frac{\partial}{\partial x_i} - \frac{\partial f}{\partial x_i} \frac{\partial}{\partial x_{i+k}} \right)$$

defines an epimorphism $\mathbb{K}P_k \rightarrow \mathbb{K}H_k$, whose kernel coincides
with this center. Thus, $\mathbb{K}H_k$ is the quotient algebra of
the Poisson algebra by its center. Further, the formula

$$f \mapsto \sum_{i=1}^{k} \left(\frac{\partial f}{\partial x_{i+k}} \frac{\partial}{\partial x_i} - \frac{\partial f}{\partial x_i} \frac{\partial}{\partial x_{i+k}} \right) + \left(\sum_{i=1}^{2k} x_i \frac{\partial f}{\partial x_i} - 2f \right) \frac{\partial}{\partial x_{2k+1}}$$

determines an isomorphism $F_{k+1} \rightarrow \mathbb{K}K_k$.

In conclusion note that, together with the algebras
$\mathbb{K}W_n$, we consider their (dense) subalgebras consisting of
polynomial vector fields. These subalgebras are denoted by
$\mathbb{K}W_n^{\text{pol}}$. We also use the notations $\mathbb{K}L_k(n)^{\text{pol}}$, $\mathbb{K}S_n^{\text{pol}}$, etc.

4. Current algebras. Suppose $\mathfrak{g}$ is a finite-dimensional Lie algebra and X is a smooth manifold. The space of
smooth maps $X \to \mathfrak{g}$, with the $\mathscr{C}^\infty$-topology and the commutator

$$[f, g] (x) = [f (x), g (x)],$$

is a topological Lie algebra known as the current algebra
and denoted by $\mathfrak{g}^X$. Suppose, for example, that $\mathfrak{g} = \mathfrak{sl} (2, \mathbb{C})$,
$X = S^1$. Choose the following basis in $\mathfrak{g}$:

$$e_{-1} = \begin{pmatrix} 0 & 1 \\ 0 & 0 \end{pmatrix}, \quad e_0 = \begin{pmatrix} 1/2 & 0 \\ 0 & -1/2 \end{pmatrix}, \quad e_1 = \begin{pmatrix} 0 & 0 \\ 1 & 0 \end{pmatrix}$$

[the commutator is defined by the same formula (5)] and de-
note by ε_i the elements of the algebra $\mathfrak{g}^X$, defined as follows:

$$\varepsilon_{3k} (t) = \exp (2\pi i k t) e_0,$$
$$\varepsilon_{3k-1} (t) = \exp (2\pi i k t) e_{-1},$$
$$\varepsilon_{3k+1} (t) = \exp (2\pi i k t) e_1.$$

Obviously, the elements ε_i constitute a topological basis
in $\mathfrak{g}^X$ and their commutators are defined by the formula

$$[\varepsilon_i, \varepsilon_j] = \alpha_{ij} \varepsilon_{i+j}, \quad \alpha_{ij} \begin{cases} = -1, \ 0 \text{ or } 1, \\ \equiv j - i \quad \mod 3 \end{cases} \qquad (12)$$

[compare with formula (5)].

In general, the case $X = S^1$ will be the most important
one for us. Together with the algebra $\mathfrak{g}^{S^1}$, we shall consider its subalgebra $(\mathfrak{g}^{S^1})^{\mathrm{pol}}$, consisting of maps described
by trigonometric polynomials. Note that for $\mathbb{K} = \mathbb{C}$ the
passage from $\mathfrak{g}$ to $(\mathfrak{g}^{S^1})^{\mathrm{pol}}$ is a particular case of a general
construction. Namely, for any commutative associative al-
gebra A , the tensor product $\mathfrak{g} \otimes A$ is a Lie algebra with
respect to the commutators $[g_1 \otimes a_1, g_2 \otimes a_2] = [g_1, g_2] \otimes a_1 a_2$, and

obviously $(\mathfrak{g}^{S^1})^{\text{pol}} = \mathfrak{g} \otimes \mathbb{C}[t, t^{-1}]$, where $\mathbb{C}[t, t^{-1}]$ denotes the Laurent polynomial algebra in the variable t.

§2. MODULES

1. Definition. Suppose $\mathfrak{g}$ is a Lie algebra. The vector space A (over the same field) is said to be module over $\mathfrak{g}$ or $\mathfrak{g}$-module if a bilinear map $\mu: \mathfrak{g} \times A \to A$ [we write ga instead of $\mu(g, a)$] such that $[g_1, g_2]a = g_1(g_2 a) - g_2(g_1 a)$ for all $a \in A$, $g_1, g_2 \in \mathfrak{g}$ is given. In other words, a $\mathfrak{g}$-module is simply a left module over the universal enveloping algebra $U(\mathfrak{g})$ of the algebra $\mathfrak{g}$. [Recall that $U(\mathfrak{g})$ is the quotient algebra of the tensor algebra $\otimes^* \mathfrak{g} = \oplus_{q=0}^{\infty} (\underbrace{\mathfrak{g} \otimes \cdots \otimes \mathfrak{g}}_{q})$ by the ideal generated by elements of the form $[g_1, g_2] - (g_1 \otimes g_2 - g_2 \otimes g_1)$. For the image of the product $g_1 \otimes \cdots \otimes g_q$ in $U(\mathfrak{g})$ we use the notation $g_1 \ldots g_q$.]

If $\mathfrak{g}$ is a topological Lie algebra then, as a rule, the space A will be a topological space here, and the map μ will be continuous (terminology: topological module). First examples: A is arbitrary, $ga \equiv 0$ ("trivial module"); $A = \mathfrak{g}$, $gg_1 = [g, g_1]$ ("adjoint representation").

For any $\mathfrak{g}$-modules A, B the following spaces $A \oplus B$, $A \otimes B$ [by definition, $g(a \otimes b) = ga \otimes b + a \otimes gb$], $S^r A$, $\Lambda^r A$, $\mathrm{Hom}(A, B)$ [by definition $g\varphi(a) = g(\varphi(a)) - \varphi(ga)$] are $\mathfrak{g}$-modules. In the topological case, Hom is automatically understood to be the space of continuous homomorphisms (in the weak topology).

A module is said to be irreducible if it has no proper submodules and is called decomposable if it is isomorphic to the sum of a finite number of irreducible modules.

2. Principal Examples. The main example of a $\mathfrak{gl}\,(n\,\Bbb{K})$ module is the space $\Bbb{K}^n$ of column-vectors on which matrices act by left multiplication. This module will be denoted by V. Another important example is the one-dimensional module E_λ, where $\lambda \in \Bbb{K}$. The structure of the module is determined by the formula $ga = -\lambda(\mathrm{Tr}\,g)\,a$, where Tr denotes trace. Obviously, $E_\lambda \otimes E_\mu = E_{\lambda+\mu}$.

Modules of the form $V \otimes \ldots \otimes V \otimes V' \otimes \ldots \otimes V'$ (the prime denotes dual space) and their submodules are called tensor modules. Modules of the form $V \otimes \ldots \otimes V \otimes V' \otimes \ldots \otimes V' \otimes E_\lambda$ and their submodules are called generalized tensor modules. An important property of tensor modules (and generalized tensor modules) is the fact that they are decomposable — see the proof in §2.1.

Since the algebra $\Bbb{K}W_n$ contains $\mathfrak{gl}\,(n,\,\Bbb{K})$ as a subalgebra (see 1.3), it is a $\mathfrak{gl}\,(n,\,\Bbb{K})$-module. For every k, the vector fields $\sum f_i \partial/\partial x_i$, where the f_i are homogeneous polynomials of degree k, constitute a submodule of this module, and this submodule is obviously isomorphic to $S^k V' \otimes V$. Thus, a $\mathfrak{gl}\,(n,\,\Bbb{K})$-module, $\Bbb{K}W_n$, is the completed direct sum (direct product)

$$\hat{\bigoplus}_{0 \leqslant k < \infty} S^k V' \otimes V.$$

The principal nontrivial modules over the algebra of smooth vector fields of a manifold X is the space of smooth functions on X, as well as the spaces of sections of vector bundles over X, associated with the tangent bundle, i.e., spaces of tensor fields of one kind or another over X. Among them are spaces of exterior forms of different degrees and the space of vector fields itself (the adjoint representa-

tion).   Similar modules exist over algebras of formal vector
fields: spaces of formal tensor fields of different types.
Both in the smooth and in the formal case, more general modules are obtained by fixing $\lambda \in \mathbb{K}$ and considering the space
of (smooth or formal) tensor fields of some type, multiplied
by the volume form raised to the power λ.   All these modules
are topological.

As an illustration, consider the algebra $\mathbb{C}W_1$. For any
$\lambda \in \mathbb{C}$ denote by F_λ the space of expressions of the form
$f(x)\, dx^{-\lambda}$,   where $f(x)$ is a formal power series in x.   The
formula

$$\left(g\, \frac{d}{dx}\right) f\, dx^{-\lambda} = (gf' - \lambda f g')\, dx^{-\lambda}$$

supplies the space F_λ with a $\mathbb{C}W_1$-module structure. For integral λ this is the module of formal tensor fields (formal
power series for $\lambda = 0$, formal differential 1-forms for $\lambda =$
-1, formal vector fields for $\lambda = 1$) and for nonintegral λ
it is a module of generalized formal tensor fields.

Exercises. 1.   For $\lambda \neq 0$ the module F_λ is irreducible.
2.   Over the algebras W_n there are no nontrivial finite-
dimensional modules.

3.   Induced and coinduced modules.   Suppose $\mathfrak{h}$ is a subalgebra of the Lie algebra $\mathfrak{g}$ and A is an arbitrary $\mathfrak{h}$-module.
The spaces

$$\mathrm{Ind}_\mathfrak{g} A = U(\mathfrak{g}) \otimes_{U(\mathfrak{h})} A,$$
$$\mathrm{Coind}_\mathfrak{g} A = \mathrm{Hom}_{U(\mathfrak{h})}(U(\mathfrak{g}), A)$$

[in the first formula $U(\mathfrak{g})$ is viewed as a right $U(\mathfrak{h})$-module,
in the second as a left $U(\mathfrak{h})$-module] possess the following
$\mathfrak{g}$-module structures:

$$g\,(g_1 \ldots g_k \otimes a) = gg_1 \ldots g_k \otimes a,$$
$$g\varphi\,(g_1 \ldots g_k) = \varphi\,(g_1 \ldots g_k g).$$

The modules $\mathrm{Ind}_{\mathfrak{g}}A$ and $\mathrm{Coind}_{\mathfrak{g}}A$ obtained from A are called induced and coinduced modules, respectively.

In order to explain the meaning of this construction, let us assume that $\dim\,(\mathfrak{g}/\mathfrak{h}) < \infty$, and fix arbitrary inclusions

$$i\colon \mathfrak{g}/\mathfrak{h} \to \mathfrak{g}, \quad j\colon S^*\,(\mathfrak{g}/\mathfrak{h}) \to \otimes^*\,(\mathfrak{g}/\mathfrak{h}),$$

which are right inverse to the projections

$$\mathfrak{g} \to \mathfrak{g}/\mathfrak{h}, \quad \otimes^*\,(\mathfrak{g}/\mathfrak{h}) \to S^*\,(\mathfrak{g}/\mathfrak{h}),$$

and consider the composition of maps

$$S^*\,(\mathfrak{g}/\mathfrak{h}) \xrightarrow{j} \otimes^*\,(\mathfrak{g}/\mathfrak{h}) \xrightarrow{\otimes^* i} \otimes^* \mathfrak{g} \xrightarrow{\mathrm{pr}} U\,(\mathfrak{g}). \tag{1}$$

This map induces the linear maps (over $\mathbb{K}$)

$$S^*\,(\mathfrak{g}/\mathfrak{h}) \otimes_{\mathbb{K}} A \xrightarrow{\cdot} \mathrm{Ind}_{\mathfrak{g}}\,A, \qquad \tag{2}$$
$$\mathrm{Hom}_{\mathbb{K}}\,(S^*\,(\mathfrak{g}/\mathfrak{h}),\,A) \leftarrow \mathrm{Coind}_{\mathfrak{g}}\,A.$$

A direct verification shows that (2) are isomorphisms. Note that in the case $\mathfrak{h} = 0$, $A = \mathbb{K}$, the first isomorphism (2) is of the form $S^* \mathfrak{g} \cong U\,(\mathfrak{g})$. This is the assertion of the well-known Poincaré–Birkhoff–Witt theorem (see [94], Chapter 1).

In the general case the isomorphisms (2) only allow us to judge the "size" of the modules $\mathrm{Ind}_{\mathfrak{g}}A$ and $\mathrm{Coind}_{\mathfrak{g}}A$: as linear spaces these modules are spaces of polynomials and formal power series on $\mathfrak{g}/\mathfrak{h}$ with coefficients in A. However, in some important particular cases it is possible to describe the action of the algebra $\mathfrak{g}$ in these modules as well. Suppose, for example, that $\mathfrak{g} = W_n$, $\mathfrak{h} = \bar{L}_0$, $A = \mathbb{K}$. The previous construction yields the linear isomorphism

$$\mathrm{Coind}_{W_n}\mathbb{K} \cong \mathbb{K}\,[[x_1, \ldots, x_n]]. \tag{3}$$

Actually, the isomorphism (3) is an isomorphism in the category of W_n-modules as well: the W_n-isomorphism $\mathbb{K}\,[[x_1, \ldots, x_n]] \to \mathrm{Hom}_{U(L_0)}(U\,(\dot{W}_n), \mathbb{K}) = \mathrm{Coind}_{W_n}\mathbb{K}$ sends the series $f \in \mathbb{K}\,[[x_1, \ldots, x_n]]$ into the $U\,(L_0)$-homomorphism $U\,(W_n) \to \mathbb{K}$, determined by the formula

$$\xi_1 \ldots \xi_k \mapsto \xi_1 \circ \ldots \circ \xi_k \, (f)|_{x_1 = \ldots = x_n = 0}$$

[this isomorphism is inverse to the isomorphism (2) constructed for i, which acts according to the formula $\mathrm{pr}\,(\partial/\partial x_i) \mapsto \partial/\partial x_i$, and for an arbitrary j].

A similar proof works for the following generalization of this last fact. Suppose A is an arbitrary tensor module or generalized tensor module over $\mathfrak{gl}\,(n, \mathbb{K})$, and suppose $\mathcal{A}$ is the W_n-module of tensor or generalized tensor fields of the corresponding type. The projection $L_0 \to L_0/L_1 = \mathfrak{gl}\,(n, \mathbb{K})$ enables us to view A as an L_0-module, which we shall do. Then we have the W_n-isomorphism

$$\mathrm{Coind}_{W_n}A = \mathcal{A}.$$

In particular,

$$\mathrm{Coind}_{W_n}V = W_n,$$
$$\mathrm{Coind}_{W_n}\Lambda^r V' = \Omega^r$$

(Ω^r denotes the space of formal exterior differential forms of degree r), and for any $\lambda \in \mathbb{K}$

$$\mathrm{Coind}_{W_1}E_\lambda = F_\lambda.$$

The modules $\mathrm{Ind}_{W_n}A$ are also easy to compute: in all the examples considered,

$$\text{Ind}_{W_n} A = (\text{Coind}_{W_n} A')'$$

(where the second prime denotes the passage to the space of continuous functionals).

§3.   COHOMOLOGY AND HOMOLOGY

1. Definitions. Suppose $\mathfrak{g}$ is a Lie algebra and A is a module over $\mathfrak{g}$. Then a q-dimensional cochain of the algebra $\mathfrak{g}$ with coefficients in A is a (continuous) skew-symmetric q-linear functional on $\mathfrak{g}$ with values in A; the space of all such cochains is denoted by $C^q(\mathfrak{g}; A)$. Thus, $C^q(\mathfrak{g}; A) = \text{Hom}(\Lambda^q \mathfrak{g}, A)$; this last representation transforms $C^q(\mathfrak{g}; A)$ into a $\mathfrak{g}$-module. The differential $d = d_q: C^q(\mathfrak{g}; A) \to C^{q+1}(\mathfrak{g}; A)$ is defined by the formula

$$dc(g_1, \ldots, g_{q+1}) = \sum_{1 \leqslant s < t \leqslant q+1} (-1)^{s+t-1} c([g_s, g_t], g_1, \ldots \hat{g}_s \ldots \hat{g}_t \ldots, g_{q+1})$$

$$+ \sum_{1 \leqslant s \leqslant q+1} (-1)^s g_s c(g_1, \ldots \hat{g}_s \ldots, g_{q+1}), \qquad (1)$$

where $c \in C^q(\mathfrak{g}; A)$, $g_1, \ldots, g_{q+1} \in \mathfrak{g}$. We complete the definitions by putting $C^q(\mathfrak{g}; A) = 0$ for $q < 0$, $d_q = 0$ for $q < 0$. As can be easily checked, $d_{q+1} \circ d_q = 0$ for all q, so that $\{C^q(\mathfrak{g}; A), d_q\}$ is an algebraic complex; this complex is denoted by $C^{\cdot}(\mathfrak{g}; A)$, while the corresponding cohomology is referred to as the cohomology of the algebra $\mathfrak{g}$ with coefficients in A and is denoted by $H^q(\mathfrak{g}; A)$. If A is the main field (examined as a trivial $\mathfrak{g}$-module), then the second sum in the right-hand side of formula (1) vanishes and may be ignored. In this case, the notations for $C^q(\mathfrak{g}; A)$, $H^q(\mathfrak{g}; A)$ are abbreviated to $C^q(\mathfrak{g})$, $H^q(\mathfrak{g})$.

Homology is simpler than cohomology (we began with cohomology only because it is more important to us). The space $C_q(\mathfrak{g}; A)$ of q-dimensional chains of the Lie algebra $\mathfrak{g}$ with

coefficients in A is defined as $A \otimes \Lambda^q \mathfrak{g}$; the differential
$\partial = \partial_q : C_q (\mathfrak{g}; A) \to C_{q-1} (\mathfrak{g}; A)$ acts in accordance with the formula

$$\partial (a \otimes (g_1 \wedge \cdots \wedge g_q)) = \sum_{1 \leqslant s < t \leqslant q} (-1)^{s+t-1} a \otimes ([g_s, g_t] \wedge g_1 \wedge \cdots \hat{g}_s \cdots$$
$$\hat{g}_t \cdots \wedge g_q) + \sum_{1 \leqslant s \leqslant q} (-1)^s g_s a \otimes (g_1 \wedge \cdots \hat{g}_s \cdots \wedge g_q); \qquad (2)$$

the homology of the complex $\{ C_q (\mathfrak{g}; A), \partial_q \}$ is referred to as
the homology of the algebra $\mathfrak{g}$ with coefficients in A and
denoted by $H_q (\mathfrak{g}; A)$; if A is the main field, the second sum
in formula (2) may be ignored and the notations $C_q (\mathfrak{g}; A)$,
$H_q (\mathfrak{g}; A)$ abbreviated to $C_q (\mathfrak{g})$, $H_q (\mathfrak{g})$. If the algebra $\mathfrak{g}$ is
finite dimensional, then obviously $C^q (\mathfrak{g}) = [C_q (\mathfrak{g})]'$ and
$H^q (\mathfrak{g}) = [H_q (\mathfrak{g})]'$. In the infinite-dimensional case the latter
is no longer true, but nevertheless an element of the space
$H^q (\mathfrak{g})$ determines a functional on $H_q (\mathfrak{g})$. In the finite-
dimensional as well as the infinite-dimensional case, an
element of the space $H^q (\mathfrak{g}; A)$ defines a linear map $H_q (\mathfrak{g}) \to A$.
Moreover, if the algebra $\mathfrak{g}$ and the module A are finite dimensional, then $H^q (\mathfrak{g}; A') = [H_q (\mathfrak{g}; A)]'$.

Now suppose $\mathfrak{h}$ is a subalgebra of the algebra $\mathfrak{g}$; A still
denotes a $\mathfrak{g}$-module. Denote by $C^q (\mathfrak{g}, \mathfrak{h}; A)$ the subspace of
the space $C^q (\mathfrak{g}; A)$, consisting of cochains c, such that
$c (g_1, \ldots, g_q) = 0$ for $g_1 \in \mathfrak{h}$ and $dc (g_1, \ldots, g_{q+1}) = 0$ for $g_1 \in \mathfrak{h}$.
Equivalent definition: $C^q (\mathfrak{g}, \mathfrak{h}; A) = \mathrm{Hom}_{\mathfrak{h}} (\Lambda^q (\mathfrak{g}/\mathfrak{h}), A)$ (the
equivalence is obvious). Elements of the space $C^q (\mathfrak{g}, \mathfrak{h}; A)$
are called relative cochains. Obviously, $dC^q (\mathfrak{g}, \mathfrak{h}; A) \subset$
$C^{q+1} (\mathfrak{g}, \mathfrak{h}; A)$, so that relative cochains constitute a subcomplex of the complex $C^{\cdot} (\mathfrak{g}; A)$. This subcomplex is denoted by
$C^{\cdot} (\mathfrak{g}, \mathfrak{h}; A)$, and its cohomology is called a (relative) cohomology of the algebra $\mathfrak{g}$ modulo $\mathfrak{h}$ with coefficients in A and

is denoted by $H^q (\mathfrak{g}, \mathfrak{h}; A)$. If A is the main field, then, instead of $C^q (\mathfrak{g}, \mathfrak{h}; A)$, $H^q (\mathfrak{g}, \mathfrak{h}; A)$, we write $C^q (\mathfrak{g}, \mathfrak{h})$. $H^q (\mathfrak{g}, \mathfrak{h})$. Note that if $\mathfrak{h}$ is an ideal in $\mathfrak{g}$, then $\Lambda^q (\mathfrak{g}/\mathfrak{h})$ is the trivial $\mathfrak{h}$-module and $\mathrm{Hom}_{\mathfrak{h}}(\Lambda^q (\mathfrak{g}/\mathfrak{h}), A) = \mathrm{Hom} (\Lambda^q (\mathfrak{g}/\mathfrak{h}), \mathrm{Inv}_{\mathfrak{h}}A) = C^q (\mathfrak{g}/\mathfrak{h}; \mathrm{Inv}_{\mathfrak{h}}A)$, where $\mathrm{Inv}_{\mathfrak{h}}A$ is the module (over $\mathfrak{g}/\mathfrak{h}$) of $\mathfrak{h}$-invariants:

$$\mathrm{Inv}_{\mathfrak{h}}A = \{a \in A \mid ha = 0 \ \text{ for all } \ h \in \mathfrak{h}\}.$$

Obviously, in this case the differentials in the complexes $C^{\cdot} (\mathfrak{g}, \mathfrak{h}; A)$ and $C^{\cdot} (\mathfrak{g}/\mathfrak{h}; \mathrm{Inv}_{\mathfrak{h}}A)$ also coincide so that $H^q (\mathfrak{g}, \mathfrak{h}; A) = H^q (\mathfrak{g}/\mathfrak{h}; \mathrm{Inv}_{\mathfrak{h}}A)$.

The definition of relative cohomology has the following generalization. Assume that $\mathfrak{h}$ is the Lie algebra of some (finite-dimensional) Lie group H and that the actions of the algebra $\mathfrak{h}$ in $\mathfrak{g}$ and in A are the differentials of certain representations of the group H, the representation of H in $\mathfrak{g}$ being the extension of the adjoint representation of H in $\mathfrak{h}$. Then, setting $C^q (\mathfrak{g}, H; A) = \mathrm{Hom}_H (\Lambda^q (\mathfrak{g}/\mathfrak{h}), A)$, we obtain one more subcomplex in $C^{\cdot} (\mathfrak{g}; A)$, whose cohomology will be denoted by $H^q (\mathfrak{g}, H; A)$. This generalization is not particularly deep: if the group H is connected, then $H^q (\mathfrak{g}, H; A) = H^q (\mathfrak{g}, \mathfrak{h}; A)$; however, it will be useful in the theory of characteristic classes of foliations (§3.1).

The definition of relative homology is similar to the definition of relative cohomology, and we shall not give it in detail, limiting ourselves to the following remark: For the spaces $C_q (\mathfrak{g}, \mathfrak{h}; A)$, $C_q (\mathfrak{g}, H; A)$ of relative chains we must take $A \otimes_{\mathfrak{h}} \Lambda^q (\mathfrak{g}/\mathfrak{h})$, $A \otimes_H \Lambda^q (\mathfrak{g}/\mathfrak{h})$.

2. Multiplicative structures. The multiplication of cochains as exterior forms transforms $C^{*} (\mathfrak{g}) = \bigoplus_q C^q (\mathfrak{g})$ into

a graded ring, and the differential d becomes the derivation

$$d\,(ab) = (da)\,b + (-1)^p a\,db \text{ for } a \in C^p\,(\mathfrak{g}), \quad b \in C^q\,(\mathfrak{g}).$$

[Note that multiplication in $\oplus\,C_q\,(\mathfrak{g})$ may be defined in the same natural way, but d is not a derivation with respect to this multiplication.] In view of this, the cohomology $H^*\,(\mathfrak{g}) = \oplus_q H^q\,(\mathfrak{g})$ also acquires a ring structure. Multiplication in cohomology is associative and commutative in the sense usual for cohomology:

$$\alpha\beta = (-1)^{pq}\beta\alpha \quad \text{for} \quad \alpha \in H^p\,(\mathfrak{g}), \quad \beta \in H^q\,(\mathfrak{g}).$$

A generalization of this multiplication is the one in $H^*\,(\mathfrak{g};\,A)$, where A is a module over $\mathfrak{g}$, which, at the same time, is an algebra whose multiplication is compatible with its $\mathfrak{g}$-module structure in the sense that

$$g\,(ab) = (ga)\,b + a\,(gb) \text{ for } g \in \mathfrak{g}, \quad a, b \in A.$$

Another generalization of multiplication is the pairing

$$H^p\,(\mathfrak{g}) \otimes H^q\,(\mathfrak{g};\,A) \to H^{p+q}\,(\mathfrak{g};\,A),$$

defined for all p, q and A; this pairing transforms $H^*\,(\mathfrak{g};\,A)$ into a graded $H^*\,(\mathfrak{g})$-module.

   As usually happens, together with multiplication and cohomology, we have the "higher multiplications," due to Massey. The simplest of these is the operator which assigns to every triple $\alpha \in H^p(\mathfrak{g})$, $\beta \in H^q\,(\mathfrak{g})$, $\gamma \in H^r\,(\mathfrak{g})$, such that $\alpha\beta = 0$, $\beta\gamma = 0$, an element $\langle\alpha, \beta, \gamma\rangle$ of the quotient space

$$H^{p+q+r-1}\,(\mathfrak{g})/[\alpha H^{q+r-1}\,(\mathfrak{g}) + H^{p+q-1}\,(\mathfrak{g})\gamma].$$

In order to construct it, it suffices to choose cocycles a, b, c which represent α, β, γ and cochains u, v, such that

$du = ab, \; dv = bc;$ the cochain $(-1)^p av - uc$ is obviously a cocycle and the cohomology class of this cocycle represents $\langle \alpha, \beta, \gamma \rangle$. The next Massey operation is defined on quadruples $\alpha \in H^p(\mathfrak{g}), \; \beta \in H^q(\mathfrak{g}), \; \gamma \in H^r(\mathfrak{g}), \; \delta \in H^s(\mathfrak{g})$, such that $\alpha\beta = 0, \, \beta\gamma = 0, \; \gamma\delta = 0$ and $\langle \alpha, \beta, \gamma \rangle = 0, \; \langle \beta, \gamma, \delta \rangle = 0$, and assumes values in the quotient space

$$H^{p+q+r+s-2}(\mathfrak{g})/[H^{p+q-1}(\mathfrak{g})\,H^{r+s-1}(\mathfrak{g}) + \alpha H^{q+r+s-2}(\mathfrak{g}) + H^{p+q+r-2}(\mathfrak{g})\,\delta].$$

Its definition, just as the definition of subsequent Massey operations, will be easily reproduced by an inventive reader.

Let us also add that ordinary multiplication, as well as Massey multiplication, can be naturally defined in $H^*(\mathfrak{g}, \mathfrak{h})$ and $H^*(\mathfrak{g}, H)$. There are also natural pairings $H^p(\mathfrak{g}) \otimes H^q(\mathfrak{g}, \mathfrak{h}) \to H^{p+q}(\mathfrak{g}, \mathfrak{h})$ and $H^p(\mathfrak{g}) \otimes H^q(\mathfrak{g}, H) \to H^{p+q}(\mathfrak{g}, H)$.

In conclusion, we shall describe a special multiplicative structure which arises in the space $H^*(\mathfrak{g}; \mathfrak{g})$ of cohomology of the algebra $\mathfrak{g}$ with coefficients in the adjoint representation. Suppose $a \in C^p(\mathfrak{g}; \mathfrak{g}), \; b \in C^q(\mathfrak{g}; \mathfrak{g})$. Define the cochain $ab \in C^{p+q-1}(\mathfrak{g}; \mathfrak{g})$ by letting

$$ab(g_1, \ldots, g_{p+q-1}) = \sum (-1)^{i_1+\ldots+i_p-p(p+1)/2} a\,(b\,(g_{i_1}, \ldots, g_{i_p}), g_{j_1}, \ldots, g_{j_{q-1}}),$$

where the sum is taken over all partitions $\{1, \ldots, p+q-1\} = \{i_1, \ldots, i_p\} \cup \{j_1, \ldots, j_{q-1}\}$ $(i_1 < \ldots < i_p, \; j_1 < \ldots < j_{q-1})$, and let $[a, b] = ab - (-1)^{(p-1)(q-1)}ba$. Obviously, for all $a \in C^p(\mathfrak{g}; \mathfrak{g})$, $b \in C^q(\mathfrak{g}; \mathfrak{g})$ we have

$$[a, b] = -(-1)^{(p-1)(q-1)}[b, a]. \tag{3}$$

A short direct computation shows that for all $a \in C^p(\mathfrak{g}; \mathfrak{g})$, $b \in C^q(\mathfrak{g}; \mathfrak{g}), \; c \in C^r(\mathfrak{g}; \mathfrak{g})$ we have

$$(-1)^{(p-1)(q-1)} [[a, b], c] + (-1)^{(q-1)(r-1)} [[b, c], a]$$
$$+ (-1)^{(r-1)(p-1)} [[c, a], b] = 0, \qquad (4)$$

and a longer, but equally direct one, shows that

$$d[a, b] = [da, b] - (-1)^{p-1} [a, db]. \qquad (5)$$

It is clear from (5) that if a, b are cocycles, then $[a, b]$ is also a cocycle, and the cohomology class of the latter is determined by the cohomology classes of a and b. Therefore, we obtain the multiplication

$$H^p(\mathfrak{g}; \mathfrak{g}) \otimes H^q(\mathfrak{g}; \mathfrak{g}) \rightarrow H^{p+q-1}(\mathfrak{g}; \mathfrak{g}) \qquad (6)$$

in cohomology, satisfying the relations (3) and (4) in which it is now necessary to assume $a \in H^p(\mathfrak{g}; \mathfrak{g})$, $b \in H^q(\mathfrak{g}; \mathfrak{g})$, $c \in H^r(\mathfrak{g}; \mathfrak{g})$. This multiplication, which will remind the topologist of the Whitehead product in homotopy groups, supplies the space $H^*(\mathfrak{g}; \mathfrak{g})$ with a "Lie superalgebra" structure; the reader who does not know what this is will have to wait until §6. Also note that this last multiplication, just as the cohomological multiplication defined previously, generates the sequence of Massey products. The definition of these multiplications repeats with obvious alterations the definition of the usual Massey product; the first of them is defined on triples $a \in H^p(\mathfrak{g}; \mathfrak{g})$, $b \in H^q(\mathfrak{g}; \mathfrak{g})$, $c \in H^r(\mathfrak{g}; \mathfrak{g})$, such that $[a, b] = 0$, $[b, c] = 0$, and assumes its values in the quotient space

$$H^{p+q+r-2}(\mathfrak{g}; \mathfrak{g})/([a, H^{q+r-1}(\mathfrak{g}; \mathfrak{g})] + [H^{p+q-1}(\mathfrak{g}; \mathfrak{g}), c]).$$

Let us also add that the multiplication (6) possesses a large number of more or less obvious generalizations, which we shall not mention since they will not be needed.

3. Motivations of the definition of cohomology. The
reader who is first familiarizing himself with the subject
would perhaps like to know where formula (1) comes from. The
true source of this formula lies in homological algebra.
Actually, $H^q (\mathfrak{g}; A)$ is Ext $(\mathbb{K}, A)$ (in the category of $\mathfrak{g}$-mod-
ules), and the definition given above gives a method for com-
puting this Ext by means of a certain standard resolvent;
the details are developed in Chapter 13 of [11]. Here we
shall consider two more particular motivations.

The first justification of formula (1) is the fact that
it does not differ in any way from the classical E. Cartan
formula for the exterior differential: if ω is a differential form of degree q on the manifold X and $\xi_1, \ldots, \xi_{q+1}$ are
vector fields on X, then

$$d\omega(\xi_1, \ldots, \xi_{q+1}) = \sum_{1 \leqslant s < t \leqslant q+1} (-1)^{s+t-1} \omega([\xi_s, \xi_t], \xi_1, \ldots \hat{\xi}_s \ldots \hat{\xi}_t \ldots, \xi_{q+1})$$

$$+ \sum_{1 \leqslant s \leqslant q+1} (-1)^s \xi_s \omega(\xi_1, \ldots \hat{\xi}_s \ldots, \xi_{q+1}).$$

It is easy to explain this coincidence by interpreting differential forms of degree q as cochains of the algebra
Vect $\dot{X}$ with coefficients in the Vect X-module $\mathscr{C}^\infty(X)$ of
smooth real-valued functions on X. The de Rham complex of
this manifold X turns out to be a subcomplex of the complex
$C^{\cdot}(\text{Vect } X; \mathscr{C}^\infty(X))$. Let us add that the special property of
differential forms which characterizes them among all co-
chains of the complex $C^{\cdot}(\text{Vect } X; \mathscr{C}^\infty(X))$ is the fact that the
maps $\Lambda^q(\text{Vect } X) \rightarrow \mathscr{C}^\infty(X)$ which they define are not only linear
but $\mathscr{C}^\infty(\dot{X})$-linear. Thus, $C^{\cdot}(\text{Vect } X; \mathscr{C}^\infty(X))$ may be viewed
as an extension of the de Rham complex. (The reader who would
like to find out how this extension influences cohomology
will have to wait until §2.4.)

The other motivation of the definition of cohomology
relates to the cohomology of the Lie algebra of a Lie group
with coefficients in the main fields. Since a skew-symmetric polylinear form on the Lie algebra $\mathfrak{g}$ of the Lie group G
becomes, under right translations, a right-invariant differential form on G, the space $C^q(\mathfrak{g})$ can be included in the
space $\Omega^q(G)$ of forms of degree q on G. The second justification of formula (1) (to be sure, without the last term)
is that this inclusion commutes with the differential, thus
yielding an inclusion of the complex $C^{\cdot}(\mathfrak{g})$ into the de Rham
complex $\Omega^{\cdot}(G)$ of the group G (as the subcomplex of right-
invariant forms). Further (in §2.1), we shall see that if
the group G is compact, then this inclusion induces an isomorphism in cohomology. In the general case, the difference
between the cohomologies of these complexes is measured by
the so-called continuous or Van Est cohomology of the group
G (see the details in §3.4).

4. An alternative description of the complexes $C^{\cdot}(\mathfrak{g}; A)$·
and $C.(\mathfrak{g}; A)$. This description will be useful in §2.1. De-
note by C the complex $C.(\mathfrak{g}; U(\mathfrak{g}))$, where the $\mathfrak{g}$-module struc-
ture in $U(\mathfrak{g})$ is defined in the usual way: $g(g_1 \ldots g_s) = gg_1 \ldots g_s.$ As is known, $U(\mathfrak{g})$ possesses another $\mathfrak{g}$-module
structure: $g(g_1 \ldots g_s) = -g_1 \ldots g_s g,$ and the second structure
supplies the spaces $C_q(\mathfrak{g}; U(\mathfrak{g}))$ with $\mathfrak{g}$-module structures such
that the differentials

$$\partial_q \colon C_q(\mathfrak{g}; U(\mathfrak{g})) \to C_{q-1}(\mathfrak{g}; U(\mathfrak{g}))$$

turn out to be $\mathfrak{g}$-homomorphisms. Having this structure in
mind, we shall view C as a complex of $\mathfrak{g}$-modules and $\mathfrak{g}$-homo-
morphisms. Obviously for any $\mathfrak{g}$-module A , the spaces $C^q(\mathfrak{g};$

A), $C_q(\mathfrak{g}; A)$ and the differentials $d_q: C^q(\mathfrak{g}; A) \to C^{q+1}(\mathfrak{g}; A)$, $\partial_q: C_q(\mathfrak{g}; A) \to C_{q-1}(\mathfrak{g}; A)$ are none other than

$$\mathrm{Hom}_{U(\mathfrak{g})}(C_q(\mathfrak{g}; U(\mathfrak{g})), A), \quad A \otimes_{U(\mathfrak{g})} C_q(\mathfrak{g}; U(\mathfrak{g})),$$
$$\mathrm{Hom}_{U(\mathfrak{g})}([\partial_q: C_q(\mathfrak{g}; U(\mathfrak{g})) \to C_{q-1}(\mathfrak{g}; U(\mathfrak{g}))], \mathrm{id}\, A),$$
$$\mathrm{id}\, A \otimes_{U(\mathfrak{g})} [\partial_q: C_q(\mathfrak{g}; U(\mathfrak{g})) \to C_{q-1}(\mathfrak{g}; U(\mathfrak{g}))].$$

Thus, the complexes $C^{\cdot}(\mathfrak{g}; A)$ and $C_{\cdot}(\mathfrak{g}; A)$ may be interpreted as $\mathrm{Hom}_{U(\mathfrak{g})}(C, A)$ and $A \otimes_{U(\mathfrak{g})} C$. The usefulness of this information is due (as we shall see in §2.1) to the following fact:

The complex C is acyclic; in more detail: $H_q(C) = 0$ for $q > 0$, while the space $H_0(C)$ is one-dimensional and is generated by the homology class of the chain $1 \in U(\mathfrak{g}) = C_0(\mathfrak{g}; U(\mathfrak{g}))$.

To prove this fact, denote by $F_p C_r(\mathfrak{g}; U(\mathfrak{g}))$ the subspace of the space $C_r(\mathfrak{g}; U(\mathfrak{g}))$, generated by chains of the form $g'_1 \ldots g'_s \otimes (g_1 \wedge \cdots \wedge g_r)$ with $s + r \leqslant p$. Obviously, the spaces $F_p C_r(\mathfrak{g}; U(\mathfrak{g}))$ constitute an increasing filtration of the complex C, and the Poincaré–Birkhoff–Witt theorem (see [94], Chapter 1, and compare with Subsection 2.3) implies that in the corresponding spectral sequence we have

$$E^0_{p,q} = S^{-q}\mathfrak{g} \otimes \Lambda^{p+q}\mathfrak{g},$$

while the differential $d^0_{p,q}$ acts according to the formula

$$g'_1 \ldots g'_{-q} \otimes (g_1 \wedge \cdots \wedge g_{p+q}) \mapsto \sum_{i=1}^{p+q} (-1)^{i-1} g'_1 \ldots g'_{-q} g_i \otimes (g_1 \wedge \cdots \hat{g}_i \cdots \wedge g_{p+q})).$$

It is easy to see that the complex E^0 is already acyclic: the formula

$$g'_1 \ldots g'_{-q} \otimes (g_1 \wedge \cdots \wedge g_{p+q}) \mapsto \sum_{i=1}^{-q} g'_1 \ldots \hat{g}'_i \ldots g'_{-q} \otimes (g'_i \wedge g_1 \wedge \cdots \wedge g_{p+q})$$

determines a certain map $D^0_{p,q}\colon E^0_{p,q} \to E_{p,q+1}$, and a brief direct computation shows that

$$d^0_{p,q+1} \circ D^0_{p,q} + D^0_{p,q-1} \circ d^0_{p,q} = p \text{ id } E^0_{p,q}.$$

Thus, for any fixed $p \neq 0$, the complex $E^0_{p,.} = \{E^0_{p,q}, \ d^0_{p,q}\}$ is acyclic while the complex $E^0_{0,.}$ reduces to $E^0_{0,0} = S^0\mathfrak{g} \otimes \Lambda^0\,\mathfrak{g} = \mathbb{K}$.

Note that the proposition proved above actually means that

$$H_q(\mathfrak{g}; U(\mathfrak{g})) = \begin{cases} \mathbb{K} & \text{for } q = 0, \\ 0 & \text{for } q \neq 0; \end{cases}$$

this fact will be considerably generalized in 5.4.

5. Digression: Weyl algebras. The considerations used in the previous subsection remind one of a construction widely used in differential geometry, and yield a multiplicative complex possessing a series of supplementary structures and called the Weyl algebra. This construction will be used later in §2.2 and I will remind the reader what it consists in. Note that Weyl algebras are usually related to finite-dimensional Lie algebras, so that in this subsection we may assume $\dim \mathfrak{g} < \infty$, although this is never used explicitly.

Let us put $W(\mathfrak{g}) = \Lambda^*\mathfrak{g}' \otimes S^*\mathfrak{g}'$ and define in the algebra $W(\mathfrak{g})$ the grading, filtration, and a differential. The grading is defined by the formula $W^q(\mathfrak{g}) = \underset{i+2j=q}{\oplus} (\Lambda^i\mathfrak{g}' \otimes S^j\mathfrak{g}')$, the filtration by the formula $F^pW(\mathfrak{g}) = \underset{2j\geqslant p}{\oplus} (\Lambda^*\mathfrak{g}' \otimes S^j\mathfrak{g}')$ [so that $W(\mathfrak{g}) = F^0W(\mathfrak{g}) \supset F^1W(\mathfrak{g}) = F^2W(\mathfrak{g}) \supset F^3\,W(\mathfrak{g}) = F^4W(\mathfrak{g}) \supset \ldots]$. The differential $d\colon W(\mathfrak{g}) \to W(\mathfrak{g})$ is defined by the formula

$$d\varphi\,(g_1 \wedge \ldots \wedge g_r \otimes h_1 \otimes \ldots \otimes h_s)$$

$$= \sum_{k=1}^{s} \varphi\,(g_1 \wedge \ldots \wedge g_r^{\varepsilon} \wedge h_i \otimes h_1 \otimes \ldots \hat{h}_i \ldots \otimes h_s)$$

$$+ \sum_{i=1}^{r} \sum_{j=1}^{s} (-1)^{i-1}\varphi\,([g_i, h_j] \wedge g_1 \wedge \ldots \hat{g}_i \ldots \wedge g_r \otimes h_1 \otimes \ldots \hat{h}_j \ldots \otimes h_s)$$

$$+ \sum_{1 \leqslant i < j \leqslant r} (-1)^{i+j-1}\varphi\,([g_i, g_j] \wedge g_1 \wedge \ldots \hat{g}_i \ldots \hat{g}_j \ldots \wedge g_r \otimes h_1 \otimes \ldots \otimes h_s).$$

Obviously, the multiplication, grading, filtration, and differential in $W\,(\mathfrak{g})$ are compatible in the appropriate way; in particular, the differential is of degree $+1$ (with respect to the grading). Notice also that the complex $W\,(\mathfrak{g})/ F^1\tilde{W}\,(\mathfrak{g})$ coincides with $C^{\cdot}(\mathfrak{g})$.

The applications of Weyl algrebras are based, for the most part, on the following two properties, the first of which is of great technical importance, while the second is the source of various constructions.

1°. The complex $W\,(\mathfrak{g})$ is acyclic [i.e., $H^0\,(W\,(\mathfrak{g})) = \mathbb{K}$ and $H^q\,(W\,(\mathfrak{g})) = 0$ for $q \neq 0$]. Indeed, the formula

$$D\varphi\,(g_1 \wedge \ldots \wedge g_r \otimes h_1 \otimes \ldots \otimes h_s)$$

$$= \sum_{i=1}^{r} (-1)^{r-i}\varphi\,(g_1 \wedge \ldots \hat{g}_i \ldots \wedge g_r \otimes g_i \otimes h_1 \otimes \ldots \otimes h_s)$$

determines a homotopy $D\colon W\,(\mathfrak{g}) \to W\,(\mathfrak{g})$, linking the identity with the zero map.

The spectral sequence associated with the filtration $\{F^pW\,(\mathfrak{g})\}$ should be noted. The limiting term of the spectral sequence is trivial, while

$$E_2^{p,\,q} = \begin{cases} 0, & \text{if } p \text{ is odd,} \\ H^q\,(\mathfrak{g};\ S^{p/2}\mathfrak{g}'), & \text{if } p \text{ is even.} \end{cases}$$

I advise the reader to return to the spectral sequence after covering §2.1 in order to prove, as an exercise, that if $\mathfrak{g}$ is the Lie algebra of a compact Lie group G, then the spec-

tral sequence is none other than the Leray—Serre spectral
sequence of the universal bundle of the group G (compare
with the remark at the end of 2.1.3).

2°. Suppose $C = \{C^q, \ \delta^q: \ C^q \to C^{q+1}\}$ is a multiplicative
complex over $\mathbb{K}$ and $\omega: \mathfrak{g}' \to C^1$ is an arbitrary linear map.
Then there exists a unique homogeneous multiplicative homo-
morphism $\Omega: W(\mathfrak{g}) \to C$, whose restriction to $\Lambda^1 \mathfrak{g}' = \mathfrak{g}' \subset W(\mathfrak{g})$
coincides with ω. The construction is obvious: the restric-
tion $\Omega \mid S^1 \mathfrak{g}$ maps $\varphi \in S^1 \mathfrak{g}' = \mathfrak{g}'$ into the difference $\Lambda^2 \omega \, (d\varphi)$
$- \delta^1 \omega \, (\varphi)$,   where $d\varphi \in \Lambda^2 \mathfrak{g}' = C^2 (\mathfrak{g}')$ is the differential
of the cochain $\varphi \in \mathfrak{g}' = C^1(\mathfrak{g})$ in the complex $C^{\cdot}(\mathfrak{g})$.

For example, if in the principal G-bundle $E \to X$ we
are given the connection $\omega \in \Omega^1 (E; \mathfrak{g})$ (G is a Lie group, $\mathfrak{g}$
the corresponding Lie algebra), then, by assigning to each
functional $\varphi \in \mathfrak{g}'$ the form $\varphi \circ \omega \in \Omega$, we obtain a linear
map $\mathfrak{g}' \to \Omega^1 L$; in view of the above, this map can be canonically extended to a homomorphism $W(\mathfrak{g}) \to \Omega^{\cdot}(E)$, which is an
important invariant of the connection; in particular, it may
be used to define characteristic classes of fiber bundles.

6. Simplest general properties. A. Coefficient
sequences and induced homomorphisms. For any short exact
sequence of $\mathfrak{g}$-modules

$$0 \to A_1 \to A_2 \to A_3 \to 0$$

we have the following exact coefficient sequences:

$$\ldots \to H^q (\mathfrak{g}; A_1) \to H^q (\mathfrak{g}; A_2) \to H^q (\mathfrak{g}; A_3) \to H^{q+1} (\mathfrak{g}; A_1) \ldots,$$
$$\ldots \to H_q (\mathfrak{g}; A_1) \to H_q (\mathfrak{g}; A_2) \to H_q (\mathfrak{g}; A_3) \to H_{q-1} (\mathfrak{g}; A_1) \ldots$$

Their construction is standard. In the same standard way,
to any homomorphism $f: \mathfrak{g} \to \mathfrak{h}$ of one Lie algebra into another
we can assign the induced homomorphisms $f^*: H^q (\mathfrak{h}; A) \to H^q (\mathfrak{g}; A)$,

$f_*: H_q(\mathfrak{g}; A) \to H_q(\mathfrak{h}; A)$ (here A is a g-module in which f introduces a g-module structure).

B. Poincaré duality. Suppose the Lie algebra g is finite dimensional and dim $\mathfrak{g} = n$. Then dim $C_n(\mathfrak{g}) = 1$ and for any nonzero $f \in C_n(\mathfrak{g})$ the formula $a, b \mapsto ab(f)$ defines a nondegenerate pairing

$$C^k(\mathfrak{g}) \times C^{k-n}(\mathfrak{g}) \to \mathbb{K} \qquad (7)$$

and, together with it, an isomorphism

$$C^k(\mathfrak{g}) \cong (C^{n-k}(\mathfrak{g}))' = C_{n-k}(\mathfrak{g}). \qquad (8)$$

In order that this isomorphism be compatible with the differentials d and ∂, it is necessary that the algebra g possess a supplementary property: it must be <u>unitary</u>. The latter means that $H_n(\mathfrak{g}) \neq 0$, i.e., f is a cycle. For the algebra g, determined by means of the "structural constants" c_{ij}^k (i.e., $[e_i, e_j] = \sum_{k=1}^n c_{ij}^k e_k$ for some basis $e_1, \ldots, e_n$ of the space g), the unitarity condition can be expressed by the formula $\sum_{j=1}^n c_{ij}^j = 0$ $(i = 1, \ldots, n)$. Examples of unitary Lie algebras are semisimple and nilpotent Lie algebras; an example of a nonunitary algebra is the Lie algebra of the Lie group of affine maps on the line. If the algebra g is unitary, then for $a \in C^{k-1}(\mathfrak{g})$, $b \in C^{n-k}(\mathfrak{g})$ we have

$$[(da) b](f) = [d(ab) + (-1)^k a(db)](f)$$
$$= ab(\partial f) + (-1)^k [a(db)](f) = (-1)^k [a(db)](f),$$

in view of which the pairing (7) and the isomorphism (8) define, respectively, the nondegenerate pairing

$$H^k(\mathfrak{g}) \times H^{n-k}(\mathfrak{g}) \to \mathbb{K} \text{'}$$

and the isomorphism

$$H^k(\mathfrak{g}) \cong (H^{n-k}(\mathfrak{g}))' = H_{n-k}(\mathfrak{g}).$$

The latter are known as <u>Poincaré duality</u> and <u>Poincaré iso-</u><u>morphism</u>. Note also that for any finite-dimensional module A over the unitary algebra $\mathfrak{g}$ a similar construction yields the nondegenerate pairing

$$H^k(\mathfrak{g}; A) \times H^{n-k}(\mathfrak{g}; A') \to \mathbb{K}$$

and the isomorphism

$$H^k(\mathfrak{g}; A) \cong (H^{n-k}(\mathfrak{g}; A'))' = H_{n-k}(\mathfrak{g}; A).$$

C. <u>Triviality of a Lie algebra's action on its homol-</u><u>ogy and cohomology.</u> As we already pointed out, $C^q(\mathfrak{g}; A)$ and $C_q(\mathfrak{g}; A)$ are modules over $\mathfrak{g}$. Obviously, the differen-tials d and ∂ are $\mathfrak{g}$-homomorphisms, so that $H^q(\mathfrak{g}; A)$ and $H_q(\mathfrak{g}; A)$ also turn out to be $\mathfrak{g}$-modules. Further, the $\mathfrak{g}$-structure in $H^{\dot{q}}(\mathfrak{g}; A)$ and $H_q(\mathfrak{g}; A)$ is trivial (i.e., $g\alpha = 0$ and $g\beta = 0$ for any $g \in \mathfrak{g}$, $\alpha \in H^q(\mathfrak{g}; A)$, $\beta \in H_q(\mathfrak{g}; A)$). To prove the cohomological part of the theorem (the proof of its homological part is similar), we must construct a homo-topy which joins the map $C^{\cdot}(\mathfrak{g}; A) \to C^{\cdot}(\mathfrak{g}; A)$, acting according to the formula $a \mapsto ga$, with the zero map, i.e., construct a sequence of homomorphisms $h_q: C^q(\mathfrak{g}; A) \to C^{q-1}(\mathfrak{g}; A)$, such that $h_q(d\dot{a}) + dh_{q-1}(a) = ga$ for any $a \in C^q(\mathfrak{g}; A)$. Such h_q may be defined by the formula

$$[h_q a](g_1, \ldots, g_{q-1}) = (-1)^q a(g_1, \ldots, g_{q-1}, g)$$

for $q > 0$ and $h_q = 0$ for $q \leqslant 0$; they possess the required properties, as may be readily checked.

7. <u>Gradings.</u> In conclusion, let us assume that our Lie algebra posssesses a grading, i.e., $\mathfrak{g}$ is a (possibly com-

pleted) direct sum of its subspaces $\mathfrak{g}_{(\lambda)}$, where the λ are integers, real or complex numbers, or, in a more general way, elements of some Abelian group; here we assume that $[\mathfrak{g}_{(\lambda)}, \mathfrak{g}_{(\mu)}] \subset \mathfrak{g}_{(\lambda+\mu)}$. The space $\mathfrak{g}_{(\lambda)}$ is said to be a <u>homogeneous component</u> of the algebra $\mathfrak{g}$. Suppose further the $\mathfrak{g}$-module A is also graded by homogeneous components $A_{(\mu)}$ in such a way that $\mathfrak{g}_{(\lambda)}A_{(\mu)} \subset A_{(\lambda+\mu)}$. (If the module A is trivial, then we usually assume that $A = A_{(0)}$.) Then gradings arise in chain and cochain spaces:

$$C_{(\lambda)}^q (\mathfrak{g}; A) = \{c \in C^q (\mathfrak{g}; A) \mid c (g_1, \ldots, g_q) \in A_{(\lambda_1 + \ldots + \lambda_q - \lambda)}$$
$$\text{for } g_i \in \mathfrak{g}_{(\lambda_i)}\};$$

$C_q^{(\lambda)} (\mathfrak{g}; A)$ is generated by the chains $a \otimes (g_1 \wedge \ldots \wedge g_q)$ with $a \in A_{(\mu)}$, $g_i \in \mathfrak{g}_{(\lambda_i)}$, $\lambda_1 + \ldots + \lambda_q + \mu = \lambda$. A direct verification shows that $d (C_{(\lambda)}^q (\mathfrak{g}; A)) \subset C_{(\lambda)}^{q+1}(\mathfrak{g}; A)$ and $\partial (C_q^{(\lambda)} (\mathfrak{g}; A))$ $\subset C_{q-1}^{(\lambda)} (\mathfrak{g}; A)$, so that both homology and cohomology acquire gradings. The multiplicative structures described in Subsection 2 are compatible with these gradings and, in particular,

$$H_{(\lambda)}^p (\mathfrak{g}) \, H_{(\mu)}^q (\mathfrak{g}) \subset H_{(\lambda+\mu)}^{p+q} (\mathfrak{g}).$$

Compatibility with these gradings, in a certain natural sense, also holds for the homomorphisms of the coefficient sequences, the inclusion homormophisms, and the Poincaré isomorphisms (see 6).

<u>Examples of gradings.</u> 1°. The spaces

$$\mathfrak{gl} (n, \mathbb{K})_{(p)} = \{\| \alpha_{ij} \| \mid \alpha_{ij} = 0, \text{ if } j - i \neq p\}$$
$$(p = - n, - n + 1, \ldots, n - 1, n)$$

give a grading of the algebra $\mathfrak{gl} (n, \mathbb{K})$.

2°. Denote by $\mathbb{K}W_{n(p)}$ the subspace of the algebra $\mathbb{K}W_n$, consisting of vector fields $\sum f_i (x_1, \ldots, x_n) \, \partial/\partial x_i$, in which

all the f_i are homogeneous polynomials of degree $p + 1$.
The spaces $\mathbb{K}W_{n\,(p)}$ $(p = -1, 0, 1\ 2, \ldots)$ yield a grading of the
algebra $\mathbb{K}W_n$.   In particular, the algebra $\mathbb{K}W_1$ acquires a
grading in which deg $e_i = i$.

§4.   PRINCIPAL ALGEBRAIC INTERPRETATIONS
       OF COHOMOLOGY

   1.   Since $C^{-1}(\mathfrak{g}; A) = 0$, we have

$$H^0(\mathfrak{g}; A) = \text{Ker } [d_0: C^0(\mathfrak{g}; A) = A \to C(\mathfrak{g}; A)]$$
$$= \{a \in A \mid ga = 0 \text{ for all } g \in \mathfrak{g}\}$$
$$= \text{Inv}_\mathfrak{g} A.$$

[In an equally obvious way one can construct the isomorphism
$H_0(\mathfrak{g}; A) = A/\mathfrak{g}A.$]

   2.   Since the differential $d_0: C^0(\mathfrak{g}) \to C^1(\mathfrak{g})$ is trivial,
we have

$$H^1(\mathfrak{g}) = \text{Ker } [d_1: C^1(\mathfrak{g}) = \mathfrak{g}' \to C^2(\mathfrak{g})]$$
$$= \{\gamma \in \mathfrak{g}' \mid \gamma([g, h]) = 0 \text{ for all } g, h \in \mathfrak{g}\}$$
$$= (\mathfrak{g}/[\mathfrak{g}, \mathfrak{g}])'.$$

(In the same obvious way one can construct the isomorphism
$H_1(\mathfrak{g}) = \mathfrak{g}/[\mathfrak{g}, \mathfrak{g}].$)

   3.   The space $H^1(\mathfrak{g}; \mathfrak{g})$ can be interpreted as the space
of "exterior derivations" of the algebra $\mathfrak{g}$. Recall that the
homomorphism $\varphi: \mathfrak{g} \to \mathfrak{g}$ is said to be a derivation if $\varphi([g,$
$h]) = [\varphi(g), h] + [g, \varphi(h)]$;   as examples note "inner derivations"
$g \mapsto [g_0, g]$, where $g_0 \in \mathfrak{g}$ is a fixed element. Exterior derivations are by definition elements of the quotient space of
the space of all derivations by the subspace of inner deriva-
tions. Cochains in $C^1(\mathfrak{g}; \mathfrak{g})$ are simply linear maps $\mathfrak{g} \to \mathfrak{g}$.
The condition $d_1(\varphi) = 0$ for $\varphi \in C^1(\mathfrak{g}; \mathfrak{g})$ means that φ is a
derivation:

$$d_1 \varphi \, (g, \, h) = \varphi \, ([g, \, h]) - g\varphi \, (h) + h\varphi \, (g)$$
$$= \varphi \, ([g, \, h]) - ([\varphi \, (g), \, h] + [g, \, \varphi \, (h)]).$$

At the same time, elements of the image of the differential $d_0 : C^0 \, (\mathfrak{g}; \, \mathfrak{g}) \to C^1 \, (\mathfrak{g}; \, \mathfrak{g})$ are inner derivations: for $g \in \mathfrak{g} = C^0 \, (\mathfrak{g}; \, \mathfrak{g})$ we have

$$d_0 \, g \, (h) = - \, gh = [-g, \cdot h].$$

4. Another interpretation of the space $H^1 \, (\mathfrak{g}; \, \mathfrak{g})$ is the following: it is the set of classes of one-dimensional "right extensions" of the algebra $\mathfrak{g}$, i.e., exact sequences

$$0 \to \mathfrak{g} \to \widetilde{\mathfrak{g}} \to \mathbb{K} \to 0$$

of Lie algebras and their homomorphisms ($\mathbb{K}$ is viewed as a Lie algebra with trivial commutator); the sequences $0 \to \mathfrak{g} \to \widetilde{\mathfrak{g}} \to \mathbb{K} \to 0$, $\quad 0 \to \mathfrak{g} \to \widetilde{\mathfrak{g}}_1 \to \mathbb{K} \to 0$ are considered equivalent if they may be included in a commutative diagram of the form

$$\begin{array}{ccccccccc} 0 \to & \mathfrak{g} & \to & \widetilde{\mathfrak{g}} & \to & \mathbb{K} & \to & 0 \\ & \downarrow \text{id} & & \downarrow & & \downarrow \text{id} & & \\ 0 \to & \mathfrak{g} & \to & \widetilde{\mathfrak{g}}_1 & \to & \mathbb{K} & \to & 0. \end{array} \qquad (1)$$

To the cohomology class of the cocycle $c \in C^1 \, (\mathfrak{g}; \, \mathfrak{g})$ corresponds the class of the extension

$$0 \to \mathfrak{g} \xrightarrow{g \, \mapsto \, (g, \, 0)} \mathfrak{g} \oplus \mathbb{K} \xrightarrow{(g, \, \lambda) \, \mapsto \lambda} \mathbb{K} \to 0,$$

where the Lie algebra structure in $\mathfrak{g} \oplus \mathbb{K}$ is defined by the formula

$$[(g_1, \, \lambda_1), \, (g_2, \, \lambda_2)] = ([g_1, \, g_2] + \lambda_2 c \, (g_1) - \lambda_1 c \, (g_2), \, 0).$$

The Jacobi identity for this commutator is equivalent to c being a cocycle: the left-hand side of this identity for $(g_1, \, \lambda_1), (g_2, \, \lambda_2), (g_3, \, \lambda_3)$, after regrouping terms, becomes

$$[[g_1, \, g_2], \, g_3] + [[g_2, \, g_3], \, g_1] + [[g_3, \, g_1], \, g_2]$$
$$+ \, \lambda_1 dc \, (g_2, \, g_3) + \lambda_2 dc \, (g_3, \, g_1) + \lambda_3 \, dc \, (g_1, \, g_2).$$

It is just as easy to check that to cohomologous cocycles
correspond equivalent extensions: sequences corresponding
to the cocycles c, $c + db$, where $b \in C^0(\mathfrak{g}; \mathfrak{g}) = \mathfrak{g}$, may be included in the diagram (1) whose middle vertical homomorphism
acts according to the formula

$$(g, \lambda) \mapsto (g + \lambda b, \lambda).$$

[To the zero of the space $H^1(\mathfrak{g}; \mathfrak{g})$ corresponds the class
of the "trivial extension," i.e., a sequence which splits.]

   Example. The formula

$$\sum_{i=1}^{2k} f_i \frac{\partial}{\partial x_i} \mapsto \sum_{i=1}^{2k} \left[\left(\sum_{j=1}^{2k} x_j \frac{\partial f_i}{\partial x_j} \right) - f_i \right] \frac{\partial}{\partial x_i}$$

determines a one-dimensional cocycle of the algebra $\mathbb{K}H_k$
with coefficients in the adjoint representation [actually
the space $H^1(\mathbb{K}H_k; \mathbb{K}H_k)$ is one-dimensional and is generated
by the class of this cocycle].   The corresponding extended
algebra is isomorphic to $\mathbb{K}\widehat{H}_k$.

   5.   Now suppose that A and B are arbitrary $\mathfrak{g}$-modules.
The space $H^1(\mathfrak{g}; \text{Hom}(B, A))$ may be interpreted as the set
of classes of exact sequences

$$0 \to A \to C \to B \to 0$$

of $\mathfrak{g}$-modules; the sequences $0 \to A \to C \to B \to 0$, $0 \to A \to C_1 \to$
$B_1 \to 0$ are assumed equivalent if they may be included in
a commutative diagram of the form

$$\begin{array}{ccccccccc} 0 \to & A & \to & C & \to & B & \to 0 \\ & \downarrow \text{id} & & \downarrow & & \downarrow \text{id} & \\ 0 \to & A & \to & C_1 & \to & B & \to 0. \end{array} \qquad (2)$$

To the cohomology class of the cocycle

$$c: \mathfrak{g} \to \text{Hom}(B, A)$$

corresponds the class of the sequence

$$0 \to A \xrightarrow{a \,\mapsto\, (a,\, 0)} A \oplus B \xrightarrow{(a,\, b) \,\mapsto\, b} B \to 0,$$

where the $\mathfrak{g}$-module structure in $A \oplus B$ is defined by the formula

$$g\,(a, \ b) = (ga + [c\,(g)]\,(b), \ gb).$$

The fact that this is indeed a $\mathfrak{g}$-module structure is equivalent to c being a cocycle:

$$[g_1, \ g_2](a, \ b) = ([g_1, \ g_2]\,a + c\,([g_1, \ g_2])(b), \ [g_1, \ g_2]\,b)$$
$$= (g_1\,(g_2 a) - g_2\,(g_1 a) + [g_1 c\,(g_2) - g_2 c\,(g_1)](b), \ g_1\,(g_2 b) - g_2\,(g_1 b))$$
$$= (g_1\,(g_2 a) + g_1\,[c\,(g_2)](b) + c\,(g_1)\,g_2\,(b), \ g_1\,(g_2 b))$$
$$- (g_2\,(g_1 a) + g_2\,[c\,(g_1)](b) + c\,(g_2)\,g_1\,(b), \ g_2\,(g_1 b))$$
$$= g_1\,(g_2\,(a, \ b)) - g_2\,(g_1\,(a, \ b)).$$

It is just as easy to check that the replacement of the cocycle c by the cocycle $c + d\varphi$, where $\varphi \in C^0\,(\mathfrak{g}; \operatorname{Hom}\,(B, \ A)) = \operatorname{Hom}\,(B, \ A)$, results in the replacement of the sequence by an equivalent one [the equivalence is described by diagram (2), in which the middle vertical homomorphism acts according to the formula $(a, \ b) \mapsto (a + \varphi\,(b), \ b)$]. [To the zero of the space $H^1\,(\mathfrak{g}; \operatorname{Hom}\,(A, \ B))$ corresponds a sequence which splits.]

6. The space $H^2\,(\mathfrak{g})$ can be interpreted as the set of classes of one-dimensional central extensions of the algebra $\mathfrak{g}$. Recall that a one-dimensional central extension of algebra $\mathfrak{g}$ is an exact sequence

$$0 \to \mathbb{K} \to \tilde{\mathfrak{g}} \to \mathfrak{g} \to 0$$

of Lie algebras and their homomorphisms, in which the image of the homomorphism $\mathbb{K} \to \tilde{\mathfrak{g}}$ is contained in the center of the algebra $\tilde{\mathfrak{g}}$. To the cohomology class of a cocycle $c \in C^2\,(\mathfrak{g})$ corresponds the extension

$$0 \to \mathbb{K} \xrightarrow{\lambda \,\mapsto\, (\lambda,\, 0)} K \oplus \mathfrak{g} \xrightarrow{(\lambda,\, g) \,\mapsto\, g} \mathfrak{g} \to 0,$$

where the commutator in the algebra $\mathbb{K} \oplus \mathfrak{g}$ is defined by the formula

$$[(\lambda, g), (\mu, h)] = (c(g, h), [g, h]).$$

It can be checked automatically that the Jacobi identity for this commutator is equivalent to c being a cocycle and that to cohomologous cocycles correspond equivalent (in an obvious sense) extensions. The trivial extension, i.e., the sequence which splits, corresponds to the zero of the space $H^2(\mathfrak{g})$.

Example. The formula

$$\sum_{i=1}^{2k} f_i \frac{\partial}{\partial x_i}, \quad \sum_{i=1}^{2k} g_i \frac{\partial}{\partial x_i} \mapsto \sum_{i=1}^{k} \begin{vmatrix} f_i(0) & g_i(0) \\ f_{i+k}(0) & g_{i+k}(0) \end{vmatrix}$$

determines a two-dimensional cocycle of the algebra $\mathbb{K}H_k$ [actually the space $H^2(\mathbb{K}H_k)$ is one-dimensional and is generated by the class of this cocycle]. The corresponding extension of the algebra is $\mathbb{K}P_k$.

A more serious example. Suppose $\mathfrak{g}$ is a semisimple Lie algebra, K is its Killing form (see [94]), and $(\mathfrak{g}^{S^1})^{pol}$ is the polynomial current algebra (see Subsection 1.4). The formula

$$\varphi, \psi \mapsto \int_{S^1} K(\varphi(t), \psi'(t)) \, dt$$

determines a cocycle in $C^2((\mathfrak{g}^{S^1})^{pol})$, which is not cohomologous to zero, and the cohomology class of this cocycle generates $H^2((\mathfrak{g}^{S^1})^{pol})$. Thus, we obtain a one-dimensional central extension of the algebra $(\mathfrak{g}^{S^1})^{pol}$, and the corresponding extension of the algebra is called a Kac—Moody algebra (actually, Kac—Moody algebras are defined axiomatically, and this axiomatic definition covers a somewhat wider class of algebras than the ones obtained from the above construction — see 2.5.3).

7. The space H^2 ($\mathfrak{g}$; $\mathfrak{g}$) can be interpreted as the set of classes of infinitesimal deformations of the algebra $\mathfrak{g}$. Recall that a <u>deformation</u> of the algebra $\mathfrak{g}$ is a smooth map

$$h: \mathfrak{g} \times \mathfrak{g} \times \mathbb{R} \to \mathfrak{g},$$

such that $h(g_1, g_2, 0) = [g_1, g_2]$ and for every $t \in \mathbb{R}$ the operation

$$g_1, g_2 \mapsto [g_1, g_2]_t = h(g_1, g_2, t)$$

determines a Lie algebra structure in $\mathfrak{g}$. Deformations h, h_1 are said to be equivalent if there exists a smooth map

$$u: \mathfrak{g} \times \mathbb{R} \to \mathfrak{g},$$

such that $u(g, 0) = g$; for every $t \in \mathbb{R}$ the map $g \mapsto u(g, t)$ is linear, and

$$h_1(g_1, g_2, t) = h(u(g_1, t), u(g_2, t), t).$$

If h is a deformation, the map

$$\eta: \mathfrak{g} \times \mathfrak{g} \to \mathfrak{g}, \quad \eta(g_1, g_2) = \frac{\partial}{\partial t} h(g_1, g_2, t)\big|_{t=0},$$

obviously satisfies the relations

$$\eta(g_1, g_2) = -\eta(g_2, g_1),$$
$$[\eta(g_1, g_2), g_3] + [\eta(g_2, g_3), g_1] + [\eta(g_3, g_1), g_2] \qquad (3)$$
$$+ \eta([g_1, g_2], g_3) + \eta([g_2, g_3], g_1) + \eta([g_3, g_1], g_2) = 0$$

(the second is obtained by taking derivatives with respect to t when $t = 0$ of the Jacobi identity for $[g_1, g_2]_t$); if the deformations h, h_1 are equivalent and $u: \mathfrak{g} \times \mathbb{R} \to \mathbb{R}$ is the map realizing this equivalence, then the maps η and η_1 which correspond to h and h_1 are related by the formula

$$\eta_1(g_1, g_2) = \eta(g_1, g_2) + [g_1, \xi(g_2)] + [\xi(g_1), g_2], \qquad (4)$$

where ξ is defined by putting $\xi(g) = \frac{\partial}{\partial t} u(g, t)\big|_{t=0}$. In this connection the bilinear map $\eta: \mathfrak{g} \times \mathfrak{g} \to \mathfrak{g}$ is called an <u>infini-</u>

<u>tesimal deformation of the algebra</u> $\mathfrak{g}$ if it satisfies relation
(3); infinitesimal deformations η, η_1 are said to be equivalent if for some linear map $\xi: \mathfrak{g} \to \mathfrak{g}$ we have relation (4).
This definition immediately implies that infinitesimal de-
formations are simply cocycles from $C^2(\mathfrak{g}; \mathfrak{g})$, while equivalent infinitestimal deformations are cohomologous cocycles, so
that the set of equivalence classes of infinitesimal deformations is actually $H^2(\mathfrak{g}; \mathfrak{g})$.

Now assume that we are given an element α of the space
$H^2(\mathfrak{g}; \mathfrak{g})$, i.e., a class of infinitesimal deformations of the
algebra $\mathfrak{g}$. Are the infinitesimal deformations of this class
derivatives with respect to some parameter of true deformations of the algebra $\mathfrak{g}$? As we shall see below (see 2.3.2G),
they are not in general; now we shall indicate some necessary
condition for this to take place. Suppose η is a cocycle
which represents α. Assume that the infinitesimal deforma-
tion η corresponds to a true deformation $h: \mathfrak{g} \times \mathfrak{g} \times \mathbb{R} \to \mathfrak{g}$,
and expand h in a Taylor series in t:

$$h(g_1, g_2, t) = [g_1, g_2] + t\eta(g_1, g_2) + t^2\zeta(g_1, g_2) + \dots$$

Further, for the commutator $[g_1, g_2]_t = h(g_1, g_2, t)$ write out
the Jacobi identity and set the coefficients of the Taylor
series of its left-hand side equal to zero:

$$
\begin{aligned}
&[[g_1, g_2], g_3] + \dots = 0 \\
&[\eta(g_1, g_2), g_3] + \eta([g_1, g_2], g_3) + \dots = 0 \\
&[\zeta(g_1, g_2), g_3] + \eta(\eta(g_1, g_2), g_3) + \zeta([g_1, g_2], g_3) + \dots = 0
\end{aligned} \tag{5}
$$

The dots in the first three lines denote summands obtained
from the ones written out by cyclic permutation of the arguments g_1, g_2, g_3. The first of the relations (5) is the Jacobi
identity for the algebra $\mathfrak{g}$, the second coincides with the
second one of the relations (3). As for the third, it shows

that the three-dimensional cocycle

$$g_1, \ g_2, \ g_3 \ \mapsto \ \eta \ (\eta \ (g_1, \ g_2), \ g_3) + \cdots$$

is cohomologous to zero: it is equal to the differential of
the cochain $g_1, \ g_2 \mapsto -\zeta \ (g_1, \ g_2)$. The cohomology class of such
a cocycle was already considered in 3.2, where it was denoted
by $[\alpha, \alpha]$. Thus, for the existence of the deformation h it
is necessary that the square $[\alpha, \alpha]$ of the class α vanish.
The next obstruction to carrying out the deformation is [as
we see from the following relation (5)] the cohomology class
of the cocycle

$$g_1, \ g_2, \ g_3 \ \mapsto \ \eta \ (\zeta \ (g_1, \ g_2), \ g_3, + \zeta \ (\eta \ (g_1, \ g_2), \ g_3) + \cdots,$$

which is precisely the "Massey cube" of the class α. Continuing the argument, we see that for the existence of the deformation it is necessary that all the Massey powers of the
class α vanish; these powers are elements of the space H^3 ($\mathfrak{g}$;
$\mathfrak{g}$), which are defined with increasing nonuniqueness, and each
is determined only if all the previous ones vanish. When
they are all equal to zero, then there exists a <u>formal</u> deformation (with respect to the parameter) of the algebra
which extends our infinitesimal deformation; it remains to
study the convergence of the series.

In conclusion, consider an interesting example. The
formula

$$(f, g) \mapsto \frac{\partial^2 f}{\partial x_1^2} \frac{\partial^2 g}{\partial x_2^2} - \frac{\partial^2 g}{\partial x_1^2} \frac{\partial^2 f}{\partial x_2^2}$$

defines, as can be shown by an automatic verification, a two-
dimensional cocycle of the Poisson àlgebra $P_1^{pol} = \mathbb{R} \ [x_1, \ x_2]$
with coefficients in the adjoint representation (this construction remains valid for the algebra P_1 and can be carried

out without difficulty to the algebras P_k for arbitrary k; we limit ourselves to considering the most instructive case). This cocycle is not cohomologous to zero and an interesting deformation of the algebra P_1^{pol} corresponds to it. In order to describe this deformation, define for $t \in \mathbb{R} \setminus 0$ the isomorphism q_t between P_1^{pol} and the space $\mathbb{R}[y, d/dy]$ of polynomial differential operators on the line by the formula

$$q_t(x_1^m x_2^n) = t |t|^{\frac{m+n}{2}} y^n \frac{d^m}{dy^m}.$$

The usual commutation of differential operators can be carried over by using the isomorphism q_t into P_1^{pol}, defining a certain operation $[\, , \,]_t$ there. Putting in addition $[\, , \,]_0 = [\, , \,]$, we obtain, as can be easily checked, a deformation of the Poisson algebra. For example,

$$[x_1^2, \ x_2^2]_t = 4x_1 x_2 + 2t;$$
$$[x_1^3, \ x_2^3]_t = 9x_1^2 x_2^2 + 18t x_1 x_2 + 3t^2.$$

Speaking informally, we have described a deformation of the Poisson algebra which transforms it into the Lie algebra of polynomial differential operators on the line with the commutation operation.

8.   Using cohomology language, we can state many other algebraic problems related to Lie algebras. Without going into detail (and in places without even giving exact definitions) we shall list certain problems of this type in this subsection. Everywhere we assume that a homomorphism f of the Lie algebra $\mathfrak{g}$ into the Lie algebra $\mathfrak{h}$ is given.

A.   Suppose $\eta \in C^2(\mathfrak{g}; \mathfrak{g})$, $\zeta \in C^2(\mathfrak{h}; \mathfrak{h})$ are infinitesimal deformations of the algebras $\mathfrak{g}, \mathfrak{h}$, and α, β are the cohomology classes of the cocycles η, ζ. Question: does there exist an infinitesimal deformation (of the homomorphism f) which extends the infinitesimal deformations η, ζ, and how many such deformations exist? Answer: the deformation exists if and only if the classes α, β have the same image under the homomorphisms $H^2(\mathfrak{g}; \mathfrak{g}) \to H^2(\mathfrak{g}; \mathfrak{h})$, $H^2(\mathfrak{h}; \mathfrak{h}) \to H^2(\mathfrak{g}; \mathfrak{h})$, determined by the homomorphism f. Under this assumption, $H^1(\mathfrak{g}; \mathfrak{h})$ acts transitively and freely in the set of classes of such deformations, so that the choice of one particular deformation induces a natural bijection between $H^1(\mathfrak{g}; \mathfrak{h})$ and a set of deformation classes.

B. Suppose

$$0 \to \mathfrak{g} \to \widetilde{\mathfrak{g}} \to \mathbb{K} \to 0, \quad 0 \to \mathfrak{h} \to \widetilde{\mathfrak{h}} \to \mathbb{K} \to 0$$

are right extensions of the algebras $\mathfrak{g}, \mathfrak{h}$ and $\alpha \in H^1(\mathfrak{g}; \mathfrak{g})$ $\beta \in H^1(\mathfrak{h}; \mathfrak{h})$ are the corresponding cohomology classes. Question: can we include these extensions into the diagram

$$
\begin{array}{ccccccccc}
0 & \to & \mathfrak{g} & \to & \widetilde{\mathfrak{g}} & \to & \mathbb{K} & \to & 0 \\
  &   & \downarrow f & & \downarrow & & \downarrow \text{id} & & \\
0 & \to & \mathfrak{h} & \to & \widetilde{\mathfrak{h}} & \to & \mathbb{K} & \to & 0
\end{array}
$$

and in how many ways? Answer: we can if and only if α, β have the same images under the homomorphisms $H^1(\mathfrak{g}; \mathfrak{g}) \to H^1(\mathfrak{g}; \mathfrak{h})$, $H^1(\mathfrak{h}; \mathfrak{h}) \to H^1(\mathfrak{g}; \mathfrak{h})$, determined by the homomorphism f; under this assumption, the choice of a specific way induces a natural bijection between the set of ways and $H^0(\mathfrak{g}; \mathfrak{h}) = \mathrm{Inv}_\mathfrak{g}\mathfrak{h} = f^{-1}(Z(\mathfrak{h}))$, where Z is the center.

C. Suppose

$$0 \to \mathbb{K} \to \widetilde{\mathfrak{g}} \to \mathfrak{g} \to 0, \quad 0 \to \mathbb{K} \to \widetilde{\mathfrak{h}} \to \mathfrak{h} \to 0$$

are one-dimensional central extensions of the algebras $\mathfrak{g}, \mathfrak{h}$ and $\alpha \in H^2(\mathfrak{g}), \beta \in H^2(\mathfrak{h})$ are the corresponding cohomology

classes. Question: can we include these extensions into
the diagram

$$0 \to \mathbb{K} \to \tilde{\mathfrak{g}} \to \mathfrak{g} \to 0$$
$$\downarrow \text{id} \quad \downarrow \quad \downarrow f$$
$$0 \to \mathbb{K} \to \tilde{\mathfrak{h}} \to \mathfrak{h} \to 0$$

and in how many ways? Answer: we can if and only if the
homomorphism $f^*: H^2(\mathfrak{h}) \to H^2(\mathfrak{g})$ maps β into α; under this assumption the choice of one way generates a natural bijection
between the set of ways and $H^1(\mathfrak{g})$.

Remark. Questions concerning the relationship of de-
formations and extensions of Lie algebra $\mathfrak{g}$ with $\mathfrak{g}$-module
structures in the space A reduce, as a rule, to those dis-
cussed previously, since a $\mathfrak{g}$-module structure in A is simply
a homomorphism of the algebra $\mathfrak{g}$ into the algebra End A of
endomorphisms of the space A.

§5. MAIN COMPUTATIONAL METHODS

1. The Serre–Hochschild spectral sequence. Suppose $\mathfrak{g}$
is a Lie algebra, $\mathfrak{h}$ its subalgebra, and A a module over $\mathfrak{g}$.

Theorem 1.5.1. There exists a spectral sequence

$$\{E_r^{p,\,q}, d_r^{p,\,q}: E_r^{p,\,q} \to E_r^{p+r,\,q-r+1}\}$$

with the following properties:

(i) $E_1^{p,\,q} = H^q(\mathfrak{h}; \text{Hom}(\Lambda^p(\mathfrak{g}/\mathfrak{h}), A))$;

(ii) $E_2^{p,\,0} = H^p(\mathfrak{g}, \mathfrak{h}; A)$;

(iii) if $\mathfrak{h}$ is an ideal, then $E_2^{p,\,q} = H^p(\mathfrak{g}/\mathfrak{h}; H^q(\mathfrak{h}; A))$;

(iv) the term E_∞ is adjoint to $H^*(\mathfrak{g}; A)$; and the na-
tural homomorphisms $H^q(\mathfrak{g};\ A) \to H^q(\mathfrak{h};\ A)$, $H^p(\mathfrak{g}, \mathfrak{h}; A) \to H^p(\mathfrak{g}; A)$
can be represented as compositions

$$H^q (\mathfrak{g}; A) \to E_\infty^{0,q} \to E_1^{0,q} = H^q (\mathfrak{h}; A),$$

$$II^p (\mathfrak{g}, \mathfrak{h}; A) = E_2^{p,0} \to E_\infty^{p,0} \to H^p (\mathfrak{g}; A),$$

where the   arrows have the obvious meaning.

(v) if $A = \mathbb{K}$ (or if A is an associative commutative algebra in which $\mathfrak{g}$ acts by means of derivations) then the spectral sequence is multiplicative; in this case, the isomorphisms (i)-(iii) are also multiplicative as well as the adjointness (iv).

Clarifications of the statement.   The space $\mathrm{Hom} (\Lambda^p (\mathfrak{g}/\mathfrak{h})$, $A)$ is a $\mathfrak{h}$-module, since such is the space $\mathfrak{g}/\mathfrak{h}$ (the latter is the quotient module of the $\mathfrak{h}$-module $\mathfrak{g}$ by the $\mathfrak{h}$-module $\mathfrak{h}$). If $\mathfrak{h}$ is an ideal, then the space $H^q (\mathfrak{h}; A)$ is a $\mathfrak{g}/\mathfrak{h}$-module: it is a $\mathfrak{g}$-module together with $\mathfrak{h}$ and is trivial as a $\mathfrak{h}$-module in view of 3.6C.

The spectral sequence of Theorem 1.5.1 was constructed in 1953 by Serre and Hochschild (see [52]) and bears their names.   There is also a homology Serre—Hochschild spectral sequence for which the precise homology analogs of statements (i)-(iv) hold.   Its construction is a repetition of the construction of the cohomology Serre—Hochschild sequence given below.

Proof of Theorem 1.5.1.   Let

$$F^p C^{p+q} (\mathfrak{g}; A) = \{c \in C^{p+q} (\mathfrak{g}; A) \,|\, c (g_1, \ldots, g_{p+q}) = 0 \text{ for } g_1, \ldots, g_{q+1} \in \mathfrak{h}\}.$$

Obviously,

$$C^r (\mathfrak{g}; A) = F^0 C^r (\mathfrak{g}; A) \supset \ldots \supset F^r C^r (\mathfrak{g}; A) \supset F^{r+1} C^r (\mathfrak{g}; A) = 0,$$

and definitions imply that $dF^p C^{p+q} (\mathfrak{g}; A) \subset F^p C^{p+q+s} (\mathfrak{g}; A)$. Thus, $\{F^p\}$ is a filtration in the complex $C^{\cdot} (\mathfrak{g}; A)$.   The corresponding spectral sequence is precisely $\{E_r^{p,q}, d_r^{p,q}\}$.

Let us compute the initial terms. Every cochain $c \in C^{p+q}(\mathfrak{g}; A)$ determines a map $\Lambda^q \mathfrak{h} \to \mathrm{Hom}\,(\Lambda^p \mathfrak{g}, A)$, sending $h_1 \wedge \ldots \wedge h_q$ into the homomorphism $g_1 \wedge \ldots \wedge g_p \mapsto c\,(h_1, \ldots, h_q, g_1, \ldots, g_p)$, and the inclusion $c \in F^p C^{p+q}(\mathfrak{g}; A)$ is equivalent to the image of this homomorphism being contained in $\mathrm{Hom}\,(\Lambda^p\,(\mathfrak{g}/\mathfrak{h}),\ A) \subset \mathrm{Hom}\,(\Lambda^p \mathfrak{g},\ A)$. We obtain the map $\varphi \colon F^p C^{p+q}\,(\mathfrak{g}; A) \to \mathrm{Hom}\,(\Lambda^p\,(\mathfrak{g}/\mathfrak{h}), A)$, which is obviously an epimorphism with kernel $F^{p+1} C^{p+q}\,(\mathfrak{g};\ A)$. Thus we get the isomorphism

$$E_0^{p,\,q} = F^p C^{p+q}\,(\mathfrak{g}; A)/F^{p+1} C^{p+q}\,(\mathfrak{g};\ A) \ \to C^q\,(\mathfrak{h};\ \mathrm{Hom}\,(\Lambda^p\,(\mathfrak{g}/\mathfrak{h}), A)).$$

This isomorphism commutes with the differentials: for $c \in F^p \bar{C}^{p+q}\,(\mathfrak{g}; A)$, $h_1, \ldots, h_{q+1} \in \mathfrak{h}$, $g_1, \ldots, g_p \in \mathfrak{g}$ we have

$$dc\,(h_1, \ldots, h_{q+1}, g_1, \ldots, g_p) = \sum_{1 \leqslant s < t \leqslant q+1} (-1)^{s+t-1} c\,([h_s, h_t], h_1, \ldots \hat{h}_s \ldots \hat{h}_t \ldots,$$

$$h_{q+1}, g_1, \ldots, g_p) + \sum_{s=1}^{q+1} \sum_{t=1}^{p} (-1)^{s+t} c\,(h_1, \ldots \hat{h}_s \ldots, h_{q+1}, [h_s, g_t], g_1, \ldots \hat{g}_t \ldots, g_p)$$

$$+ \sum_{s=1}^{q+1} (-1)^s h_s c\,(h_1, \ldots \hat{h}_s \ldots, h_{q+1}, g_1, \ldots\ g_p)$$

$$= \sum_{1 \leqslant s < t \leqslant q+1} (-1)^{s+t-1} [\varphi\,(c)\,([h_s, h_t], h_1, \ldots \hat{h}_s \ldots \hat{h}_t \ldots, h_{q+1})]\,(\bar{g}_1, \ldots, \bar{g}_p)$$

$$+ \sum_{s=1}^{q+1} (-1)^s \{h_s\,[\varphi\,(c)\,(h_1, \ldots \hat{h}_s \ldots, h_{q+1})]\}\,(\bar{g}_1, \ldots, \bar{g}_p)$$

$$= [d\varphi\,(c)\,(h_1, \ldots h_{q+1})]\,(\bar{g}_1, \ldots, \bar{g}_p),$$

where $\bar{g}_1, \ldots, \bar{g}_p$ are the classes of the elements $g_1, \ldots, g_p$ in $\mathfrak{g}/\mathfrak{h}$. Thus,

$$\text{(i)} \qquad E_1^{p,\,q} = H^q\,(\mathfrak{h};\ \mathrm{Hom}\,(\Lambda^p\,(\mathfrak{g}/\mathfrak{h}),\ A)).$$

The differential

$$d_1^{p,\,q} \colon E_1^{p,\,q} = H^q\,(\mathfrak{h};\ \mathrm{Hom}\,(\Lambda^p\,(\mathfrak{g}/\mathfrak{h}),\ A))$$

$$\to E_1^{p+1,\,q} = H^q\,(\mathfrak{h};\ \mathrm{Hom}\,(\Lambda^{p+1}\,(\mathfrak{g}/\mathfrak{h}),\ A)).$$

is not induced by any natural homomorphism $\mathrm{Hom}\,(\Lambda^p\,(\mathfrak{g}/\mathfrak{h}),\ A) \to \mathrm{Hom}\,(\Lambda^{p+1}\,(\mathfrak{g}/\mathfrak{h}),\ A)$, and in general no nontrivial natural

$\mathfrak{h}$-homomorphism Hom $(\Lambda^p(\mathfrak{g}/\mathfrak{h}),\ A) \to$ Hom $(\Lambda^{p+1}(\mathfrak{g}/\mathfrak{h}),\ A)$ exists. Of course, we do have the differential $d: C^p(\mathfrak{g},\ \mathfrak{h};\ A) \to C^{p+1}(\mathfrak{g},\ \mathfrak{h};\ A)$, which is a natural homomorphism

$$\text{Hom}_{\mathfrak{h}}(\Lambda^p(\mathfrak{g}/\mathfrak{h}),\ A) \to \text{Hom}_{\mathfrak{h}}(\Lambda^{p+1}(\mathfrak{g}/\mathfrak{h}),\ A),$$

and an obvious verification shows that the diagram

$$
\begin{array}{ccc}
H^q(\mathfrak{h}) \otimes C^p(\mathfrak{g},\ \mathfrak{h};\ A) & \xrightarrow{\ \text{id} \odot d\ } & H^q(\mathfrak{h}) \otimes C^{p+1}(\mathfrak{g},\ \mathfrak{h};\ A) \\
\| & & \| \\
H^q(\mathfrak{h};\ \text{Hom}_{\mathfrak{h}}(\Lambda^p(\mathfrak{g}/\mathfrak{h}),\ A)) & & H^q(\mathfrak{h};\ \text{Hom}_{\mathfrak{h}}(\Lambda^{p+1}(\mathfrak{g}/\mathfrak{h}),\ A)) \\
\downarrow & & \downarrow \\
H^q(\mathfrak{h};\ \text{Hom}(\Lambda^p(\mathfrak{g}/\mathfrak{h}),\ A)) & \xrightarrow{\ d_1^{p,\,q}\ } & H^q(\mathfrak{h};\ \text{Hom}(\Lambda^{p+1}(\mathfrak{g}/\mathfrak{h}),\ A)),
\end{array}
\tag{1}
$$

whose vertical arrows are induced by the inclusions $\text{Hom}_{\mathfrak{h}}(\Lambda^r(\mathfrak{g}/\mathfrak{h}),\ A) \to$ Hom $(\Lambda^r(\mathfrak{g}/\mathfrak{h}),\ A)$, is commutative. As we shall see in Chapter 2, in certain important cases these inclusions turn out to be isomorphisms, and diagram (1) satisfactorily described the differential $d_1^{p,\,q}$ as well as the term $E_2^{p,\,q}$. In the general case, however, these arrows are isomorphisms for $q=0$ (see 4.1), so that $E_1^{p,0} = C^p(\mathfrak{g},\mathfrak{h};\ A)$, $d_1^{p,0} = [d: C^p(\mathfrak{g}, \mathfrak{h};\ A) \to C^{p+1}(\mathfrak{g},\ \mathfrak{h};\ A)$ and

$$\text{(ii)} \qquad\qquad E_2^{p,0} = H^p(\mathfrak{g},\ \mathfrak{h};\ A).$$

Finally, if $\mathfrak{h}$ is an ideal, then the $\mathfrak{h}$-module $\mathfrak{g}/\mathfrak{h}$ is trivial and

$$
\begin{aligned}
H^q(\mathfrak{h};\ \text{Hom}(\Lambda^p(\mathfrak{g}/\mathfrak{h}),\ A)) &= H^q(\mathfrak{h};\ \Lambda^p(\mathfrak{g}/\mathfrak{h})' \otimes A) \\
= \Lambda^p(\mathfrak{g}/\mathfrak{h})' \otimes H^q(\mathfrak{h};\ A) &= \text{Hom}(\Lambda^q(\mathfrak{g}/\mathfrak{h}),\ H^q(\mathfrak{h};\ A)) = C^p(\mathfrak{g}/\mathfrak{h};\ H^q(\mathfrak{h};\ A)).
\end{aligned}
$$

The differential $d_1^{p,\,q}$ coincides in this case with the differential $d: C^p(\mathfrak{g}/\mathfrak{h};\ H^q(\mathfrak{h};\ A)) \to C^{p+1}(\mathfrak{g}/\mathfrak{h};\ H^q(\mathfrak{h};\ A))$ and we have

$$\text{(iii)} \qquad\qquad E_2^{p,\,q} = H^p(\mathfrak{g}/\mathfrak{h};\ H^q(\mathfrak{h};\ A)).$$

Finally, statement (iv) is obvious, while (v) follows from the obvious multiplicativity of the filtration

$$F^p C^{p+q} (\mathfrak{g}) F^{p'} C^{p'+q'} (\mathfrak{g}) \subset F^{p+p'} C^{p+p'+q+q'} (\mathfrak{g}).$$

The theorem is proved.

This theorem may be generalized to relative cases. Namely, if $\mathfrak{k}$ is a subalgebra of the algebra $\mathfrak{h}$, there is a spectral sequence for which

(i) $E_1^{p,\,q} = H^q (\mathfrak{h}, \mathfrak{k}; \mathrm{Hom} (\Lambda^p (\mathfrak{g}/\mathfrak{h}), A))$;

(ii) $E_2^{p,\,0} = H^p (\mathfrak{g}, \mathfrak{h}; A)$;

(iii) if $\mathfrak{h}$ is an ideal, then $E_2^{p,\,q} = H^p (\mathfrak{g}/\mathfrak{h}; H^q (\mathfrak{h}, \mathfrak{k}; A))$;

(iv) the term E_∞ is adjoint to $H^* (\mathfrak{g}, \mathfrak{k}; A)$.

There also exists a spectral sequence which converges to $H^* (\mathfrak{g}, K; A)$, where K is a Lie group with Lie algebra $\mathfrak{k}$, which acts appropriately in $\mathfrak{g}$, $\mathfrak{h}$ and A (see the end of 3.1); for this spectral sequence we have $E_1^{p,\,q} = H^q (\mathfrak{h}, K; \mathrm{Hom} (\Lambda^p (\mathfrak{g}/\mathfrak{h}), A))$. Both of these spectral sequences have analogs in homology.

Let us add that in the case of gradings (see 3.7) all the Serre–Hochschild spectral sequences obtain a supplementary grading, i.e., they split up into a sum of spectral sequences corresponding to different homogeneity indices.

2. **Inner gradings.** Assume that the Lie algebra $\mathfrak{g}$ contains an element g_0, such that $\mathfrak{g}$ possesses a (topological) basis constituted by the eigenvectors of the operator $g \mapsto [g_0, g]$. Then the algebra $\mathfrak{g}$ possesses a natural grading with the following homogeneous components:

$$\mathfrak{g}_{(\lambda)} = \{g \in \mathfrak{g} \mid [g_0, g] = \lambda g\};$$

indeed, if $g \in \mathfrak{g}_{(\lambda)}$, $h \in \mathfrak{g}_{(\mu)}$, then

$$[g_0, [g, h]] = [[g_0, g], h] + [g, [g_0, h]]$$
$$= [\lambda g, h] + [g, \mu h] = (\lambda + \mu) [g, h].$$

Such gradings are called <u>inner</u>. According to 3.7, the complex $C^{\cdot}(\mathfrak{g})$ splits up into a (completed) direct sum of subcomplexes $C_{(\lambda)}(\mathfrak{g})$.

Theorem 1.5.2. The inclusion $C_{(0)}^{\cdot}(\mathfrak{g}) \to C^{\cdot}(\mathfrak{g})$ induces an isomorphism in homology.

Proof. We must show that the complexes $C_{(\lambda)}^{\cdot}(\mathfrak{g})$ with $\lambda \neq 0$ are acyclic. Define the map $D_{(\lambda)}^q \colon C_{(\lambda)}^q(\mathfrak{g}) \to C_{(\lambda)}^{q-1}(\mathfrak{g})$ by letting

$$[D_{(\lambda)}^q (c)](g_1, \ldots, g_{q-1}) = c(g_0, g_1, \ldots, g_{q-1})$$

(obviously, $g_0 \in \mathfrak{g}_{(0)}$). Then for $c \in C_{(\lambda)}^q(\mathfrak{g})$, $g_1 \in \mathfrak{g}_{(\lambda_1)}, \ldots, g_q \in \mathfrak{g}_{(\lambda_q)}$, $\lambda_1 + \ldots + \lambda_q = \lambda$, we have

$$[D_{(\lambda)}^{q+1}(dc)](g_1, \ldots, g_q) = dc(g_0, g_1, \ldots, g_q)$$
$$= \sum_{s=1}^{q} (-1)^{s-1} c([g_0, g_s], g_1, \ldots \hat{g}_s \ldots, g_q)$$
$$- \sum_{1 \leqslant s < t \leqslant q} (-1)^{s+t-1} c(g_0, [g_s, g_t], g_1, \ldots \hat{g}_s \ldots \hat{g}_t \ldots, g_q)$$
$$= \sum_{s=1}^{q} (-1)^{s-1} \lambda_s c(g_s, g_1, \ldots \hat{g}_s \ldots, g_q)$$
$$- \sum_{1 \leqslant s < t \leqslant q} (-1)^{s+t-1} [D_{(\lambda)}^q (c)]([g_s, g_t], g_1, \ldots \hat{g}_s \ldots \hat{g}_t \ldots, g_q)$$
$$= [\lambda c - d D_{(\lambda)}^q (c)](g_1, \ldots, g_q),$$

i.e., $\left\{ \frac{1}{\lambda} D_{(\lambda)}^q \right\}$ is (for $\lambda \neq 0$) a homotopy joining the identity of the complex $C_{(\lambda)}(\mathfrak{g})$ with the zero map.

Let us indicate two generalizations of this theorem. First, if the $\mathfrak{g}$-module A possesses a topological basis constituted by the eigenvectors of the transformation $a \mapsto g_0 a$, then, by letting $A_{(\lambda)} = \{a \in A \mid g_0 a = \lambda a\}$, we obtain a grading corresponding to the scheme in 3.7 of the complex $C^{\cdot}(\mathfrak{g}; A)$.

Similarly to Theorem 1.5.2, we can prove the following statement.

Theorem 1.5.2a.   The inclusion $C_{(0)}^{\cdot}(\mathfrak{g}; A) \to C^{\cdot}(\mathfrak{g}; A)$ induces an isomorphism in homology.

Second generalization: suppose $g_1, \ldots, g_r \in \mathfrak{g}$ are pairwise commuting elements such that $\mathfrak{g}$ possesses a topological
basis consisting of the vectors which are eigenvectors for
all the operators $g \mapsto [g_i, g]$.   Assume further that A is a
$\mathfrak{g}$-module possessing a topological basis consisting of the
vectors which are eigenvectors for all the operators $a \mapsto g_i a$.
Then the complex $C^{\cdot}(\mathfrak{g}; A)$ acquires the natural grading
$$\{C_{(\lambda_1,\ldots,\lambda_r)}^{\cdot}(\mathfrak{g}; A)\}.$$

Theorem 1.5.2b.   The inclusion $C_{(0\ldots0)}^{\cdot}(\mathfrak{g}; A) \to C^{\cdot}(\mathfrak{g}; A)$
induces an isomorphism in cohomology.

Finally, the homology analogs of Theorems 1.5.2, 1.5.2a,
and 1.5.2b can be stated and proved just as their cohomology
duals except, and this is an important difference, that the
algebra $\mathfrak{g}$ and the module A no longer need be assumed topological and must possess not a topological but a real basis,
constituted by eigenvectors of all the transformations $g \mapsto$
$[g_i, g]$ and $a \mapsto g_i a$.

3. The Laplace operator.   The following is a considerably simplified finite-dimensional analog of the Hodge—
de Rham theory.

Assume that the Lie algebra $\mathfrak{g}$ is finite-dimensional or
graded in such a way that all the spaces $C_{(\lambda)}^q(\mathfrak{g})$ are finite
dimensional (for example, this property is possessed by all
gradings of the form $\mathfrak{g} = \hat{\bigoplus}_{\lambda \in \mathbb{Z}^n} \mathfrak{g}_{(\lambda)}$ with finite-dimensional $\mathfrak{g}_{(\lambda)}$,
for which there exists an N, such that $\mathfrak{g}_{(k_1,\ldots,k_n)} = 0$ whenever

$\min(k_1, \ldots, k_n) < -N$). Assume further that in the space $\mathfrak{g}$ or in every space $\mathfrak{g}_{(\lambda)}$ a Hermitian or Euclidean metric has been chosen. Then all the spaces $C^q_{(\lambda)}(\mathfrak{g})$ acquire a metric and we can identify $C^q_{(\lambda)}(\mathfrak{g})$ with $(C^q_{(\lambda)}(\mathfrak{g}))'$, i.e., with $C^{(\lambda)}_q(\mathfrak{g})$. The operator $\partial \colon C^{(\lambda)}_q(\mathfrak{g}) \to C^{(\lambda)}_{q-1}(\mathfrak{g})$ is then transformed into the operator $C^q_{(\lambda)}(\mathfrak{g}) \to C^{q-1}_{(\lambda)}(\mathfrak{g})$, while the operator

$$\Delta = \Delta^q_{(\lambda)} = d \circ \partial + \partial \circ d \colon C^q_{(\lambda)}(\mathfrak{g}) \to C^q_{(\lambda)}(\mathfrak{g})$$

is called the <u>Laplace operator</u>. This is a self-adjoint operator which commutes with d and ∂. The formula

$$\langle c, \Delta c \rangle = \langle c, d\partial c + \partial dc \rangle = \langle \partial c, \partial c \rangle + \langle dc, dc \rangle,$$

where the angle brackets denote the inner product, show that the operator Δ is positive definite. It also shows that if $\Delta c = 0$, then $dc = 0$ and $\partial c = 0$. Cochains c satisfying $\Delta c = 0$ are called <u>harmonic</u>.

<u>Theorem 1.5.3</u>. Every element of the space $H^q_{(\lambda)}(\mathfrak{g})$ can be represented by a unique harmonic cocycle from $C^q_{(\lambda)}(\mathfrak{g})$. In other words, there is a natural isomorphism

$$\mathrm{Ker}\ \Delta^q_{(\lambda)} = H^q_{(\lambda)}(\mathfrak{g}).$$

<u>Proof</u>. Let

$$L_{(\mu)} C^q_{(\lambda)}(\mathfrak{g}) = \{c \in C^q_{(\lambda)}(\mathfrak{g}) \mid \Delta c = \mu c\}.$$

Obviously, $dL_{(\mu)} C^q_{(\lambda)}(\mathfrak{g}) \subset L_{(\mu)} C^{q+1}_{(\lambda)}(\mathfrak{g})$, so that the complex $C_{(\lambda)}(\mathfrak{g})$ acquires a supplementary grading $\{L_{(\mu)} C_{(\lambda)}(\mathfrak{g})\}$. We must prove that for $\mu \neq 0$ the complex $L_{(\mu)} C_{(\lambda)}(\mathfrak{g})$ is acyclic and for $\mu = 0$ the differential of this complex is trivial. But we already know the second assertion, while the first follows from the fact that the homomorphisms

$$\frac{1}{\mu} \partial \colon L_{(\mu)} C^q_{(\lambda)}(\mathfrak{g}) \to L_{(\mu)} C^{q-1}_{(\lambda)}(\mathfrak{g})$$

constitute (for $\mu \neq 0$) homotopy which joins the identity of the complex $L_{(\mu)} C_{(\lambda)}^{'}(\mathfrak{g})$ with the trivial map.

4. Relationship with induced and coinduced modules. This relationship is expressed by the following theorem.

Theorem 1.5.4. For any Lie algebra $\mathfrak{g}$, its subalgebra $\mathfrak{h}$, and $\mathfrak{h}$-module A there are natural isomorphisms (with respect to $\mathfrak{g}$, $\mathfrak{h}$ and A)

$$H^q(\mathfrak{g};\ \mathrm{Coind}_\mathfrak{g}A) = H^q(\mathfrak{h};\ A);$$
$$H_q(\mathfrak{g};\ \mathrm{Ind}_\mathfrak{g}A) = H_q(\mathfrak{h};\ A).$$

For example, if $\mathfrak{h} = 0$, then the second of these relations becomes the formula established in 3.4:

$$H_q(\mathfrak{g}; U(\mathfrak{g})) = \begin{cases} \mathbb{K} & \text{for } q = 0, \\ 0 & \text{for } q \neq 0. \end{cases}$$

The proof of Theorem 1.5.4 generalizes the corresponding proof from 3.4. Introduce a filtration into the complex $C(\mathfrak{g};\ \mathrm{Ind}_\mathfrak{g}A)$ by taking for $F^p C_u(\mathfrak{g};\ \mathrm{Ind}_\mathfrak{g}A)$ the subspace of the space $C_u(\mathfrak{g};\ \mathrm{Ind}_\mathfrak{g}A)$, generated by chains of the form

$$(a \otimes g_1' \ldots g_r') \otimes (g_1 \wedge \cdots \wedge g_s \wedge h_1 \wedge \cdots \wedge h_t)$$
$$\text{for } s + t = u,\ r + s \leqslant p,$$

where $a \in A$, $g_1', \ldots, g_r'$, $g_1, \ldots, g_s \in \mathfrak{g}$, $h_1, \ldots, h_t \in \mathfrak{h}$. Obviously, in the spectral sequence which then arises, we have $E_{p,q}^0 = \oplus_r E_{p,q}^0(r)$, where

$$E_{p,q}^0(r) = A \otimes S^{p-r}(\mathfrak{g}/\mathfrak{h}) \otimes \Lambda^r(\mathfrak{g}/\mathfrak{h}) \otimes \Lambda^{p+q-r}\mathfrak{h},$$

and the differential $d_{p,q}^0$ acts in accordance with the formula

$$d_{p,q}^0[a \otimes \bar{g}_1' \ldots \bar{g}_{p-r}' \otimes (\bar{g}_1 \wedge \cdots \wedge \bar{g}_r) \otimes (h_1 \wedge \cdots \wedge h_{p+q-r})]$$
$$= \sum_{i=1}^r (-1)^i a \otimes \bar{g}_1' \ldots \bar{g}_{p-r}' \bar{g}_i \otimes (\bar{g}_1 \wedge \cdots \hat{\bar{g}}_i \ldots \wedge \bar{g}_r) \otimes (h_1 \wedge \cdots \wedge h_{p+q-r})$$
$$(\bar{g}_j',\ \bar{g}_k \in \mathfrak{g}/\mathfrak{h},\ h_l \in \mathfrak{h}).$$

The formula

$$a \otimes \bar{g}_1' \ldots \bar{g}_{p-r}' \otimes (\bar{g}_1 \wedge \cdots \wedge \bar{g}_r) \otimes (h_1 \wedge \cdots \wedge h_{p+q-r})$$

$$\mapsto \sum_{j=1}^{p-r} a \otimes \bar{g}_1' \ldots \tilde{\bar{g}}_j \ldots \bar{g}_{p-r}' \otimes (\bar{g}_j' \wedge \bar{g}_1 \wedge \cdots \wedge \bar{g}_r) \otimes (h_1 \wedge \cdots \wedge h_{p+q-r})$$

determines a homotopy in the complex E_p^0, which joins 0 with p id. Thus,

$$E_{p,q}^1 = 0 = E_{p,q}^\infty \quad \text{for} \quad p \neq 0.$$

At the same time, it is obvious that $E_{0,q}^0 = E_{0,q}^0(0) = A \otimes \Lambda^q \mathfrak{h} = C_q(\mathfrak{h}; A)$, $d_{0,q}^0 = 0$, $d_{0,q}^1 = [\partial: C_q(\mathfrak{h}; A) \to C_{q-1}(\mathfrak{h}; A)]$ so that

$$E_{0,q}^2 = H_q(\mathfrak{h}; A) = E_{0,q}^\infty.$$

The homology part of the theorem is proved; the proof of its cohomology part is similar.

§6. LIE SUPERALGEBRAS

1. Main definitions. Almost all of the contents of this chapter remain valid when we pass on to the so-called "supercase." This passage consists in assuming that all the objects under consideration possess a $\mathbb{Z}_2$-grading, in writing minus one to the appropriate power in the appropriate places of all definitions, and adding the prefix "super" to all the terms. Thus, a superspace is simply a vector space with a $\mathbb{Z}_2$-grading: $V = V_0 \oplus V_1$. (Usually elements of the space V_0 are called even, and elements of the space V_1, odd; the indices 0 and 1 are sometimes viewed as integers and sometimes as residues modulo 2.) A Lie superalgebra is a superspace $\mathfrak{g} = \mathfrak{g}_0 \oplus \mathfrak{g}_1$, with a commutation operation (or "supercommutation") $[,]$, which satisfies the conditions

$$[g_1, g_2] = -(-1)^{p_1 p_2}[g_2, g_1],$$

$$(-1)^{p_1 p_3}[[g_1, g_2], g_3] + (-1)^{p_2 p_3}[[g_2, g_3], g_1] + (-1)^{p_3 p_1}[[g_3, g_1], g_2] = 0$$

$$(g_1 \in \mathfrak{g}_{p_1}, \ \ g_2 \in \mathfrak{g}_{p_2}, \ \ g_3 \in \mathfrak{g}_{p_3})$$

[compare with formulas (3), (4) from 3.2]. Thus, $\mathfrak{g}_0$ is a Lie algebra and $\mathfrak{g}_1$ is a module over $\mathfrak{g}_0$; the Lie superalgebra structure also contains the symmetric pairing $S^2 \mathfrak{g}_1 \rightarrow \mathfrak{g}_0$, which is a $\mathfrak{g}_0$-homomorphism and satisfies the Jacobi identity applied to three elements of the space $\mathfrak{g}_1$.

A <u>subsuperalgebra</u> of the Lie algebra $\mathfrak{g}$ is a subspace $\mathfrak{h} \subset \mathfrak{g}$, closed with respect to commutation if it is also a "subsuperspace," i.e., satisfies the relation $\mathfrak{h} = \mathfrak{h}_0 \oplus \mathfrak{h}_1$, where $\mathfrak{h}_0 = \mathfrak{h} \cap \mathfrak{g}_0$, $\mathfrak{h}_1 = \mathfrak{h} \cap \mathfrak{g}_1$.

The superspace $A = A_0 \oplus A_1$ is said to be a <u>module over the Lie superalgebra</u> $\mathfrak{g} = \mathfrak{g}_0 \oplus \mathfrak{g}_1$, if it is supplied with a bilinear map $\mathfrak{g} \times A \rightarrow A$, $(g, a) \mapsto ga$, such that $\mathfrak{g}_p A_q \subset A_{p+q}$ and

$$[g_1, g_2] a = g_1 (g_2 a) - (-1)^{p_1 p_2} g_2 (g_1 a)$$

for $g_i \in \mathfrak{g}_{p_i}$, $a \in A$. As examples, let us mention the trivial module $(ga = 0)$ and the adjoint representation $(A = \mathfrak{g}, ga = [g, a])$.

2. <u>Principal examples of Lie superalgebras</u>. The first example is the Lie superalgebra $\mathfrak{gl}(m, n)$ of endomorphisms of (m, n)-dimensional superspace $\mathbb{K}^{m, n} = V_0 \oplus V_1$, where $V_0 = \mathbb{K}^m$, $V_1 = \mathbb{K}^n$. In $\mathfrak{gl}(m, n)$ there is a natural grading

$$\mathfrak{gl}(m, n)_0 = \mathrm{Hom}(V_0, V_0) \oplus \mathrm{Hom}(V_1, V_1),$$
$$\mathfrak{gl}(m, n)_1 = \mathrm{Hom}(V_0, V_1) \oplus \mathrm{Hom}(V_1, V_0),$$

while the supercommutator is defined by the formula

$$[g, h] = g \circ h - (-1)^{pq} h \circ g \quad \text{for} \quad g \in \mathfrak{gl}(m, n)_p, \ h \in \mathfrak{gl}(m, n)_q.$$

In other words, $\mathfrak{gl}(m, n)_0 = \mathfrak{gl}(m, \mathbb{K}) \oplus \mathfrak{gl}(n, \mathbb{K})$, the $\mathfrak{gl}(m, n)_0$-module structure in $\mathfrak{gl}(m, n)_1 = \mathrm{Hom}(V_0, V_1) \oplus \mathrm{Hom}(V_1, V_0)$ is standard, while the supercommutator $S^2 \mathfrak{gl}(m, n)_1 \rightarrow \mathfrak{gl}(m, n)_0$

acts according to the formula

$$[(\varphi_1, \ \psi_1), \ (\varphi_2, \ \psi_2)] = (\psi_2 \circ \varphi_1 + \psi_1 \circ \varphi_2, \ \varphi_2 \circ \psi_1 + \varphi_1 \circ \psi_2),$$

where $\varphi_1, \ \varphi_2 \in \mathrm{Hom} \ (V_0, \ V_1), \ \psi_1, \ \psi_2 \in \mathrm{Hom} \ (V_1, \ V_0)$.

The <u>supertrace</u> $\mathrm{Str} \ (\alpha)$ of the element $\alpha = (\gamma, \delta, \varphi, \psi)$ of the superalgebra $\mathfrak{gl} \ (m, \ n) = \mathrm{Hom} \ (V_0, \ V_0) \oplus \mathrm{Hom} \ (V_1, \ V_1) \oplus$ $\mathrm{Hom} \ (V_0, \ V_1) \oplus \mathrm{Hom} \ (V_1, \ V_0)$ is defined by the formula $\mathrm{Str} \ (\alpha) = \mathrm{Tr} \ \gamma - \mathrm{Tr} \ \delta$. The superalgebra

$$\mathfrak{sl} \ (m, n) = [\mathfrak{gl} \ (m, n), \mathfrak{gl} \ (m, n)] = \mathrm{Ker} \ \mathrm{Str}$$

is simple for $m \neq n$, and for $m = n$ possesses a one-dimensional center which is generated by the identity endomorphism .id $\mathbb{K}^{\hat{m}, n}$.

Denote by ω the bilinear form on $\mathbb{R}^{m, \, 2n}$, defined by the formula

$$\omega ((x_1, \ldots, x_{m+2n}), (y_1, \ldots, y_{m+2n})) = \sum_{i=1}^{m} x_i y_i + \sum_{j=1}^{n} \begin{vmatrix} x_{m+j} & x_{m+n+j} \\ y_{m+j} & y_{m+n+j} \end{vmatrix},$$

and by $\mathfrak{osp} \ (m, n)$ the subspace of the space $\mathfrak{gl} \ (m, 2n)$, consisting of the endomorphisms which annihilate ω. From the relation $\omega (V_0, V_1) = 0$. it is obviously follows that $\mathfrak{osp} \ (m, n)$ is a subsuperalgebra of the Lie superalgebra $\mathfrak{gl} \ (m, 2n)$; it is also clear that

$$\mathfrak{osp} \ (m, n)_0 = \mathfrak{o} \ (m) + \mathfrak{sp} \ (n, \mathbb{R}),$$

$$\mathfrak{osp} \ (m, n)_1 = V_0 \otimes V_1,$$

where the Lie algebra structure in $\mathfrak{osp} \ (m, \ n)_0$ and the action of $\mathfrak{osp} \ (m, \ n)_0$ in $\mathfrak{osp} \ (m, \ n)_1$ are standard ($V_{0,}$ and V_1 possess the Euclidean and symplectic structures defined by the restrictions $\omega \, | \, V_0$ and $\omega \, | \, V_1$), while the supercommutator $S^2 \ (V_0 \otimes V_1) \to \mathfrak{o} \ (m) \oplus \mathfrak{sp} \ (n, \ \mathbb{R})$ acts in accordance with the formula

$$(a_0 \otimes a_1, b_0 \otimes b_1) \mapsto (\omega (a_1, b_1) a_0 \wedge b_0, \quad \omega (a_0, b_0) a_1 b_1)$$
$$\in \Lambda^2 V_0 \oplus S^2 V_1 = \mathfrak{o} (m) \oplus \mathfrak{sp} (n, \mathbb{R})$$

(we identify $\Lambda^2 V_0$ and $S^2 V_1$ with $\mathfrak{o} (m) \subset \text{End } V_0$ and $\mathfrak{sp} (n, \mathbb{R}) \subset \text{End } V_1$ in the usual way, i.e., according to the formulas $a \wedge b \mapsto \{c \mapsto \omega (a, c) b - \omega (b, c) a\}$, $ab \mapsto \{c \mapsto \omega (a, c) b + \omega (b, c) a\}$, in the first of which $a, b, c \in V_0$, while in the second $a, b, c \in V_1$).

The next example: the superalgebra $P (n)$ is defined by the formulas

$$P (n)_0 = \mathfrak{gl} (n, \mathbb{K}),$$
$$P (n)_1 = \Lambda^2 V \oplus S^2 V', \text{ where } V = \mathbb{K}^n,$$

the action $\mathfrak{gl} (n, \mathbb{K})$ in $\Lambda^2 V \oplus S^2 V'$ is standard, the supercommutator $S^2 (\Lambda^2 V \oplus S^2 V') \to V \oplus V' = \mathfrak{gl} (n, \mathbb{K})$ is defined by the formula

$$(a_{11} \wedge a_{12}, b_{11} b_{12}), (a_{21} \wedge a_{22}, b_{21} b_{22}) \mapsto \sum_{i, j, k=1}^{2} (- 1)^k b_{ij} (a_{i'k}) a_{i'k'} \otimes b_{ij'},$$

where $a_{st} \in V$, $b_{st} \in V'$, and i', j', k' denote numbers which are equal to 1 or 2 and are not equal, respectively, to i, j, k.

Finally, the Lie superalgebra $Q (n)$ is defined in a very simple way:

$$Q (n)_0 = Q (n)_1 = \mathfrak{gl} (n, \mathbb{K}),$$

the commutator in $Q (n)_0$, and the action of $Q (n)_0$ in $Q (n)_1$ are standard, the supercommutator $S^2 Q (n)_1 \to Q (n)_0$ is defined by the formula

$$(a, b) \mapsto ab + ba.$$

The Lie superalgebras [$\mathfrak{gl} (m, n)$, $\mathfrak{sl} (m, n)$, $\mathfrak{osp} (m, 2n)$, $P (n)$, $Q (n)$], listed above, are more or less similar to finite-dimensional simple Lie algebras and they are sometimes called classical. The following Lie superalgebras are related to the infinite-dimensional Lie algebras considered in 1.3 (although some of them are actually finite dimensional). Suppose $x_1, \ldots, x_m; y_1, \ldots, y_n$ are coordinates in the (m, n)-dimensional superspace $\mathbb{K}^{m,n} = V_0 \oplus V_1$. Let

$$F^{m,\,n} = \hat{S}*V_0' \otimes \Lambda*V_1' = \mathbb{K}\,[[x_1, \ldots, x_m]] \otimes \Lambda\,(y_1, \ldots, y_m),$$
$$F_0^{m,\,n} = \hat{S}*V_0' \otimes [\bigoplus_{q \equiv 0 \bmod 2} \Lambda^q V_1'], \quad F_1^{m,\,n} = \hat{S}*V_0' \otimes [\bigoplus_{q \equiv 1 \bmod 2} \Lambda^q V_1']$$

and consider the set of all continuous "superderivations" of the "superalgebra" $F^{m,n}$, i.e., of continuous endomorphisms $\varphi: F^{m,n} \to F^{m,n}$, such that

$$\varphi\,(xy) = \varphi\,(x)\, y + (-1)^p x \varphi\,(y)$$

for $x \in F_p^{m,n}$, $y \in F^{m,n}$. The superderivation φ is said to be <u>even</u> or <u>odd</u> if $\varphi\,(F_p^{m,n}) \subset F_p^{m,n}$ or $\varphi\,(F_0^{m,n}) \subset F_1^{m,n}$, $\varphi\,(F_1^{m,n}) \subset F_0^{m,n}$. Denote by $W\,(m,\,n)$, $W\,(m,\,n)_0$, $W\,(m,\,n)_1$, respectively, the space of all superderivations, the space of even superderivations, and the space of odd superderivations of the space $F^{m,n}$, and define the supercommutator of superderivations $\alpha \in W\,(m,\,n)_p$, $\beta \in W\,(m,\,n)_q$ by the formula

$$[\alpha, \beta] = \alpha \circ \beta - (-1)^{pq} \beta \circ \alpha.$$

It is obvious that we obtain a Lie superalgebra in this way. For $m = 0$ it is finite dimensional, while for $m > 0$ it is infinite dimensional; for $n = 0$ it becomes the Lie algebra W_m. Its elements may be written "in coordinates" in the form

$$\sum_{i=1}^{m} f_i\, \frac{\partial}{\partial x_i} + \sum_{j=1}^{n} g_j\, \frac{\partial}{\partial y_j}, \qquad (1)$$

where f_i, g_j are power series in $x_1, \ldots, x_m$, whose coefficients are elements of the Grassmann algebra in $y_1, \ldots, y_n$. When taking the commutator of such "vector fields," one should keep in mind that signs should be changed in changing places of the y_i with each other or with $\partial/\partial y_j$ (for example, $\frac{\partial}{\partial y_2}(y_1 \wedge y_2) = -y_1$).

If the field (1) belongs to $W(m, n)_p$, then its divergence is, by definition, the expression $\sum (\partial f_i/\partial x_i) - (-1)^p \sum (\partial g_j/\partial y_j)$; by linearity the definition of divergence can be carried over to the entire space $W(m, n)$. Fields with trivial divergence constitute a subsuperalgebra of the Lie superalgebra $W(m, n)$, which is denoted by $S(m, n)$.

Vector fields of the form

$$\sum_{i=1}^{k} \left(\frac{\partial f}{\partial x_{i+k}} \frac{\partial}{\partial x_i} - \frac{\partial f}{\partial x_i} \frac{\partial}{\partial x_{i+k}} \right) + (-1)^p \sum_{j=1}^{k} \frac{\partial f}{\partial y_j} \frac{\partial}{\partial y_j},$$

where $f \in F_p^{2k, n}$, constitute a subspace $H(k, n)_p$ of the space $W(2k, n)_p$, and the sum $H(k, n) = H(k, n)_0 \oplus H(k, n)_1$ is a subsuperalgebra of the Lie superalgebra $W(2k, n)$. There exists a definition of this superalgebra similar to the definition of the Lie algebra H_k given in 1.3. There exist also super-analogs of contact vector field algebras. For details see [1].

In the following Subsections 3-5, we will be concerned with the cohomology of Lie superalgebras. To make our exposition more succinct, we shall not mention the homology of Lie superalgebras. The corresponding theory is similar to the cohomology one, and the reader will have no difficulty in recovering the omitted definitions.

3. Definition of cohomology. By definition, the super-
space of q-dimensional cocycles of the Lie superalgebra $\mathfrak{g} =
\mathfrak{g}_0 \oplus \mathfrak{g}_1$ with coefficients in the $\mathfrak{g}$-module $A = A_0 \oplus A_1$ is
given by the formulas

$$C^q(\mathfrak{g}; A) = \bigoplus_{q_0 + q_1 = q} \text{Hom}(\Lambda^{q_0}\mathfrak{g}_0 \otimes S^{q_1}\mathfrak{g}_1, A),$$

$$C_p^q(\mathfrak{g}; A) = \bigoplus_{\substack{q_0 + q_1 = q, \\ q_1 + r \equiv p \bmod 2}} \text{Hom}(\Lambda^{q_0}\mathfrak{g}_0 \otimes S^{q_1}\mathfrak{g}_1, A_r).$$

The differential $d: C^q(\mathfrak{g}; A) \to C^{q+1}(\mathfrak{g}; A)$ is defined by the
formula

$$dc(g_1, \ldots, g_{q_0}, h_1, \ldots, h_{q_1})$$

$$= \sum_{1 \leqslant s < t \leqslant q_0} (-1)^{s+t-1} c([g_s, g_t], g_1, \ldots \hat{g}_s \ldots \hat{g}_t \ldots, g_{q_0}, h_1, \ldots, h_{q_1})$$

$$+ \sum_{s=1}^{q_0} \sum_{t=1}^{q_1} (-1)^{s-1} c(g_1, \ldots \hat{g}_s \ldots, g_{q_0}, [g_s, h_t], h_1, \ldots \hat{h}_t \ldots, h_{q_1})$$

$$+ \sum_{1 \leqslant s < t \leqslant q_1} c([h_s, h_t], g_1, \ldots, g_{q_0}, h_1, \ldots \hat{h}_s \ldots \hat{h}_t \ldots, h_{q_1})$$

$$+ \sum_{s=1}^{q_0} (-1)^s g_s c(g_1, \ldots \hat{g}_s \ldots, g_{q_0}, h_1, \ldots, h_{q_1})$$

$$+ (-1)^{q_0-1} \sum_{s=1}^{q_1} h_s c(g_1, \ldots, g_{q_0}, h_1, \ldots \hat{h}_s \ldots, h_{q_1}),$$

where $c \in C^q(\mathfrak{g}; A)$, $g_1, \ldots, g_{q_0} \in \mathfrak{g}_0$, $h_1, \ldots, h_{q_1} \in \mathfrak{g}_1$. Obviously, $d \circ d = 0$ and $d(C_p^q(\mathfrak{g}; A)) \subset C_p^{q+1}(\mathfrak{g}; A)$, so that we obtain
the cohomology $H_p^q(\mathfrak{g}; A)$ ($q = 0, 1, 2, \ldots; p = 0, 1$).

The contents of 3.2 and 3.4 may be carried over to the
supercase without any difficulty. In particular, we have
bilinear multiplications

$$H_{p_1}^{q_1}(\mathfrak{g}) \times H_{p_2}^{q_2}(\mathfrak{g}) \to H_{p_1+p_2}^{q_1+q_2}(\mathfrak{g}),$$

$$H_{p_1}^{q_1}(\mathfrak{g}; \mathfrak{g}) \times H_{p_2}^{q_2}(\mathfrak{g}; \mathfrak{g}) \to H_{p_1+p_2}^{q_1+q_2-1}(\mathfrak{g}; \mathfrak{g})$$

(the corresponding products are denoted by $\alpha\beta$ and $[\alpha, \beta]$),
satisfying the conditions

$$\beta\alpha = (-1)^{(q_1+p_1)(q_2+p_2)}\alpha\beta, \quad \alpha\,(\beta\gamma) = (\alpha\beta)\,\gamma;$$

$$[\eta, \ \xi] = -(-1)^{(q_1+p_1-1)(q_2+p_2-1)}\,[\xi, \ \eta],$$

$$(-1)^{(q_1+p_1-1)(q_3+p_3-1)}\,[[\xi, \ \eta], \ \zeta] + (-1)^{(q_2+p_2-1)(q_3+p_3-1)}\,[[\eta, \ \zeta], \ \xi]$$

$$+ (-1)^{(q_3+p_3-1)(q_1+p_1-1)}\,[[\zeta, \ \xi], \ \eta] = 0,$$

where $\alpha \in H_{p_1}^{q_1}(\mathfrak{g})$, $\beta \in H_{p_2}^{q_2}(\mathfrak{g})$, $\gamma \in H_{p_3}^{q_3}(\mathfrak{g})$, $\xi \in H_{p_1}^{q_1}(\mathfrak{g}; \ \mathfrak{g})$, $\eta \in H_{p_2}^{q_2}(\mathfrak{g}; \ \mathfrak{g})$, $\zeta \in H_{p_3}^{q_3}(\mathfrak{g}; \ \mathfrak{g})$; the corresponding Massey products are also defined.

4. Algebraic interpretation of cohomology. The principal change which must be made in §4 when passing to the supercase is that only homogeneous, i.e., even or odd, cohomology classes acquire an interpretation. For example, elements of the space $H_p^2(\mathfrak{g})$ are interpreted as classes of central extensions

$$0 \to \mathbb{K}\,(p) \to \bar{\mathfrak{g}} \to \mathfrak{g} \to 0,$$

where $\mathbb{K}\,(p)$ is a one-dimensional Lie superalgebra with trivial supercommutator and with the superspace structure described by the formula $\mathbb{K}\,(p)_\nu \cong \mathbb{K}$.

Elements of the space $H_0^2(\mathfrak{g}; \ \mathfrak{g})$ are interpreted as infinitesimal deformations of the superalgebra $\mathfrak{g}$, and the relationship between infinitesimal and real deformations is described exactly as the corresponding relationship in the case of Lie algebras (the supercommutator is deformed, not the superspace structure). Elements of the space $H_1^2(\mathfrak{g}; \ \mathfrak{g})$ are naturally called "odd infinitesimal deformations"; note that "odd real deformations" do not exist, while the squares and higher Massey powers of elements of the space $H_1^2(\mathfrak{g}; \ \mathfrak{g})$ are all trivial.

Elements of the space $H_0^1(\mathfrak{g}; \ \mathfrak{g})$ are interpreted as classes of one-dimensional right extensions

$$0 \to \mathfrak{g} \to \tilde{\mathfrak{g}} \to \mathbb{K}(0) \to 0;$$

as for one-dimensional right extensions of the form

$$0 \to \mathfrak{g} \to \tilde{\mathfrak{g}} \to \mathbb{K}(1) \to 0, \tag{2}$$

their relationship to cohomology is more involved. In this case, we may assume that $\tilde{\mathfrak{g}}_0 = \mathfrak{g}_0$, $\tilde{\mathfrak{g}}_1 = \mathfrak{g}_1 \oplus \mathbb{K}$, and, in order to complete the definition of the supercommutator in $\tilde{\mathfrak{g}}$, we must define the operations $[c, \ ]: \mathfrak{q}_1 \to \mathfrak{g}_0$ and $[c, \ ]: \mathfrak{g}_0 \to \mathfrak{g}_1$ and the element $[c, c] \in \tilde{\mathfrak{g}}_0$, where c denotes the generator of the summand $\mathbb{K}$ in $\mathfrak{g}_1$. In other words, it is necessary to choose cochains $\varphi \in C_1^1(\mathfrak{g}; \mathfrak{g})$ and $\gamma \in \mathfrak{g}_0 = C_0^0(\mathfrak{g}; \mathfrak{g})$. The verification of the Jacobi identity must be carried out in four cases: when zero, one, two, or three of the three given elements of the superalgebra coincide with c, while the others are contained in $\mathfrak{g}$. In the first case, the identity reduces to the Jacobi identity for $\mathfrak{g}$, and in the other three cases it gives the relations

$$d\varphi = 0, \quad d\gamma = 2\varphi \circ \varphi, \quad \varphi(\gamma) = 0. \tag{3}$$

It is easy to see, further, that the pairs (φ, γ), (φ', γ') define equivalent extensions if and only if there exists a $\beta \in \mathfrak{g}_1 = C_1^0(\mathfrak{g}; \mathfrak{g})$ such that

$$\varphi' = \varphi + d\beta, \quad \gamma' = \gamma + \varphi(\beta) + [\beta, \beta]. \tag{4}$$

The first of the relations (3) shows that φ is a cocycle, i.e., defines a certain class $\lambda \in H_1^1(\mathfrak{g}; \mathfrak{g})$. The second of these relations shows that the square $[\lambda, \lambda] \in H_0^1(\mathfrak{g}; \mathfrak{g})$ is trivial, the third that the corresponding Massey cube [with values in $H_1^0(\mathfrak{g}; \mathfrak{g}) = \mathfrak{g}_1$] vanishes. Note that subsequent Massey powers of class λ vanish because of dimension. Finally, the first of the relations (4) shows that in order to de-

fine the class of extensions we need only consider the cohomology class of the cocycle φ, while the second of the relations (3) shows that γ is chosen from φ uniquely, up to any
summand which is a cocycle belonging to $C_0^0(\mathfrak{g}; \mathfrak{g})$, i.e., an
element of $\mathfrak{g}_0$, belonging to the center of the superalgebra $\mathfrak{g}$.
Thus, if the superalgebra $\mathfrak{g}$ has no center, or if its center
does not contain any even element, then the classes of extensions of the form (2) correspond bijectively to elements of
the space $H_1^1(\mathfrak{g}; \mathfrak{g})$, whose Massey square and cube vanish. [If
$\mathfrak{g}$ has a center, an element of the space $H_1^1(\mathfrak{g}; \mathfrak{g}')$ may correspond to several right extensions; for example, if $\mathfrak{g} = \mathbb{K}(0)$,
then $H_1^1(\mathfrak{g}; \mathfrak{g}) = 0$, but there is a nontrivial right extension

$$0 \to \mathbb{K}(0) \to \tilde{\mathfrak{g}} \to \mathbb{K}(1) \to 0:$$

$\tilde{\mathfrak{g}}_0 = \mathbb{K}$, $\tilde{\mathfrak{g}}_1 = \mathbb{K}$, and if b, c are the generators of $\tilde{\mathfrak{g}}_0$, $\tilde{\mathfrak{g}}_1$,
then $[b, c] = 0$, $[c, c] = b$.]

5.   The Serre—Hochschild spectral sequence.   The construction of the Serre—Hochschild spectral sequence and the
list of main properties in the supercase is a repetition,
with obvious alterations, of 5.1.   It should be noted that
the entire spectral sequence turns out to be $\mathbb{Z}_2$-graded,
i.e., is the sum of two spectral sequences which converge
respectively to even and odd cohomology. We shall dwell in
more detail on the most useful spectral sequence, correspond-
ing to the pair $(\mathfrak{g} = \mathfrak{g}_0 \oplus \mathfrak{g}_1, \mathfrak{g}_0)$. If for the coefficients
we take the trivial module $\mathbb{K} = \mathbb{K}^{1,0}$, then for this spectral
sequence we have

$$E_1^{p,q} = H^q(\mathfrak{g}_0; S^p \mathfrak{g}_1'),$$

which implies that the space $E_1^{p,q}$ consists of even elements
for even q and odd elements for odd q. Since the differen-

tials of the spectral sequence are homogeneous with respect
to the $\mathbb{Z}_2$-grading, it follows that the differentials with
odd numbers are trivial and the decomposition of the spectral
sequence into the sum of even and odd parts described above
reduces to the following:

$$E_r^{p,q} = [\bigoplus_{q \equiv 0 \bmod 2} E_r^{p,q}] \oplus [\bigoplus_{q \equiv 1 \bmod 2} E_r^{p,q}].$$

In the general case, when for the coefficients we take the
module $A = A_0 \oplus A_1$, we have

$$E_1^{p,q} = H^q (\mathfrak{g}_0; \text{Hom} (S^p \mathfrak{g}_1, A)),$$

and a very simple calculation shows that the differential
$d_1^{p,q}$: $E_1^{p,q} \rightarrow E_1^{p+1,q}$ is induced by the map

$$\text{Hom} (S^p \mathfrak{g}_1, A) \rightarrow \text{Hom} (S^{p+1} \mathfrak{g}_1, A),$$

$$\varphi \mapsto \left\{ (g_1, \ldots, g_{p+1}) \mapsto \sum_{i=1}^{p+1} g_i \varphi (g_1, \ldots \hat{g}_i \ldots, g_{p+1}) \right\}, \tag{5}$$

which, as can be easily checked, is a $\mathfrak{g}_0$-homomorphism.

Remark. This description of the differential d_1 deserves two comments. First of all, one should not be surprised that we succeeded in advancing in the computation of
the initial part of the Serre–Hochschild spectral sequence
for Lie superalgebras further than we did in the similar problem
for Lie algebras (see 5.1). A similar description may
be given also for the first differential of the spectral sequence corresponding to the pair $(\mathfrak{g}, \mathfrak{h})$ of Lie algebras
under the condition that $\mathfrak{g}$ possesses a subspace L, such that
$\mathfrak{g} = \mathfrak{h} \oplus L$, $[\mathfrak{h}, L] \subset L$, $[L, L] \subset \mathfrak{h}$. Secondly, the triviality of
the composition $d_1^{p+1,q} \circ d_1^{p,q}$: $E_1^{p,q} \rightarrow E_1^{p+2,q}$ is not clear directly
from the description; moreover, the composition $\text{Hom} (S^p \mathfrak{g}_1,
A) \rightarrow \text{Hom} (S^{p+2} \mathfrak{g}_1, A)$ of two homomorphisms of the form (5) is
not trivial in general. As an exercise for the reader, we

leave the verification, carried out by means of a direct calculation, that, in view of our description of d_1, the composition $d_1^{p+1,q} \circ d_1^{p,q}$ is actually trivial; this calculation
will remind him of the triviality proof of the action of Lie
algebras in their cohomology (see 3.6C).

Chapter 2. Computations

§1. COMPUTATIONS FOR FINITE-DIMENSIONAL
     LIE ALGEBRAS

This section is mainly devoted to the homology and co-homology of the algebra $\mathfrak{gl}\,(n,\;\mathbb{K})$. The case of other finite-dimensional algebras is briefly discussed in the last subsection.

1. The ring $H^*\,(\mathfrak{gl}\,(n,\;\mathbb{K}))$. We shall begin with the proof of a fact which was already mentioned in 1.3.3.

Lemma. If G is a connected compact lie group and $\mathfrak{g}$ is its Lie algebra, then the canonical inclusion $C^{\cdot}\,(\mathfrak{g}) \to \Omega^{\cdot}\,(G)$ (see 1.3.3) induces an isomorphism in cohomology. Thus, in this case, $H^*\,(\mathfrak{g}) = H^*\,(G;\,\mathbb{K})$.†

Proof. According to 1.3.3, our statement is equivalent to the homology equivalence of the complex $\Omega^{\cdot}G$ with its subcomplex $\Omega^{\cdot}_{\mathrm{inv}}(G)$ consisting of right-invariant forms. De-

---

†Generalization: if G is a compact connected Lie group, $\mathfrak{g}$ its Lie algebra, H is closed subgroup, and $H_0, \mathfrak{h}$ the component of unity and the Lie algebra of the group H , then $H^*\,(\mathfrak{g},\ \mathfrak{h}) = H^*\,(G/H_0;\ \mathbb{K})$, $H^*\,(\mathfrak{g},\ H) = H^*\,(G/H;\ \mathbb{K})$.

note by n the dimension of the group G and fix a bilateral-
ly invariant volume form ω on G [existence follows from the
fact that the algebra $\mathfrak{g}$ is unitary (see the definition in
1.3.6B and the proof in [94]): the unitarity guarantees the
$\mathfrak{g}$-invariance of the generator c of the space $C^n(\mathfrak{g})$, which
in turn implies the left G-invariance of the form obtained
by spreading c over G by right translations; the right G-
invariance of this form is obvious]; let us normalize ω by
the condition $\int_G \omega = 0$. For the form $\eta \in \Omega^q(G)$ denote by $\bar\eta$
the form obtained from η by integrating over ω, i.e., the
form

$$e_1, \ldots, e_q \mapsto \int_G \eta(e_1 g, \ldots, e_q g)\, \omega$$

[$e_1, \ldots, e_q$ are tangent vectors to G, with the same initial
point]. It is clear that $\bar\eta \in \Omega^q_{\text{inv}}(G)$, that $\bar\eta = \eta$ for $\eta \in$
$\Omega^q_{\text{inv}}(G)$, and that the map $\eta \to \bar\eta$ commutes with the differential. It remains to show that this map induces the identity
in $H^*(G;\ \mathbb{K})$. To do this, note that the passage from η to $\bar\eta$
may be represented as the composition of three operations:
the map $\mu^*\colon \Omega^q(G) \to \Omega^q(G \times G)$, where μ is the multiplication
in G, multiplication by the form $p_2^*\omega$, where p_2 is the projection of $G \times G$ on the second factor, and integration
along the fibers of the projection p_1 on the first factor.
Obviously, the composition of second and third maps induces
a homomorphism in cohomology which equals $i_1^*\colon H^q(G \times G;\ \mathbb{K}) \to$
$H^q(G;\ \mathbb{K})$, where i_1 is defined by the formula $i_1(g) = (g,\ e)$,
so that the composition of all three maps induces the coho-
mology homomorphism $i_1^* \circ \mu^* = (\mu \circ i_1)^* = \text{id}^* = \text{id}$.

Theorem 2.1.1. The ring $H^*(\mathfrak{gl}(n,\ \mathbb{K}))$ is the exterior
algebra in n generators of degrees $1, 3, 5, \ldots, 2n - 1$. The

homomorphism $H^q (\mathfrak{gl}\, (n,\ \mathbb{K})) \to H^q (\mathfrak{gl}\, (n-1, \mathbb{K}))$, induced by the inclusion $\mathfrak{gl}\, (n-1,\ \mathbb{K}) \to \mathfrak{gl}\, (n,\ \mathbb{K})$, is an isomorphism for $q < n$
and has a one-dimensional kernel for $q = n$.

Proof. The relation $\mathbb{C}\mathfrak{u}\, (n) = \mathbb{C}\mathfrak{gl}\,(n, \mathbb{R})$ (see 1.1.1) shows
that $H^* (\mathfrak{gl}\, (n, \mathbb{R})) \cong H^* (\mathfrak{u}\, (n))$, while the lemma proved above
shows that $H^*(\mathfrak{u}\, (n)) \cong H^*\, (U(n); \mathbb{R})$. The isomorphism $H^* (\mathfrak{gl}\,(n,$
$\mathbb{R})) \cong H^* (U\, (n); \mathbb{R})$ thus obtained is obviously compatible with
the standard inclusions $\mathfrak{gl}\, (n-1, \mathbb{R}) \to \mathfrak{gl}\, (n, \mathbb{R})$ and $U\, (n-$
$1) \to U\, (n)$. This proves the theorem in the case $\mathbb{K} = \mathbb{R}$, while
the case $\mathbb{K} = \mathbb{C}$ reduces to the previous one in view of the
relation $\mathbb{C}\mathfrak{gl}\, (n, \mathbb{R}) = \mathfrak{gl}\, (n, \mathbb{C})$.

This proof of Theorem 2.1.1 allows us to make the following specification in its statement.

Theorem 2.1.1'. The homomorphism

$$C^{\cdot} (\mathfrak{gl}\, (n, \mathbb{C})) \to \Omega^{\cdot} (\mathrm{GL}\, (n, \mathbb{C}); \mathbb{C}),$$

under which to each cochain corresponds a complex-valued
right-invariant differential form, induces an isomorphism

$$H^* (\mathfrak{gl}\, (n, \mathbb{C})) \xrightarrow{\cong} H^* (\mathrm{GL}\, (n, \mathbb{C}); \mathbb{C}).$$

Further (in Subsection 4), we shall explicitly indicate
the generators of the ring $H^* (\mathfrak{gl}\, (n,\ \mathbb{K}))$.

2. The case of nontrivial coefficients. The symbol e_j^i
further denotes the matrix whose entries are all zero, except the one standing on the intersection of the i-th line
and the j-th column, which equals one. By V denote the
space $\mathbb{K}^n$ with the standard $\mathfrak{gl}\, (n, \mathbb{K})$-module structure. The
tensor product

$$\underbrace{V' \otimes \ldots \otimes V'}_{k} \otimes \underbrace{V \otimes \ldots \otimes V}_{l}$$

is denoted by $T_l^k(n)$ or T_l^k. The elements of the standard basis in T_l^k are denoted by $e_{j_1 \ldots j_l}^{i_1 \ldots i_k}$. Thus,

$$e_j^i e_{j_1 \ldots j_l}^{i_1 \ldots i_k} = \sum_{i_s=j} e_{j_1 \ldots j_l}^{i_1 \ldots i_{s-1} i i_{s+1} \ldots i_k} - \sum_{j_t=i} e_{j_1 \ldots j_{t-1} j j_{t+1} \ldots j_l}^{i_1 \ldots i_k}.$$

Theorem 2.1.2. For any tensor $\mathfrak{gl}(n, \mathbb{K})$-module A the inclusion $\mathrm{Inv}\, A \to A$ induces isomorphisms

$$H_q(\mathfrak{gl}(n, \mathbb{K}); \mathrm{Inv}\, A) \to H_q(\mathfrak{gl}(n, \mathbb{K}); A),$$
$$H^q(\mathfrak{gl}(n, \mathbb{K}); \mathrm{Inv}\, A) \to H^q(\mathfrak{gl}(n, \mathbb{K}); A).$$

Thus,

$$H_q(\mathfrak{gl}(n, \mathbb{K}); A) = \mathrm{Inv}\, A \otimes H_q(\mathfrak{gl}(n, \mathbb{K})),$$
$$H^q(\mathfrak{gl}(n, \mathbb{K}); A) = \mathrm{Inv}\, A \otimes H^q(\mathfrak{gl}(n, \mathbb{K})).$$

The central role in the proof of this theorem is played by the element

$$\Delta = \sum_{i,\, j=1}^n e_j^i e_i^j$$

of the enveloping algebra $U(\mathfrak{gl}(n, \mathbb{K}))$. Since the $\mathfrak{gl}(n, \mathbb{K})$-module is a $U(\mathfrak{gl}(n, \mathbb{K}))$-module, Δ defines a certain endomorphism of any $\mathfrak{gl}(n, \mathbb{K})$-module; this endomorphism is known as the Casimir operator.

Suppose the proposed tensor module A is a submodule of the module T_l^k. Fix in T_l^k a Euclidean or Hermitian structure with respect to which the basis $e_{j_1 \ldots j_l}^{i_1 \ldots i_k}$ is orthonormed.

Lemma 1. The transformations of the space T_l^k, carried out by elements e_j^i, e_i^j of the algebra $\mathfrak{gl}(n, \mathbb{K})$, are adjoint to each other.

This is obvious.

Lemma 2. The Casimir operator T_l^k can be diagonalized. The kernel of this operator coincides with Inv T_l^k.

Indeed, it follows from Lemma 1 that the operator Δ is self-adjoint and therefore can be diagonalized. It is also clear that if $a \in$ Inv T_l^k, then $\Delta a = 0$. Finally, $\Delta a = 0$ implies

$$0 = \langle \Delta a, a \rangle = \langle \sum e_j^i e_i^j a, a \rangle = \sum \langle e_j^i e_i^j a, a \rangle = \sum \langle e_i^j a, e_i^j a \rangle,$$

i.e., $\langle e_i^j a, e_i^j a \rangle = 0$, and, therefore, $e_i^j a = 0$ for any i, j.

Lemma 3. Δ belongs to the center of the enveloping algebra $U(\mathfrak{gl}(n, \mathbb{K}))$.

Proof. Direct verification.

Corollaries. 1°. In any $\mathfrak{gl}(n, \mathbb{K})$-module the Casimir operator is a $\mathfrak{gl}(n, \mathbb{K})$-endomorphism. 2°. The subspaces of the finite-dimensional $\mathfrak{gl}(n, \mathbb{K})$-module corresponding to the eigenvalues of the Casimir operator are submodules.

Lemma 4. For any $\mathfrak{gl}(n, \mathbb{K})$-module B the homomorphisms

$$\Delta_*: H_q(\mathfrak{gl}(n, \mathbb{K}); B) \to H_q(\mathfrak{gl}(n, \mathbb{K}); B),$$
$$\Delta_*: H^q(\mathfrak{gl}(n, \mathbb{K}); B) \to H^q(\mathfrak{gl}(n, \mathbb{K}); B),$$

induced by the Casimir operator are trivial.

Proof. Recall that the complexes $C.(\mathfrak{gl}(n, \mathbb{K}); B)$ and $C^{\cdot}(\mathfrak{gl}(n, \mathbb{K}); B)$ can be interpreted as $C \otimes_{U(\mathfrak{gl}(n, \mathbb{K}))} B$ and $\mathrm{Hom}_{U(\mathfrak{gl}(n, \mathbb{K}))}(C, B)$, where C is the complex $C.(\mathfrak{gl}(n, \mathbb{K}); U(\mathfrak{gl}(n, \mathbb{K})))$, considered as a complex of $U(\mathfrak{gl}(n, \mathbb{K}))$-modules and $U(\mathfrak{gl}(n, \mathbb{K}))$-homomorphisms (see 1.3.4). In view of this, for the proof of the lemma it suffices to show that the homomorphism $\Delta_*: C \to C$ is $U(\mathfrak{gl}(n, \mathbb{K}))$-homotopic to zero. Since

the $U(\mathfrak{gl}(n, \mathbb{K}))$-modules which constitute C are free, this
is equivalent to showing the triviality of the homomorphism
$\Delta_*: H_*(C) \to H_*(C)$. But, as was shown in 1.3.4, $H_*(C)$ is
generated by a unique element, the homology class of the
chain $1 \in C_0(\mathfrak{gl}(n, \mathbb{K}); U(\mathfrak{gl}(n, \mathbb{K})))$, and, clearly, Δ_* sends this
class into 0.

Lemma 5. Suppose B is a finite-dimensional $\mathfrak{gl}(n, \mathbb{K})$-
module. If the Casimir operator in B is nondegenerate,
then $H_q(\mathfrak{gl}(n, \mathbb{K}); B) = 0$ and $H^q(\mathfrak{gl}(n, \mathbb{K}); B) = 0$ for all q.

Proof. We can limit ourselves to the case when $\Delta: B \to$
B is the multiplication on $\lambda \neq 0$ (for $\mathbb{K} = \mathbb{C}$ the general
case reduces to this one by means of coefficient sequences —
see 1.3.6A; the case $\mathbb{K} = \mathbb{R}$ reduces to the case $\mathbb{K} = \mathbb{C}$ by
complexification). In this case the operators

$$\Delta_*: H_q(\mathfrak{gl}(n, \mathbb{K}); B) \to H_q(\mathfrak{gl}(n, \mathbb{K}); B),$$
$$\Delta_*: H^q(\mathfrak{gl}(n, \mathbb{K}); B) \to H^q(\mathfrak{gl}(n, \mathbb{K}); B)$$

are also multiplications by λ, but these operators are trivial in view of Lemma 4.

Theorem 2.1.2 follows from Lemmas 2 and 5: by Lemma 2,
$A = \text{Inv } A \oplus \Delta(A)$ and on $\Delta(A)$ the Casimir operator is nondegenerate; by Lemma 5, the homology and cohomology of the
algebra $\mathfrak{gl}(n, \mathbb{K})$ with coefficients in $\Delta(A)$ are trivial.

Theorem 2.1.3. Any tensor $\mathfrak{gl}(n, \mathbb{K})$-module is completely reducible. In more detail: for any submodule B of tensor module A there exists a submodule C of the module A,
such that the projection $A \to A/B$ is an isomorphism of C
onto A/B.

Indeed, by Lemma 1, the orthogonal complement to B in A is a submodule of the module A.

3. **The theorem on invariants.** We now study the space Inv $T_l^k(n)$. The necessity of this study is motivated by Theorem 2.1.2, but it shall become even clearer in the sequel.

Theorem 2.1.4. If $k \neq l$, then Inv $T_l^k = 0$. The space Inv T_k^k is generated by the tensors

$$c(\sigma) = \sum_{i_1, \ldots, i_k = 1}^{n} e_{\sigma(i_1) \ldots \sigma(i_k)}^{i_1 \ldots i_k},$$

where $\sigma \in \mathrm{Symm}(k)$.

Remarks. 1°. Under the identification of T_k^k with $(T_k^k)'$ and with $\mathrm{Hom}(V \otimes \ldots \otimes V, V \otimes \ldots \otimes V)$, the tensor $c(\sigma)$ becomes, respectively, the functional and the homomorphism described by the formulas

$$\beta_1 \otimes \ldots \otimes \beta_k \otimes \alpha_1 \otimes \ldots \otimes \alpha_k \mapsto \beta_1(\alpha_{\sigma(1)}) \ldots \beta_k(\alpha_{\sigma(k)}),$$
$$\alpha_1 \otimes \ldots \otimes \alpha_k \mapsto \alpha_{\sigma(1)} \otimes \ldots \otimes \alpha_{\sigma(k)},$$

where $\alpha_1, \ldots, \alpha_k \in V$, $\beta_1, \ldots, \beta_k \in V'$.

2°. If $k \leqslant n$, then the elements $c(\sigma)$ of the space $T_k^k(n)$ are linearly independent; for $k = n + 1$ they are subjected to only one relation $\sum \mathrm{sgn}(\sigma) c(\sigma) = 0$; for $k > n + 1$ all relations between the $c(\sigma)$ are algebraic consequences of this relation. For the details and proofs, see Weyl's book [93].

Proof of Theorem 2.1.4. The element $c = \sum e_i^i$ of the algebra $\mathfrak{gl}(n, \mathbb{K})$ determines the multiplication by $l - k$ in T_l^k. This already shows that Inv $T_l^k = 0$ for $k \neq l$. Let $I(k, n) = \mathrm{Inv}\, T_k^k(n)$ and denote by $J(k, n)$ the subspace of the space $I(k, n)$ generated by the tensors $c(\sigma)$ [the fact that the $c(\sigma)$ are invariants is obvious]. We must prove that $J(k, n) = I(k, n)$. This is obvious for $k = 0$ and $n = 0$, and, applying induction over k and n, we can assume that $J(k - 1, n) = I(k - 1, n)$, $J(k, n - 1) = I(k, n - 1)$.

Define the maps r_l^s: $T_k^k(n) \to T_{k-1}^{k-1}(n)$, R_l^s: $T_{k-1}^{k-1}(n) \to T_k^k(n)$, π: $T_k^k(n) \to T_k^k(n - 1)$ by setting

$$r_l^s(e_{j_1\ldots j_k}^{i_1\ldots i_k}) = \begin{cases} 0 & \text{for } i_s \neq j_l, \\ e_{j_1\ldots j_{l-1}j_{l+1}\ldots j_k}^{i_1\ldots i_{s-1}i_{s+1}\ldots i_k} & \text{for } i_s = j_l; \end{cases}$$

$$R_l^s(e_{j_1\ldots j_{k-1}}^{i_1\ldots i_{k-1}}) = \sum_{u=1}^{n} e_{j_1\ldots j_{l-1}uj_l\ldots j_{k-1}}^{i_1\ldots i_{s-1}ui_s\ldots i_{k-1}},$$

$$\pi(e_{j_1\ldots j_k}^{i_1\ldots i_k}) = \begin{cases} 0, & \text{if at least one of} \\ & \quad \text{the indices } i, j \text{ equals } n, \\ e_{j_1\ldots j_k}^{i_1\ldots i_k} & \text{otherwise.} \end{cases}$$

Obviously, $r_l^s(I(k, n)) \subset I(k - 1, n)$, $r_l^s(J(k, n)) \subset J(k - 1, n)$, $R_l^s(I(k - 1, n)) \subset I(k, n)$, $R_l^s(J(k - 1, n)) \subset J(k, n)$, $\pi(I(k, n)) \subset I(k, n - 1)$ and $\pi(c(\sigma)) = c(\sigma)$. It is also clear that r_l^s and R_l^s are adjoint operators [with respect to the Euclidean and Hermitian structures in $T_k^k(n)$, $T_{k-1}^{k-1}(n)$, introduced in Subsection 2]. Let us put

$$\nabla = \sum_{s,t} R_l^s \circ r_l^s, \quad K(k, n) = I(k, n) \cap \text{Ker } \pi.$$

Lemma 1. Suppose $\sum \alpha_{j_1\ldots j_k}^{i_1\ldots i_k} e_{j_1\ldots j_k}^{i_1\ldots i_k} \in I(k, n)$. If $\alpha_{j_1\ldots j_k}^{i_1\ldots i_k} \neq 0$, then for $s = 1, \ldots, n$ we have

$$\rho_s(i_1, \ldots, i_k) = \rho_s(j_1, \ldots, j_k),$$

where $\rho_s(m_1, \ldots, m_k)$ is the number of times s appears among the $m_1, \ldots, m_k$.

This follows from the obvious relation

$$e_s^s e_{j_1\ldots j_k}^{i_1\ldots i_k} = (\rho_s(j_1, \ldots, j_k) - \rho_s(i_1, \ldots, i_k))e_{j_1\ldots j_k}^{i_1\ldots i_k}.$$

Lemma 2. $\nabla (I (k, n)) \subset J (k, n)$.

Proof. $\nabla (I (k, n)) \subset \oplus R_i^s (I (k - 1, n)) = \oplus R_i^s (J (k - 1, n)) \subset J (k, n)$.

Lemma 3. $\operatorname{Ker} \nabla = \bigcap\limits_{s,t} \operatorname{Ker} r_t^s$.

Proof. If $\nabla x = 0$, then $0 = \langle \nabla x, x \rangle = \langle \sum R_i^s \circ r_t^s x, x \rangle = \sum \langle R_i^s \circ r_t^s x, x \rangle = \sum \langle r_t^s x, r_t^s x \rangle$; hence $r_t^s x = 0$ for all s, t. Thus, $\operatorname{Ker} \nabla \subset \bigcap\limits_{s,t} \operatorname{Ker} r_t^s$, while the inverse inclusion is obvious.

Lemma 4. $I (k, n) = J (k, n) + K (k, n)$.

Proof. Suppose $x \in I (k, n)$. Since $I (k, n - 1) = J (k, n - 1)$, it follows that $\pi (x)$ is a linear combination of the tensors $c (\sigma)$. Denote by y the same linear combination of the elements $c (\sigma)$ in the space $I (k, n)$. Then $x = y + (x - y)$ and $y \in J (\kappa, n)$, $x - y \in K (k, n)$.

Lemma 5. $\operatorname{Ker} \nabla \cap K (k, n) = 0$.

Proof. Suppose $x \in \operatorname{Ker} \nabla \cap K (k, n)$ and $x = \sum \alpha_{j_1 \ldots j_k}^{i_1 \ldots i_k} e_{j_1 \ldots j_k}^{i_1 \ldots i_k}$. Lemma 1 allows us to assume that this sum contains only $e_{j_1 \ldots j_k}^{i_1 \ldots i_k}$ with $\rho_s (i_1, \ldots, i_k) = \rho_s (j_1, \ldots, j_k)$; set $m = \rho_n (i_1, \ldots, i_k) [= \rho_n (j_1, \ldots, j_k)]$. The inclusion $x \in K (k, n)$ shows that $m = 0$ implies $\alpha_{j_1 \ldots j_k}^{i_1 \ldots i_k} = 0$; assume that $\alpha_{j_1 \ldots j_k}^{i_1 \ldots i_k} = 0$ for $m < m_0$, and consider $\alpha_{j_1 \ldots j_k}^{i_1 \ldots i_k}$ for $m = m_0$. Suppose $i_s = j_t = n$. Since $r_t^s (x) = 0$ (see Lemma 3) it follows that the coefficient in $r_t^s (x)$ of $e_{j_1 \ldots j_{t-1} j_{t+1} \ldots j_k}^{i_1 \ldots i_{s-1} i_{s+1} \ldots i_k}$ vanishes:

$$\sum_{u=1}^{n} \alpha_{j_1 \ldots j_{t-1} u j_{t+1} \ldots j_k}^{i_1 \ldots i_{s-1} u i_{s+1} \ldots i_k} = 0.$$

But, if $u < n$, then $\alpha_{j_1 \ldots j_{t-1} u j_{t+1} \ldots j_k}^{i_1 \ldots i_{s-1} u i_{s+1} \ldots i_k} = 0$ by assumption. Therefore, $\alpha_{j_1 \ldots j_{t-1} n j_{t+1} \ldots j_k}^{i_1 \ldots i_{s-1} n i_{s+1} \ldots i_k} = \alpha_{j_1 \ldots j_k}^{i_1 \ldots i_k} = 0$.

<u>Lemma 6</u>. $\nabla(K\,(k,\,n)) \subset K\,(k,\,n)$.

<u>Proof</u>. As can be shown by a short direct calculation, if n does not appear in the sets $i_1, \ldots, i_k; j_1, \ldots, j_k$, the parts of our decomposition with respect to the basis of tensors $\Delta e_{j_1 \ldots j_k}^{i_1 \ldots i_k}$ and $\Delta e_{j_1 \ldots j_k}^{i_1 \ldots i_k}$ (where Δ is the Casimir operator – see Subsection 1), consisting of terms with such $e_{j_1' \ldots j_k'}^{i_1' \ldots i_k'}$ that the number n appears among the $i_1', \ldots, i_k'; j_1', \ldots, j_k'$, are respectively equal to

$$-2 \sum_{i_s = j_t} e_{j_1 \ldots j_{t-1} n j_{t+1} \ldots j_k}^{i_1 \ldots i_{s-1} n i_{s+1} \ldots i_k}, \qquad \sum_{i_s = j_t} e_{j_1 \ldots j_{t-1} n j_{t+1} \ldots j_k}^{i_1 \ldots i_{s-1} n i_{s+1} \ldots i_k}.$$

Thus, the space $\mathrm{Ker}\,\pi$ is invariant with respect to the operator $\nabla + \frac{1}{2}\Delta$. But the restriction $\Delta \mid I\,(k,\,n)$ is trivial; hence the intersection $\mathrm{Ker}\,\pi \cap I\,(k,\,n) = K\,(k,\,n)$ is invariant with respect to ∇.

Our theorem obviously follows from the lemmas proved above: by Lemmas 5 and 6, $\nabla\,(K\,(k,\,n)) = K\,(k,\,n)$; by Lemma 2 this implies $K\,(k,\,n) \subset J\,(k,\,n)$, after which Lemma 4 yields the relation $I\,(k,\,n) = J\,(k,\,n)$.

As an application of Theorem 2.1.4 we shall carry out a computation (useful for the sequel) of the ring of "symmetric invariants" of the algebra $\mathfrak{gl}\,(n, \mathbb{K})$, i.e., of the ring

$$\mathrm{Inv}\,S^*\,(\mathfrak{gl}\,(n, \mathbb{K})) = \mathrm{Inv}\,S^*\,(\mathfrak{gl}\,(n, \mathbb{K})').$$

<u>Theorem 2.1.5</u>. $\mathrm{Inv}\,S^*\,(\mathfrak{gl}\,(n, \mathbb{K})')$ is the polynomial ring in the variables $\zeta_k \in \mathrm{Inv}\,S^k\,(\mathfrak{gl}\,(n, \mathbb{K})')$, $k = 1, 2, \ldots, n$, determined by the formula

$$\zeta_k\,(g_1, \ldots, g_k) = \sum_{\sigma \equiv \mathrm{Symm}\,(k)} \mathrm{Tr}\,(g_{\sigma(1)} \ldots g_{\sigma(k)}). \tag{1}$$

[Note that formula (1) determines ζ_k for all natural k.]

Proof. By Theorem 2.1.4, an arbitrary invariant $\xi \in$ Inv $\otimes^k (\mathfrak{gl}(n, \mathbb{K})')$ is of the form

$$\sum_{\sigma \in \mathrm{Symm}\,(k)} a(\sigma) \prod_i \beta_i(\alpha_{\sigma(i)}) \quad [\alpha_j \in V, \beta_j \in V'],$$

where a is some $\mathbb{K}$-valued function on Symm (k). It is clear that if $(i_{11}, \ldots, i_{1k_1}), \ldots, (i_{s1}, \ldots, i_{sk_s})$ are cycles of the permutation σ (we assume that $k_1 \leqslant \ldots \leqslant k_s$), then

$$\prod_i \beta_i(\alpha_{\sigma(i)}) = \mathrm{Tr}\,(g_{i_{11}} \ldots g_{i_{1k_1}}) \ldots \mathrm{Tr}\,(g_{i_{s1}} \ldots g_{i_{sk_s}}),$$

where $g_j = \alpha_j \otimes \beta_j$. Furthermore, it is clear that the fact that the invariant ξ is symmetric is equivalent to the function a being constant on classes of conjugate elements, i.e., $a(\sigma)$ depends only on $k_1, \ldots, k_s$. Therefore,

$$\xi(g_1, \ldots, g_k) = \sum_{\substack{k_1 \leqslant \ldots \leqslant k_s,\ k_1 + \ldots + k_s = k, \\ \tau \in \mathrm{Symm}\,(k)}} a(k_1, \ldots, k_s)\, \mathrm{Tr}\,(g_{\tau(1)} \ldots g_{\tau(k_1)})$$

$$\ldots \mathrm{Tr}\,(g_{\tau(k_1 + \ldots + k_{s-1} + 1)} \ldots g_{\tau(k)})$$

for some $a(k_1, \ldots, k_s)$, i.e.,

$$\xi = \sum_{k_1, \ldots, k_s} a(k_1, \ldots, k_s)\, \zeta_{k_1} \ldots \zeta_{k_s}.$$

It remains to prove that $\zeta_{n+1}, \zeta_{n+2}, \ldots$ can be expressed in terms of $\zeta_1, \ldots, \zeta_n$, while $\zeta_1, \ldots, \zeta_n$ are not subject to any relations. But this is obvious: symmetric functions in $g_1, \ldots, g_k$ are defined by their values on sets of coinciding elements, while

$$\zeta_k(g, \ldots, g) = k!\,\mathrm{Tr}\,(g^k) = k!\,(\lambda_1(g)^k + \ldots + \lambda_n(g)^k),$$

where $\lambda_1(g), \ldots, \lambda_n(g)$ are the eigenvalues of the matrix g; the required statement now becomes a well-known proposition from the theory of symmetric polynomials.

It is interesting to compare Theorems 2.1.1, 2.1.2, and 2.1.5 with the spectral sequence from 1.3.5, which is defined for an arbitrary Lie algebra $\mathfrak{g}$ and for which $E_\infty = E_\infty^{0,0}$, while $E_2^{p,q} = H^p(\mathfrak{g}; S^{q/2}\,\mathfrak{g}')$. We see that if $\mathfrak{g} = \mathfrak{gl}(n, \mathbb{K})$. then the ring E_2 is the tensor product of the exterior algebra in n variables contained in $E_2^{0,1}, E_2^{0,3}, \ldots, E_2^{0,2n-1}$, and the ring of polynomials in n generators contained in $E_2^{2,0}, E_2^{1,0}, \ldots,$ $E_2^{2n,0}$. The structure of the differentials of such a spectral sequence is well known: the exterior generators are transgressive and their images under transgression represent the polynomial generators. In other words, the spectral sequence, beginning with the second term, coincides with the cohomology spectral sequence of the universal bundle of the group $U(n)$ (compare with the Remark in 1.3.5, 1°).

4. <u>Refinement of Theorem 2.1.1</u>. Define the cochain $\Phi_{q,n} = \Phi_q \in C^{2q-1}(\mathfrak{gl}(n, \mathbb{K}))$ by the formula

$$\Phi_q(g_1, \ldots, g_{2q-1}) = \sum_{\sigma \in \mathrm{Symm}(2q-1)} \mathrm{sgn}\,\sigma\,\mathrm{Tr}(g_{\sigma(1)} \ldots g_{\sigma(2q-1)}).$$

Obviously, the homomorphism induced by the standard inclusion $\mathfrak{gl}(n-1, \mathbb{K}) \to \mathfrak{gl}(n, \mathbb{K})$ sends $\Phi_{q,n}$ into $\Phi_{q,n-1}$.

<u>Lemma 1</u>. Φ_q is a cocycle.

<u>Proof</u>: Direct verification.

Denote the cohomology class of the cocycle $\Phi_{q,n}$ by $\varphi_{q,n}$ or φ_q.

<u>Theorem 2.1.6</u>. $H^*(\mathfrak{gl}(n, \mathbb{K}))$ is the exterior algebra in $\varphi_{1,n}, \ldots, \varphi_{n,n}$.

<u>Lemma 2</u>. Let $C_{\mathrm{inv}}^q(\mathfrak{gl}(n, \mathbb{K})) = \mathrm{Inv}\,C^q(\mathfrak{gl}(n, \mathbb{K}))$. Then (i) $d(C_{\mathrm{inv}}^q(\mathfrak{gl}(n, \mathbb{K}))) \subset C_{\mathrm{inv}}^{q+1}(\mathfrak{gl}(n, \mathbb{K}))$, so that the spaces $C_{\mathrm{inv}}^q(\mathfrak{gl}(n, \mathbb{K}))$ constitute the subcomplex $C_{\mathrm{inv}}(\mathfrak{gl}(n, \mathbb{K}))$ of the complex

$C^{\cdot}$ ($\mathfrak{gl}(n, \mathbb{K})$),   and (ii) the inclusion C_{inv} ($\mathfrak{gl}(n, \mathbb{K})$) $\to$ $C^{\cdot}$ ($\mathfrak{gl}(n, \mathbb{K})$) induces an isomorphism in cohomology.

Proof. The statement (i) follows from the fact that d is a $\mathfrak{gl}(n, \mathbb{K})$-homomorphism; let us prove (ii). The algebra $\mathfrak{gl}(n, \mathbb{K})$ as a $\mathfrak{gl}(n, \mathbb{K})$-module is $V' \otimes V$. Therefore, C^q ($\mathfrak{gl}(n, \mathbb{K})$)   is a tensor module. Applying Theorem 2.1.3 to the module Ker $d \subset C^q$ ($\mathfrak{gl}(n, \mathbb{K})$) and to its submodule Im d,   we obtain the realization of the $\mathfrak{gl}(n, \mathbb{K})$-module H^q ($\mathfrak{gl}(n, \mathbb{K})$) in Ker d.   Since the $\mathfrak{gl}(n, \mathbb{K})$-module H^q ($\mathfrak{gl}(n, \mathbb{K})$) is trivial (see 1.3.6C), we conclude that every cohomology class is represented by a $\mathfrak{gl}(n, \mathbb{K})$-invariant cocycle. Further, applying Theorem 2.1.3 to the module d^{-1} (C^q_{inv} ($\mathfrak{gl}(n, \mathbb{K})$))   and to its submodule Ker $d \subset C^{q-1}$ ($\mathfrak{gl}(n, \mathbb{K})$), we can find a submodule of C^{q-1} ($\mathfrak{gl}(n, \mathbb{K})$)   mapped ismorphically by the differential d onto C^q_{inv} ($\mathfrak{gl}(n, \mathbb{K})$) $\cap$ Im d.   Thus, if an invariant cocycle belongs to the image of the differential, it is the differential of an invariant cochain.

Lemma 3. C^*_{inv} ($\mathfrak{gl}(n, \mathbb{K})$)   is generated multiplicatively by the cocycles Φ_q.

This immediately follows from Theorem 2.1.4: if r_1, ..., r_s are the lengths of the cycles constituting the permutation $\sigma \in \mathrm{Symm}(q)$, then

$$\mathrm{Alt}\, c\,(\sigma) \in C^q(\mathfrak{gl}(n, \mathbb{K})) \subset T^q_q$$

is, up to a sign, the product of the cochains $\tilde{\Phi}_{r_i} \in C^{r_i}$ ($\mathfrak{gl}(n, \mathbb{K})$),   where

$$\tilde{\Phi}_r (g_1, \ldots, g_q) = \sum_{\tau \in \mathrm{Symm}(r)} \mathrm{sgn}\,(\tau)\, \mathrm{Tr}\,(g_{\tau(1)} \ldots g_{\tau(r)}).$$

But, obviously, $\tilde{\Phi}_r = 0$ for even r and $\tilde{\Phi}_{2q-1} = \Phi_q$.

<u>Corollary</u>. The differential in the complex $C_{\text{inv}}^{\cdot}$ ($\mathfrak{gl}\,(n,\,\mathbb{K})$) is trivial; in particular, C_{inv}^{q} ($\mathfrak{gl}\,(n,\,\mathbb{K})$) $= H^{q}$ ($\mathfrak{gl}\,(n,\,\mathbb{K})$).

<u>Lemma 4</u>. $\Phi_{n,\,n}$ cannot be represented in the form of a polynomial in $\Phi_{1,\,n},\,\ldots,\,\Phi_{n-1,\,n}$.

<u>Proof</u>. As shown by a direct calculation,

$$\Phi_{n,\,n}\,(e_n^1,\,e_1^n,\,\ldots,\,e_n^{n-1},\,e_{n-1}^n,\,e_n^n) = (n-1)!\,(2n-1).$$

At the same time it is obvious that $\Phi_i\,(g_1,\,\ldots,\,g_{2i-1}) = 0$ for any subset $g_1,\,\ldots,\,g_{2i-1}$ of the set $e_n^1,\,e_1^n,\,\ldots,\,e_n^{n-1},\,e_{n-1}^n,\,e_n^n$ which does not include e_n^n. Therefore,

$$\Phi_{i_1,\,n}\ldots\Phi_{i_s,\,n}\,(e_n^1,\,e_1^n,\,\ldots,\,e_n^{n-1},\,e_{n-1}^n,\,e_n^n) = 0$$

for $s \geqslant 2$.

It is now quite easy to prove Theorem 2.1.5: using induction over n, we can assume that $H^*\,(\mathfrak{gl}\,(n-1,\,\mathbb{K}))$ is the exterior algebra with generators $\varphi_{1,\,n-1},\,\ldots,\,\varphi_{n-1,\,n-1}$ and, in view of Theorem 2.1.1, we need only show that $\varphi_{n,\,n}$ cannot be represented as a polynomial in $\varphi_{1,\,n},\,\ldots,\,\varphi_{n-1,\,n}$. But this follows immediately from Lemma 4 and the corollary to Lemma 3.

<u>5. Generalizations</u>. The theory developed in this section is a special case of a considerably more general theory, which includes the cohomology of a wide class of finite-dimensional Lie algebras. We shall limit ourselves to the statement of two main results, which generalize Theorems 2.1.2 and 2.1.3. Proofs and further statements can be found by the reader in [93] and [94].

<u>Theorem 2.1.7</u>. Any finite-dimensional module over a semisimple Lie algebra is completely reducible.

<u>Theorem 2.1.8</u>. For any finite-dimensional module A over a semisimple Lie algebra $\mathfrak{g}$ we have

$$H_q \, (\mathfrak{g}; \, A) = H_q \, (\mathfrak{g}; \, \mathrm{Inv}_\mathfrak{g} \, A),$$
$$H^q \, (\mathfrak{g}; \, A) = H^q \, (\mathfrak{g}; \, \mathrm{Inv}_\mathfrak{g} \, A).$$

These theorems cannot be applied literally to the algebra $\mathfrak{gl} \, (n, \, \mathbb{K})$, since it is not semisimple, although it is the sum of a simple algebra $\mathfrak{sl} \, (n, \, \mathbb{K})$ and a one-dimensional commutative algebra. The latter fact allows us to carry over both theorems to the case $\mathfrak{g} = \mathfrak{gl} \, (n, \, \mathbb{K})$. Then, in the first of them, it is necessary to make the supplementary assumption that the transformation defined in the proposed module by the unit matrix $E \in \mathfrak{gl} \, (n, \, \mathbb{K})$ can be diagonalized; the second theorem can be carried over to the case $\mathfrak{g} = \mathfrak{gl} \, (n, \, \mathbb{K})$ without any restrictions.

§2.   COMPUTATIONS FOR LIE ALGEBRAS
      OF FORMAL VECTOR FIELDS.
      GENERAL RESULTS

In this section we shall be concerned with the continuous cohomology of the algebras W_n of formal vector fields and some of their subalgebras. The homology of the algebras W_n appears to be boundless and will not be considered. A more reasonable domain for the application of homology theory are subalgebras of the algebra W_n^{pol} of polynomial vector fields. Thus, obviously, $[H_q \, (W_n^{\mathrm{pol}})]' = H^q \, (W_n)$, and similar relations are valid for other vector-field algebras and for more complicated coefficient modules.

The case $n = 1$ is exceptional, if only because it allows one to move ahead in the computation of cohomology much further than in the general case. This case will be considered in a separate section (the next one).

1. Finite dimensionality. (The results of this subsection are the contents of a joint paper by I. M. Gelfand and

the author [32].) The spaces of continuous cohomology of formal vector-field Lie algebras are often finite dimensional, and this can sometimes be proved by using the results of 1.5.2. Their application is based on the following observation. Suppose $c_1, \ldots, c_n \in \mathbb{K}$, and let B be a subset of the field $\mathbb{K}$ consisting of elements of the form $k_1 c_1 + \ldots + k_n c_n - c_i$ ($1 \leqslant i \leqslant n$, $k_1, \ldots, k_n$ nonnegative integers). Let

$$g_0 = \sum_{i=1}^{n} c_i x_i \frac{\partial}{\partial x_i}$$

and, for $b \in B$, let

$$S_b = \{\xi \in W_n \,|\, [g_0, \xi] = b\xi\} = \left\{ \sum_{k_1 c_1 + \ldots + k_n c_n - c_i = b} a_{k_1, \ldots, k_n; \, i} \, x_1^{k_1} \ldots x_n^{k_n} \frac{\partial}{\partial x_i} \right\}.$$

For example, if $c_1 = \ldots = c_n = 1$, then $B = \{m \in \mathbb{Z} \,|\, m \geqslant -1\}$ and S_b consists of the vector fields $\sum p_i \partial / \partial x_i$ with homogeneous p_i of degrees $b + 1$ (thus, L_k is the completed direct sum $\widehat{\bigoplus}_{b \geqslant k} S_b$); if we also have $n = 1$, then $S_b = \mathbb{K} e_b$.

Lemma. If $\mathfrak{g}$ is a subalgebra of the algebra W_n, containing g_0, then

$$\mathfrak{g} = \bigoplus_{b \in B} (\mathfrak{g} \cap S_b).$$

Proof. We must check that if $\xi_1 \in S_{b_1}, \ldots, \xi_m \in S_{b_m}$, where $b_1, \ldots, b_m$ are pairwise distinct elements of B and $\xi_1 + \ldots + \xi_m \in \mathfrak{g}$, then we also have $\xi_1 \in \mathfrak{g}, \ldots, \xi_m \in \mathfrak{g}$. But this follows from the inclusion

$$\underbrace{[g_0, [g_0, \ldots, [g_0, \xi_1 + \ldots + \xi_m] \ldots]]}_{s}$$

$$= b_1^s \xi_1 + \ldots + b_m^s \xi_m \in \mathfrak{g} \qquad (s = 0, 1, 2, \ldots).$$

Thus, for $g_0 \in \mathfrak{g}$ we are in the situation of 1.5.2 and can apply Theorems 1.5.2 and 1.5.2A. The first claims that the complex $C^{\cdot}(\mathfrak{g})$ is homology equivalent to its part, $C_0^{\cdot}(\mathfrak{g})$, where

$$C_0^q(\mathfrak{g}) = \bigoplus_{\substack{b_1, \ldots, b_q \in B, \\ b_1 + \ldots + b_q = 0}} (\mathfrak{g}_{b_1} \wedge \cdots \wedge \mathfrak{g}_{b_q})' \subset \Lambda^q \mathfrak{g}' = C^q(\mathfrak{g})$$

$(\mathfrak{g}_b = \mathfrak{g} \cap S_b)$; the second assumes given the $\mathfrak{g}$-module A with a basis consisting of eigenvectors of the transformation $a \mapsto g_0 a$, and claims that the complex $C^{\cdot}(\mathfrak{g}; A)$ is homology equivalent to the complex $C_0^{\cdot}(\mathfrak{g}; A)$, where

$$C_0^q(\mathfrak{g}; A) = \bigoplus_{\substack{b_1, \ldots, b_q \in B, \, d \in \mathbb{K}, \\ b_1 + \ldots + b_q = d}} \mathrm{Hom}\,(\mathfrak{g}_{b_1} \wedge \cdots \wedge \mathfrak{g}_{b_q}, A_d) \subset C^q(\mathfrak{g}; A)$$

$(A_d = \{a \in A \,|\, ga = da\})$. Most important is the case when the $c_1, \ldots, c_n$ are positive real numbers while the module A is finite dimensional. In this case, for any $b \in B$, the equation $k_1 c_1 + \ldots + k_n c_n - c_i = b$ has a finite number of solutions in nonnegative integers $k_1, \ldots, k_n$, so that all the spaces S_b, and with them all the spaces $\mathfrak{g}_b$, are finite dimensional. Furthermore, in the case considered, B consists of real numbers among which only a finite amount can be less than any number given in advance. In view of this, the spaces $C_0^q(\mathfrak{g}; A)$ are finite dimensional and may be nontrivial only for a finite number of different q. We reach the following conclusion concerning cohomology.

Theorem 2.2.1. If the algebra $\mathfrak{g} \subset W_n$ contains the element $\sum c_i x_i \partial / \partial x_i$ with positive $c_1, \ldots, c_n$, then for any finite-dimensional $\mathfrak{g}$-module A the space $H^q(\mathfrak{g}; A)$ is finite dimensional.

Indeed, if A possesses a basis consisting of eigenvectors of the transformations $\sum c_i x_i \partial / \partial x_i$, then this fact follows immediately from the previous arguments. For $\mathbb{K} = \mathbb{C}$,

the general case reduces to the special case by means of coefficient sequences (see 1.3.6A). The case $\mathbb{K} = \mathbb{R}$ reduces
to the case $\mathbb{K} = \mathbb{C}$ by complexification.

Corollary 2.2.2. The following spaces are finite dimen-
sional: $H^*(\mathbb{K}W_n)$, $H^*(\mathbb{K}\hat{S}_n)$, $H^*(\mathbb{K}\hat{H}_k)$, $H^*(\mathbb{K}K_k)$.

Indeed, the algebras $\mathbb{K}W_n$, $\mathbb{K}\hat{S}_n$, $\mathbb{K}\hat{H}_k$ contain the field
$\sum x_i \frac{\partial}{\partial x_i}$, while the algebra $\mathbb{K}K_k$ contains the field $x_1 \frac{\partial}{\partial x_1} +$
$\ldots + x_{2k} \frac{\partial}{\partial x_{2k}} + 2x_{2k+1} \frac{\partial}{\partial x_{2k+1}}$. (Note that the grading determined
by the field $\sum x_i \frac{\partial}{\partial x_i}$ in $\mathbb{K}W_n$ does not differ from the grad-
ing described directly in Example 2° in 1.3.7.)

Theorem 2.2.1 cannot be applied to the algebras $\mathbb{K}S_n$,
$\mathbb{K}H_k$ and it is not even known at present whether the spaces
$H^q(\mathbb{K}S_n)$, $H^q(\mathbb{K}H_k)$ are finite dimensional or not (some details
are developed in Subsection 7). However, it may be applied
to the intersections of algebras from Corollary 2.2.2 with
the algebra L_0 which thus also have finite-dimensional cohomology with coefficients in finite-dimensional modules.
Comparing this with Theorem 1.5.4 and with 1.2.3, we obtain:

Corollary 2.2.3. If $\mathfrak{g}$ is one of the algebras $\mathbb{K}W_n, \mathbb{K}\hat{S}_n$,
$\mathbb{K}\hat{H}_k$, $\mathbb{K}K_k$, then the cohomology of the algebra $\mathfrak{g}$ with coefficients in the module $\text{Coind}_{\mathfrak{g}}A$ are finite dimensional for any
finite-dimensional $(L_0 \cap \mathfrak{g})$-module A. In particular, the
cohomology of the algebra $\mathbb{K}W_n$ with coefficients in the module of formal tensor fields or generalized formal tensor
fields of any type is finite dimensional.

2. The ring $H^*(W_n)$. (The computation was carried
out by I. M. Gelfand and myself in [29]; an exposition of

this paper is contained in [37] and [44].) Denote by X the
inverse image of the (Schubert) $2n$-dimensional skeleton
$\mathrm{sk}_{2n}\mathbb{C}\mathrm{G}\,(\infty,\,n)$ of the complex Grassmann space in the total
space $\mathbb{C}\mathrm{V}\,(\infty,\,n)$ of the classical universal $\mathrm{U}\,(n)$-bundle.

Theorem 2.2.4. For every q

$$H^q\,(\mathbb{K}W_n) \cong H^q\,(X_n;\,\mathbb{K}).$$

The multiplicative structure in $H^*\,(W_n)$ is trivial (the
product of any elements of positive dimension is 0). The
Massey product in $H^*\,(W_n)$ is also trivial.

Proof. Just as in §1.1, we identify the subalgebra of
the algebra W_n, consisting of linear fields (i.e., of fields
of the form $\sum a_{ij}x_i \frac{\partial}{\partial x_j}$), with $\mathfrak{gl}\,(n,\,\mathbb{K})$. This identification
transforms the algebra W_n and the spaces $C^q\,(W_n)$ into
$\mathfrak{gl}\,(n,\,\mathbb{K})$-modules and the differentials $d\colon C^q\,(W_n) \to C^{q+1}\,(W_n)$
become $\mathfrak{gl}\,(n,\,\mathbb{K})$-homomorphisms; the $\mathfrak{gl}\,(n,\,\mathbb{K})$-structure which
thus arises in $H^q\,(W_n)$ will be trivial in accordance with
1.3.6C. [Note that this action of the algebra $\mathfrak{gl}\,(n,\,\mathbb{K})$
in W_n and $C^q\,(W_n)$ may be described in more direct fashion:
the group $\mathrm{GL}\,(n,\,\mathbb{K})$ acts in $\mathbb{K}^n$ and, therefore, acts in W_n
and $C^q\,(W_n)$, while the passage to the action of the algebra
$\mathfrak{gl}\,(n,\,\dot{\mathbb{K}})$ is carried out as usual by means of derivation.]

Consider the Serre–Hochschild spectral sequence corres-
ponding to the subalgebra $\mathfrak{gl}\,(n,\,\mathbb{K})$ of the algebra W_n. For
this spectral sequence we have

$$E_1^{p,\,q} = H^q\,(\mathfrak{gl}\,(n,\,\mathbb{K});\,\Lambda^p\,(W_n/\mathfrak{gl}\,(n,\,\mathbb{K}))').$$

Let us see what the coefficients in the last cohomology are.
As a $\mathfrak{gl}\,(n,\,\mathbb{K})$-module, W_n equals

$$W_n = (\widehat{S}*V') \otimes V = \widehat{\oplus}_{j=0}^{\infty} (S^j V') \otimes V$$

(see 1.2.2). In the last sum the summand corresponding to $j = 1$ is $\mathfrak{gl}(n, \mathbb{K})$. Therefore,

$$W_n/\mathfrak{gl}(n, \mathbb{K}) = \widehat{\bigoplus_{j=0,2,3,\ldots}} [(S^j V') \otimes V],$$

$$\Lambda^p (W_n/\mathfrak{gl}(n, \mathbb{K}))' = \Lambda^p (\widehat{\bigoplus_{j=0,2,3,\ldots}} [(S^j V') \otimes V])$$

$$= \bigoplus_{p_0+p_2+p_3+\ldots=p} (\bigotimes_{j=0,2,3,\ldots} \Lambda^{p_j} [(S^j V') \otimes V]).$$

It is important to note that all the summands of this last sum are finite-dimensional spaces and even tensor modules. We continue:

$$E_1^{p,q} = H^q (\mathfrak{gl}(n, \mathbb{K}); \Lambda^p (W_n/\mathfrak{gl}(n, \mathbb{K}))')$$

$$= \bigoplus_{p_0+p_2+p_3+\ldots=p} H^q (\mathfrak{gl}(n, \mathbb{K}); \bigotimes_{j=0,2,3,\ldots} \Lambda^{p_j} [(S^j V) \otimes V'])$$

$$= \bigoplus_{p_0+p_2+p_3+\ldots=p} H^q (\mathfrak{gl}(n, \mathbb{K})) \otimes \mathrm{Inv} \bigotimes_{j=0,2,3,\ldots} \Lambda^{p_j} [(S^j V) \otimes V'])$$

$$= H^q (\mathfrak{gl}(n, \mathbb{K})) \otimes [\bigoplus_{p_0+p_2+p_3+\ldots=p} \mathrm{Inv} \bigotimes_{j=0,2,3,\ldots} \Lambda^{p_j} [(S^j V) \otimes V'])]$$

(we have used Theorem 2.1.4). Note that the second factor in the last product is $E_1^{p,0}$; for the sequel it is important that $\oplus_p E_1^{p,0}$ is a ring. Elements of the space $\oplus_{j \neq 1} \Lambda^{p_j} \times [(S^j V) \otimes V']$ may be viewed as functionals in the variables

$$\alpha_1; \ldots; \alpha_{p_0};$$
$$\beta_{p_0+1}^1, \beta_{p_0+1}^2, \alpha_{p_0+1}; \ldots; \beta_{p_0+p_2}^1, \beta_{p_0+p_2}^2, \alpha_{p_0+p_2};$$
$$\beta_{p_0+p_2+1}^1, \beta_{p_0+p_2+1}^2, \beta_{p_0+p_2+1}^3, \alpha_{p_0+p_2+1}; \ldots, \alpha_{p_0+p_2+p_3};$$

where all the α belong to V and all the β belong to V'. Here the functional is not changed by interchanging any two β with the same lower index and is multiplied by -1 in each of the following cases: (1) when α_i and α_j are interchanged and $1 \leqslant i < j \leqslant p_0$; (2) when the group of variables $\beta_i^1, \beta_i^2, \alpha_i$ is interchanged with the group of variables $\beta_j^1, \beta_j^2, \alpha_j$, and $p_0 < i < j \leqslant p_0 + p_2$; (3) when similar groups of variables from the third line separated by semicolons are interchanged; etc.

Lemma 1. The ring

$$\bigoplus_p E_1^{p,\,0} = \bigoplus_{p_0,\,p_2,\,p_3,\dots} \mathrm{Inv}\Big[\bigotimes_{j=0,\,2,\,3,\dots} \Lambda^{p_j}[(S^j V)\otimes V']\Big]$$

is generated by the elements

$$\Psi_r \in \Lambda^r V' \otimes \Lambda^r[(S^2 V)\otimes V'],$$

defined by the formula

$$\Psi_r(\alpha_1,\dots,\alpha_r;\ \beta_{r+1}^1,\ \beta_{r+1}^2,\ \alpha_{r+1};\dots;\ \beta_{2r}^1,\ \beta_{2r}^2,\ \alpha_{2r}) =$$

$$\sum_{\substack{\sigma,\ \tau\in\mathrm{Symm}\,(r),\\ \nu_1,\dots,\nu_r\in\mathrm{Symm}\,(2)}} \Big[\ \mathrm{sgn}\,(\sigma)\,\mathrm{sgn}\,(\tau)\prod_{j=1}^r \beta_{r+\tau(j)}^{\nu_j(1)}\,(\alpha_{\sigma(j)})\,\beta_{r+\tau(j)}^{\nu_j(2)}\,(\alpha_{r+\tau(j-1)})\ \Big]_j,$$

in which $\tau(0)$ is assumed equal to $\tau(r)$. Here $\Psi_{r_1}^{m_1}\dots\Psi_{r_k}^{m_k} = 0$ if $m_1 r_1 + \dots + m_k r_k > n$ (in particular, $\Psi_r = 0$ when $r > n$), and these relations are the defining relations between the Ψ_r.

Remarks. 1°. Recall that the identifications that we have made allow us to consider the Ψ_r as cochains (as will be seen in the sequel, cocycles) belonging to $C^{2r}(W_n, \mathfrak{gl}\,(n, \mathbb{K})) \subset C^{2r}(W_n)$.

2°. The formula for Ψ_r looks so complicated because of the symmetry conditions. We could have written, more briefly,

$$\Psi_r(\alpha,\beta) = \beta_{r+1}^1(\alpha_1)\dots\beta_{2r}^1(\alpha_r)\,\beta_{r+1}^2(\alpha_{2r})\,\beta_{r+2}^2(\alpha_{r+1})\dots\beta_{2r}^2(\alpha_{2r-1}) +\dots,$$

where the last ellipsis denotes summands obtained from the one written out by all permissible permutations of the variables and supplied with the appropriate signs. Perhaps the readers will find it easier to understand this formula by means of the pictures which I. M. Gelfand and I used to represent the functionals Ψ_r and other similar functionals. Figure 1 represents the picture corresponding to Ψ_r (for

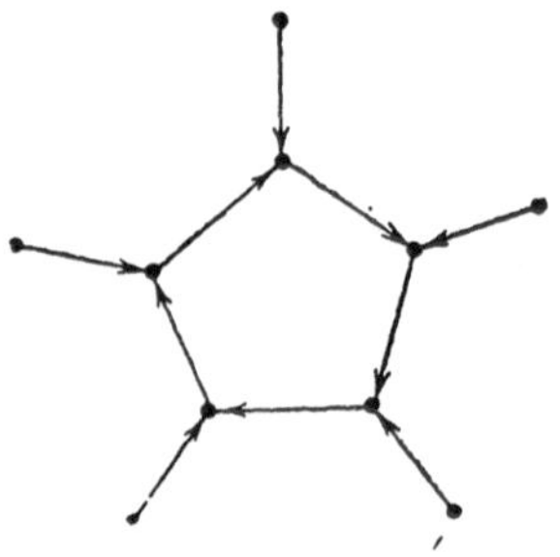

Fig. 1

$r = 5$) — "the r-th hedgehog." In order to recover Ψ_r from this picture, we must number its $2r$ points, assigning (in any way) the numbers $1, \ldots, r$, to the exterior points, and the numbers $r + 1, \ldots, 2r$ to the interior points. After that, for any arrow going from the k-th point to the l-th one, we write $\beta_l(\alpha_k)$, assign the superscripts 1 and 2 to β (so as not to get β's with identical indices), form the product of all the expressions written out, and take the sum of such products (with appropriate signs) over all possible numerations and indices.

<u>Proof of Lemma 1.</u> Since

$$\Lambda^{p_0} V' \otimes \Lambda^{p_2}[(S^2 V) \otimes V'] \otimes \Lambda^{p_3}[(S^3 V) \otimes V'] \otimes \ldots \subset T^{p_0 + p_2 + p_3 + \ldots}_{2p_2 + 3p_3 + \ldots},$$

the nonzero invariants may arise only when $p_0 + p_2 + p_3 + \ldots = 2p_2 + 3p_3 + \ldots$, i.e., if $p_0 = p_2 + 2p_3 + \ldots$ (see Theorem 2.1.4). By this same theorem, all these invariants are linear combinations of functionals of the form

$$(\alpha, \beta) \mapsto \prod \beta_l^j(\alpha_k),$$

where in the last product each α and each β must appear exactly once. Some of these functionals are forbidden by our symmetry conditions: namely, the product cannot simultaneously contain the factors $\beta_{l_1}^{j_1}(\alpha_{k_1})$ and $\beta_{l_2}^{j_2}(\alpha_{k_2})$, for which

$i_1 = i_2$, $j_1 \neq j_2$, $k_1 \leqslant p_0$, $k_2 \leqslant p_0$; the transpositions of $\beta_{i_1}^{j_1}$ with $\beta_{i_2}^{j_2}$ and of α_{k_1} with α_{k_2} do not change the product, but the resulting functional must change its sign. Thus, at least p_0 variables β must possess distinct subscripts, i.e., we must have $p_0 \leqslant p_2 + p_3 + \ldots$ Together with the relation $p_0 = p_2 + 2p_3 + \ldots$ established above, this shows that $p_3 = p_4' = \ldots = 0$ and $p_0 = p_2$; let us put $p_0 = p_2 = r$. Thus, our invariant is contained in the space $\Lambda^r V' \otimes \Lambda^r [(S^2 V) \otimes V']$ and is the linear combination of functionals of the form

$$(\alpha_1; \ldots; \alpha_r; \beta_{r+1}^1, \beta_{r+1}^2, \alpha_{r+1}; \ldots; \beta_{2r}^1, \beta_{2r}^2, \alpha_{2r})$$

$$\mapsto \beta_{r+1}^1 (\alpha_{\sigma_1(1)}) \ldots \beta_{2r}^1 (\alpha_{\sigma_1(r)}) \beta_{r+1}^2 (\alpha_{\sigma_2(1)}) \ldots \beta_{2r}^2 (\alpha_{\sigma_2(r)}),$$

where all the numbers $\sigma_1 (1), \ldots, \sigma_1 (r), \sigma_2 (1), \ldots, \sigma_2 (r)$ are pairwise distinct and, for each j, one of the numbers $\sigma_1 (j)$, $\sigma_2 (j)$ is less than or equal to r, while the other is greater than r. By permissible permutation of its variables, the last functional can be transformed into a functional of the form

$$(\alpha_1; \ldots; \alpha_r; \beta_{r+1}^1, \beta_{r+1}^2, \alpha_{r+1}; \ldots; \beta_{2r}^1, \beta_{2r}^2, \alpha_{2r})$$

$$\mapsto \beta_{r+1}^1 (\alpha_1) \ldots \beta_{2r}^1 (\alpha_r) \beta_{r+1}^2 (\alpha_{r+\sigma(1)}) \ldots \beta_{2r}^2 (\alpha_{r+\sigma(r)}),$$

where $\sigma \in \text{Symm} (r)$. The element of the space $\Lambda^r V' \otimes \Lambda^r \times [(S^2 V) \otimes V']$, obtained from the functional constructed above by (skew-) symmetrization, will be denoted by Ψ_σ. Obviously, Ψ_σ is determined by the conjugacy class of elements of the group $\text{Symm} (r)$, to which the element σ belongs and $\Psi_\sigma = \Psi_{\sigma_1} \Psi_{\sigma_2}$ (multiplication in the ring $\oplus_p E_1^{p, 0}$), if the permutation σ is formed by the permutations $\sigma_1 \in \text{Symm} (r_1)$, $\sigma_2 \in \text{Symm} (r_2)$ for $r_1 + r_2 = r$. Thus, the ring $\oplus_p E_1^{p, 0}$ is generated by elements $\Psi_r [\in E_1^{2r, 0}]$, corresponding to elementary cyclic permutations, so that it remains to check that these

elements are related precisely by the relations indicated
in the statement of the lemma. These relations may be writ-
ten as follows: $\Psi_\sigma = 0$ for $\sigma \in \mathrm{Symm}\,(r)$ when $r > n$.

The fact that these relations actually hold is obvious:
$\Psi_\sigma \in \Lambda^r V' \otimes \Lambda^r\,[(S^2 V) \otimes V']$, while $\Lambda^r V' = 0$ for $r > n$. In
order to verify that there are no other relations, we must
fix a permutation $\tau \in \mathrm{Symm}\,(r)$ with $r \leqslant n$ and set

$$\dot{\alpha}_1 = e_1, \ldots, \alpha_r = e_r; \quad \alpha_{r+1} = e_{\tau(1)}, \ldots, \alpha_{2r} = e_{\tau(r)};$$
$$\beta^1_{r+1} = \beta^2_{r+1} = e^1, \ldots, \beta^1_{2r} = \beta^2_{2r} = e^r.$$

Obviously Ψ_σ assumes a nonzero value on this set of variables if and only if σ and τ are conjugate in $\mathrm{Symm}\,(r)$.

The lemma is proved.

Thus, E_1 is the tensor product of the ring $E_1^{0,*} = H^* \times$
$(\mathfrak{gl}\,(n, \mathbb{K}))$ and the ring $E_1^{*,0}$, which is the quotient ring of
the polynomial ring in $\Psi_1, \ldots, \Psi_n$ (with $\Psi_r \in E_1^{2r,0}$) by the
ideal consisting of polynomials of degree $> 2n$ (we assume
that $\deg \Psi_r = 2r$). Since $E_1^{p,q} = 0$ for any odd p, the differential d_1 is trivial and $E_2 = E_1$.

Now we are in a position to note that the Leray–Serre
cohomology spectral sequence of the bundle $X_n \to \mathrm{sk}_{2n}\mathbb{C}G\,(\infty, n)$
has precisely the same term E_2. We shall show that the dif-
ferentials of these two spectral sequences (beginning from
the second one) also act identically. Recall that in the
spectral sequence of the bundle $X_n \to \mathrm{sk}_{2n}\mathbb{C}G\,(\infty, n)$ the ex-
ternal generators of the ring $E_2^{0,*} = H^*\,(U\,(n);\,\mathbb{K})$ are transgressive, and their images under transgression are the poly-
nomial generators of the ring $H^*\,(\mathrm{sk}_{2n}\mathbb{C}G\,(\infty, n);\,\mathbb{K})$.

<u>Lemma 2</u>. $H^q\,(W_n) = 0$ for $0 < q \leqslant n$.

Remarks. 1°. Actually $H^q(W_n) = 0$ for $0 < q \leqslant 2n$, but this is difficult to prove directly. 2°. The proof of Lemma 2 proposed here was taken from Guillemin's lectures [44].

Proof of Lemma 2. Consider the map $\eta_i \colon C^q(W_n) \to C^q(W_n)$, defined by the formula

$$[\eta_i c](\xi_1, \ldots, \xi_q) = \sum_{r=1}^{q} c\left(\xi_1, \ldots, \xi_{r-1}, \left[\frac{\partial}{\partial x_i}, \xi_r\right], \xi_{r+1}, \ldots, \xi_q\right)$$

$(i = 1, \ldots, n)$. Clearly the maps η_i commute among themselves and with the differential d and we have $\eta_i(C_s^q(W_n)) \subset C_{s+1}^q(W_n)$ (the subscript relates to the grading determined in $C^{\cdot}(W_n)$ by the element $\sum x_i \partial/\partial x_i$ of the algebra W_n — see Subsections 1 and 1.3.7C). Here are two more properties of the maps η_i: (i) they are monomorphisms; (ii) if $\{c_{i_1 \ldots i_m} \in C^q(W_n)\}_{i_1, \ldots, i_m = 1}^n$ is a skew-symmetric (with respect to indices) set of cochains with $m < n$ and if for all $i_1, \ldots, i_{m+1}$

$$\sum_{r=1}^{m+1} (-1)^r \eta_{i_r} (c_{i_1 \ldots \hat{i}_r \ldots i_{m+1}}) = 0,$$

then there exists a skew-symmetric family $\{b_{j_1 \ldots j_{m-1}} \in C^q(W_n)\}$ such that for all $i_1, \ldots, i_m$ we have

$$c_{i_1 \ldots i_m} = \sum_{r=1}^{m} (-1)^r \eta_{i_r} (b_{i_1 \ldots \hat{i}_r \ldots i_m}).$$

In order to prove (i) and (ii), let $S = \mathbb{K}[\eta_1, \ldots, \eta_n]$; the homomorphisms η_i supply $C^q(W_n)$ with an S-module structure. With this structure, $C^q(W_n)$ becomes a direct summand in the S-module $W_n' \otimes \ldots \otimes W_n'$ (q factors) in which the endomorphisms η_i are now defined by the same formulas as in $C^q(W_n)$;

thus we need only prove (i) and (ii) for $W_n' \otimes \ldots \otimes W_\pi'$
rather than for $C^q(W_n)$. Furthermore, W_n' is a free S-
module $(W_n' \cong S \otimes V \cong S \oplus \ldots \oplus S)$, so that $W_n'' \otimes \ldots \otimes W_n'$
is a free $S \otimes \ldots \otimes S$-module, i.e., $W_n' \otimes \ldots \otimes W_n'$ is the
sum of several S-modules isomorphic to $S \otimes \ldots \otimes S$, and we
can prove properties (i) and (ii) for $S \otimes \ldots \otimes S$. But,
for $S \otimes \ldots \otimes S$ they are obvious.

Now suppose $\gamma \in H^q(W_n)$, where $0 < q \leqslant n$, and let $c \in$
$C_0^q(W_n)$ be a cocycle representing γ. Since the $\eta_l(c)$ are
cocycles from $C_1^q(W_n)$, while the complex $C_1^{\cdot}(W_n)$ is acyclic
(Theorem 1.5.2), there exist elements $c_1, \ldots, c_n \in C_1^{q-1}(W_n)$
satisfying $dc_i = \eta_i(c)$. But for the same reasons we can find
elements $c_{ij} \in C_2^{q-2}(W_n)$ satisfying $dc_{ij} = \eta_j(c_i) - \eta_i(c_j)$, as
well as the supplementary condition $c_{ji} = -c_{ij}$. Continuing
in the same way, we shall obtain (for $s = 1, \ldots, q$) skew-symmetric families of cocycles $\{c_{i_1 \ldots i_s} \in C_s^{q-s}(W_n)\}$ such that

$$dc_{i_1 \ldots i_s} = \sum_{r=1}^{s} (-1)^r \eta_{i_r}(c_{i_1 \ldots \hat{i}_r \ldots i_s}).$$

Since $C_q^0(W_n) = 0$, we have $c_{i_1 \ldots i_q} \equiv 0$ and, therefore,

$$\sum_{r=1}^{q} (-1)^r \eta_{i_r}(c_{i_1 \ldots \hat{i}_r \ldots i_q}) = 0.$$

By (ii) there exists a skew-symmetric family $\{b_{j_1 \ldots j_{q-2}} \in$
$C_{q-2}^1(W_n)\}$ satisfying

$$c_{i_1 \ldots i_{q-1}} = \sum_{r=1}^{q-1} (-1)^r \eta_{i_r}(b_{i_1 \ldots \hat{i}_r \ldots i_{q-1}}).$$

Replacing the cochains $c_{i_1 \ldots i_{q-2}}$ by the cochains $c_{i_1 \ldots i_{q-2}}' =$
$c_{i_1 \ldots i_{q-2}} - db_{i_1 \ldots i_{q-2}}$, we still obtain

$$dc'_{i_1\ldots i_{q-2}} = \sum_{r=1}^{q-2} (-1)^r \eta_{i_r} (c_{i_1\ldots \hat{i}_r \ldots i_{q-2}}),$$

and, at the same time,

$$\sum_{r=1}^{q-1} (-1)^r \eta_{i_r} (c'_{i_1\ldots \hat{i}_r \ldots i_{q-1}}) = 0.$$

Applying (ii) once more, we can choose cochains $b_{j_1\ldots j_{q-3}} \in C_{q-3}^2 (W_n)$ so that

$$c_{i_1\ldots i_{q-2}} = \sum_{r=1}^{q-2} (-1)^r \eta_{i_r} (b_{i_1\ldots \hat{i}_r \ldots i_{q-2}}),$$

etc. Finally, we shall get cochains c'_i satisfying $dc'_i = \eta_i(c)$ and $\eta_j c'_i - \eta_i c'_j = 0$; choosing b so as to have $\eta_i(b) = c'_i$, we see that $d\eta_i(b) = \eta_i(c)$ (for any i) and $db = c$ by (i).

Lemma 2 is proved. From it we see that in our Serre–Hochschild spectral sequence we have $E_\infty^{p,q} = 0$ for $0 < p + q \leqslant n$, from which it follows in the standard way that the exterior generators φ_i of the ring $E_2^{0,*} = H^*(\mathfrak{gl}(n, \mathbb{K}))$ of dimension $< n$ are transgressive and are mapped by transgression into the multiplicative generators of the ring $E_2^{*,0}$. In order to verify that the same is true for the other generators of the ring $E_2^{*,0}$, it suffices to compare our Serre–Hochschild spectral sequence with a similar spectral sequence for W_N when $N \geqslant 2n$. These two spectral sequences are related by the homomorphism induced by the inclusion $W_n \to W_N$, and this homomorphism sends the standard generators of the rings $E_2^{0,*}$, $E_2^{*,0}$ from the spectral sequence for W_N into the corresponding generators for W_n. But in the spectral sequence for W_N the exterior generators of dimension $< 2n$ of the ring $E_2^{0,*}$ were transgressive and

were mapped by transgressions into the generators of the corresponding part of the ring $E_2^{*,0}$. Therefore, the same is
true of the spectral sequence for W_n.

The additive part of Theorem 2.2.4 is proved. In order
to prove its multiplicative part, we shall need another lemma.

Lemma 3. In our Serre—Hochschild spectral sequence
$E_\infty^{p,q} = 0$ for $(p, q) \neq (0, 0)$, $p \leqslant n$.

Proof. The part $\bigoplus_{p \geqslant n} E_2^{p,q}$ of the term E_2 is additively generated by elements of the form $\varphi_{i_1} \ldots \varphi_{i_s}$, of the form
$\varphi_{i_1} \ldots \varphi_{i_s} \Psi_{j_1}^{m_1} \ldots \Psi_{j_t}^{m_t}$, and of the form $\Psi_{j_1}^{m_1} \ldots \Psi_{j_t}^{m_t}$, where $i_1 <$
$\ldots < i_s$, $j_1 < \ldots < j_t$, $2(j_1 m_1 + \ldots + j_t m_t) \leqslant n$, $s \neq 0$, $t \neq 0$,
$m_1 \neq 0, \ldots, m_t \neq 0$. An element of the first form does not
belong to the kernel of the differential $d_{2 i_1}$, an element
of the third form belongs to the image of the differential
$d_{2 j_1}$, an element of the second form belongs to the image of
the differential $d_{2 j_1}$, if $i_1 \geqslant j_1$, and does not belong to the
kernel of the differential $d_{2 i_1}$, if $i_1 < j_1$. [The condition
$p \leqslant n$ is needed for the last statement: if $i_1 < j_1$, then

$$d_{2 i_1} (\varphi_{i_1} \ldots \varphi_{i_s} \Psi_{j_1}^{m_1} \ldots \Psi_{j_t}^{m_t}) \in E_{2 i_1}^{2 i_1 + 2(j_1 m_1 + \ldots + j_t m_t), \, 2(i_2 + \ldots + i_s) + s - 1}$$

and it is necessary to have $2 i_1 + 2(j_1 m_1 + \ldots + j_t m_t)$ less
than or equal to $2n$.]

This lemma shows that every element of positive degree
of the ring $H^*(W_n)$ is of filtration $> n$, while the product of any two such elements must have filtration $> 2n$, i.e.,
must vanish. These same arguments are applicable to Massey
products as well.

3. Commentary and additions to Theorem 2.2.4. A. The
cohomology of the space X_n may be computed by the standard

methods of algebraic topology. This computation shows, in particular, that the multiplicative structure of the ring $H^*(X_n; \mathbb{K})$ is also trivial, so that the entire content of Theorem 2.2.4 reduces to the multiplicative isomorphism

$$\varPi^*(W_n) \cong H^*(X_n; \mathbb{K}).$$

Here are some more corollaries of this computation and of Theorem 2.2.4:

1°. $\quad\quad\quad\quad \dim H^*(W_n) < \infty;$

2°. $H^q(W_n) = 0$ for $0 < q < 2n + 1$ and for $q > n(n+2);$

3°. $\quad\quad H^q(W_1) = \begin{cases} \mathbb{K} & \text{for } q = 0, 3, \\ 0 & \text{for all other } q \end{cases}$

(actually, $X_1 = S^3$);

4°. $\quad\quad H^q(W_2) = \begin{cases} \mathbb{K} \oplus \mathbb{K} & \text{for } q = 5, 8, \\ \mathbb{K} & \text{for } q = 0, 7, \\ 0 & \text{for all other } q; \end{cases}$

5°. $\dim \varPi^{2n+1}(W_n) = p(n+1) - 1$, $\dim H^{n(n+2)}(W_n) = p(n)$,

where p denotes the number of decompositions into a sum of positive integers.

B. Now we shall describe a convenient (although not very canonical) method for choosing an additive basis in $\oplus_{q>0} H^q(W_n)$; this basis first appeared in unpublished notes by J. Vey and is usually referred to as the Vey basis.

Since $\Psi_1, \ldots, \Psi_n$ are cohomologically trivial $\mathfrak{gl}(n, \mathbb{K})$-invariant cocycles in $C^*(W_n)$, there exist $\mathfrak{gl}(n, \mathbb{K})$-invariant cochains $\tilde{\Phi}_1, \ldots, \tilde{\Phi}_n \in C^*(W_n)$ satisfying $d\tilde{\Phi}_i = \Psi_i$; they may be chosen so as to satisfy the condition $\tilde{\Phi}_i \mid \mathfrak{gl}(n, \mathbb{K}) = \Phi_i$. Obviously the cochains

$$\tilde{\Phi}_{p_1} \ldots \tilde{\Phi}_{p_l} \Psi_{r_1} \ldots \Psi_{r_m} \in C^{2(p_1 + \ldots + p_l + r_1 + \ldots + r_m) - 1}(W_n)$$

for $1 \leqslant p_1 < \ldots < p_l \leqslant n$, $1 \leqslant r_1 \leqslant \ldots \leqslant r_m \leqslant n$, $p_1 \geqslant r_1$, $r_1 + \ldots + r_m \leqslant n$, $p_1 + r_1 + \ldots + r_m > n$ are cocycles and their cohomology classes $c_{p_1 \ldots p_l; r_1 \ldots r_m}$ constitute an additive basis in $\oplus_{q>0} H^q(W_n)$. This is the Vey basis. The nonuniqueness of this construction is due to the fact that the choice of cochains $\tilde{\Phi}_i$ is not unique. (Note that actually the classes $c_{p_1 \ldots p_l; r_1 \ldots r_m}$ with $r_1 + \ldots + r_m = n$ are well defined.)

<u>C</u>. The statement as well as the proof of Theorem 2.2.4 may be improved by using the Weyl algebras defined in 1.3.5. Namely, the linear map $W_n \rightarrow \mathfrak{gl}(n, \mathbb{K})$, defined by the formula

$$\sum_i \left(a_i + \sum_j a_{ij} x_j + \ldots \right) \frac{\partial}{\partial x_i} \mapsto \| a_{ij} \|, \text{ determines a linear map}$$

$\mathfrak{gl}(n, \mathbb{K})' \rightarrow W_n' = C^1(W_n)$, which according to statement 2° from 1.3.5 can be canonically extended to a homomorphism $W(\mathfrak{gl}(n, \mathbb{K})) \rightarrow C^\cdot(W_n)$. The restriction of this homomorphism to $S^1(\mathfrak{gl}(n, \mathbb{K})')$ can easily be computed (using 1.3.5): this is the homomorphism $\mathfrak{gl}(n, \mathbb{K})' \rightarrow C^2(W_n)$, sending the functional $\varphi \in \mathfrak{gl}(n, \mathbb{K})'$ into the two-dimensional cochain

$$\left[\sum_i \left(a_i + \sum_j a_{ij} x_j + \sum_{j,k} a_{ijk} x_j x_k + \ldots \right) \frac{\partial}{\partial x_i} \right]$$

$$\wedge \left[\sum_i \left(b_i + \sum_j b_{ij} x_j + \sum_{j,k} b_{ijk} x_j x_k + \ldots \right) \frac{\partial}{\partial x_i} \right] \mapsto \varphi(\| c_{ij} \|),$$

where $c_{ij} = \sum_k (a_k b_{jik} - b_k a_{jik})$. Obviously this cochain vanishes if both its arguments belong to $L_0(n) \subset W_n$, which implies (in view of the fact that $\operatorname{codim} L_0(n) = n$) that the product of cochains from the image of this homomorphism, if there are more than n of them, vanishes. This means in its turn that the restriction of our homomorphism $W(\mathfrak{gl}(n, \mathbb{K})) \rightarrow C^\cdot(W_n)$ to $F_{2n+1} W(\mathfrak{gl}(n, \mathbb{K}))$ vanishes, since we are in fact dealing with the homomorphism

$$W\left(\mathfrak{gl}\left(n,\,\mathbb{K}\right)\right)_{2n}=W\left(\mathfrak{gl}\left(n,\mathbb{K}\right)\right)/F_{2n+1}W\left(\mathfrak{gl}\left(n,\,\mathbb{K}\right)\right)\to C^{\cdot}\left(W_{n}\right). \qquad (1)$$

[The quotient complex $W\left(\mathfrak{gl}\left(n,\,\mathbb{K}\right)\right)_{2n}$ is usually known as the truncated Weyl algebra.]

Theorem 2.2.4'. The homomorphism (1) induces an isomorphism in cohomology.

In order to deduce Theorem 2.2.4 from this theorem, it suffices to compare the spectral sequence of the bundle $X_n \to \mathrm{sk}_{2n}\mathbb{C}G\left(\infty,\,n\right)$ with the spectral sequence associated with the filtration

$$\{F_iW\left(\mathfrak{gl}\left(n,\,\mathbb{K}\right)\right)/F_{2n+1}W\left(\mathfrak{gl}\left(n,\,\mathbb{K}\right)\right)\} \qquad (2)$$

in $W\left(\mathfrak{gl}\left(n,\,\mathbb{K}\right)\right)_{2n}$. This latter spectral sequence satisfies

$$E_2^{p,\,q}=\begin{cases} H^q\left(\mathfrak{gl}\left(n,\,\mathbb{K}\right);\,S^{p/2}\left(\mathfrak{gl}\left(n,\,\mathbb{K}\right)'\right)\right). \\ \quad \text{if } p \text{ is even and } p\leqslant 2n, \\ 0 \text{ otherwise} \end{cases}$$

(see 1.3.5 again) and, at the same time,

$$H^q\left(\mathfrak{gl}\left(n,\,\mathbb{K}\right);\,S^{p/2}\left(\mathfrak{gl}\left(n,\,\mathbb{K}\right)'\right)\right) =H^q\left(\mathfrak{gl}\left(n,\,\mathbb{K}\right)\right)\otimes \mathrm{Inv}\,S^{p/2}\left(\mathfrak{gl}\left(n,\,\mathbb{K}\right)'\right)$$

(see Theorem 2.1.2). The ring $\mathrm{Inv}\,S^*\left(\mathfrak{gl}\left(n,\,\mathbb{K}\right)'\right)$ is the polynomial ring in the generators

$$\zeta_k \in S^k\left(\mathfrak{gl}\left(n,\,\mathbb{K}\right)'\right),\quad k=1,\,\ldots,\,n,$$

$$\zeta_k\left(g_1,\,\ldots,\,g_k\right)= \sum_{\sigma\in \mathrm{Symin}\,(k)} \mathrm{Tr}\left(g_{\sigma(1)} \cdots g_{\sigma(k)}\right);$$

see Theorem 2.1.5. This already shows that the second terms of the spectral sequences under consideration coincide and the acyclicity of the complex $W\left(\mathfrak{gl}\left(n,\,\mathbb{K}\right)\right)$ (see 1.3.5) implies that, in the spectral sequence associated with the filtration (2), the exterior generators of the algebra $E_2^{0,\,*}$ are transgressive and are mapped by transgression into the mul-

tiplicative generators of the algebra E_2^*, i.e., the differentials of the two spectral sequences act in the same way.

The proof of Theorem 2.2.4' also uses a spectral sequence associated with the filtration (2), only now we compare this spectral sequence with the Serre—Hochschild spectral sequence from Subsection 2. To do this, note that the
homomorphism (1) is compatible with the filtration (2) and
with the Serre—Hochschild filtration in $C^{\cdot}(W_n)$, so that we
obtain a homomorphism of one spectral sequence to another.
It suffices to verify that this homomorphism establishes an
isomorphism between the terms E_2 and even (in view of its
multiplicativity) between the parts $E_2^{0,*}$, $E_2^{*,0}$ of these terms.
For $E_2^{0,*}$ this is obvious (on $E_2^{0,*}$ the homomorphism actually
is the identity) and for $E_2^{*,0}$ it follows from what we have
already pointed out in this subsection and in Lemma 1 of Subsection 2, as well as from the directly verifiable fact that,
up to a nonzero factor, the homomorphism sends ξ_r to Ψ_r.

Note that this proof is based on Lemma 1 from Subsection
2 but does not use Lemma 2, so that overall it is shorter
than the proof of Theorem 2.2.4. Note also that the statement of 2.2.4', in a certain sense, is more convenient than
the statement of 2.2.4; for example, it implies the following
useful statement:

Corollary 2.2.5. Any cohomology class of the algebra
W_n is represented by a cocycle which depends only on 2-jets
of its arguments.

Indeed, this property is possessed by all the cochains
in the image of the homomorphism (1).

(For another proof of this corollary see [31].)

D.   Possibly the reader will find the following direct computation of the ring $H^*(W_1)$ of interest.   Denote by the symbol $y_1^{m_1} \ldots y_q^{m_q}$ the q-linear functional on W_1, which acts in accordance with the formula

$$\left(\sum_m a_{1m} x^m \frac{\partial}{\partial x}, \ldots, \sum_m a_{qm} x^m \frac{\partial}{\partial x} \right) \mapsto m_1! \ldots m_q! \, a_{1m_1} \ldots a_{qm_q}.$$

This notation allows us to identify the q-th tensor power of the space W_1' with $\mathbb{K}[y_1, \ldots, y_q]$, and $C^q(W_1)$ with the space of skew-symmetric polynomials in q variables.   After this identification, the differential $d \colon C^q(W_1) \to C^{q+1}(W_1)$ can be written in the form

$$dP(y_1, \ldots, y_{q+1}) = \sum_{1 \leqslant s < t \leqslant q+1} (-1)^{s+t-1} (y_t - y_s) P(y_s + y_t, y_1, \ldots \hat{y}_s \ldots \hat{y}_t \ldots, y_{q+1}).$$

It is clear from this formula that the differential maps homogeneous polynomials into homogeneous ones and increases their degree by 1; therefore, the homogeneous components of any cocycle are also cocycles.   Suppose P is a homogeneous cocycle of degree m .   Set

$$R(y_1, \ldots, y_{q-1}) = P_q'(y_1, \ldots, y_{q-1}, 0)$$

(P_s' is the derivative of the polynomial P with respect to the s-th variable).   Taking the derivative of the relation $0 = dP(y_1, \ldots, y_{q+1})$ with respect to y_{q+1} and then setting $y_{q+1} = 0$,   we obtain

$$0 = \sum_{1 \leqslant s < t \leqslant q} (-1)^{s+t-1} (y_t - y_s) P_q'(y_s + y_t, y_1, \ldots \hat{y}_s \ldots \hat{y}_t \ldots, y_q, 0)$$
$$+ \sum_{1 \leqslant s \leqslant q} (-1)^{s+q} P(y_s, y_1, \ldots \hat{y}_s \ldots, y_q)$$
$$+ \sum_{1 \leqslant s \leqslant q} (-1)^{s+q} (-y_s) P_1'(y_s, y_1, \ldots \hat{y}_s \ldots, y_q).$$

The first of these sums equals $dR(y_1, \ldots, y_q)$; also,

$$P(y_s, y_1, \ldots \hat{y}_s \ldots, y_q) = (-1)^{s-1} P(y_1, \ldots, y_q),$$
$$P'_1(y_s, y_1, \ldots \hat{y}_s \ldots, y_q) = (-1)^{s-1} P'_s(y_1, \ldots, y_q)$$

and $\sum_{|s} y_s P'_s = mP$ ("Euler's formula"). Thus,

$$0 = dR + (-1)^{q-1} qP + (-1)^q mP,$$

i.e., $(-1)^q (q - m) P = dR$. If $q \neq m$, this shows that the cocycle P is cohomologous to 0. Thus, in computing the cohomology, we can limit ourselves to polynomials with a number of variables equal to the degree. But such polynomials appear rarely: skew-symmetric polynomials are divisible by the product of all possible differences of their variables and are of degree $\geqslant q(q-1)/2$, where q is the number of variables. To be more precise, there are only four (up to constant factors) homogeneous skew-symmetric polynomials of degree equal to the number of variables: $P_0 = 1$, $P_1(y) = y$, $P_2(y_1, y_2) = y_2^2 - y_1^2$, and $P_3(y_1, y_2, y_3) = (y_2 - y_1)(y_3 - y_2)(y_1 - y_3)$. Further, $dP_0 = 0$, $dP_1 = P_2$, $dP_3 = 0$; therefore, $H^0(W_1) \cong \mathbb{K}$, $H^3(W_1) \cong \mathbb{K}$ and $H^q(W_1) = 0$ for $q \neq 0, 3$.

E. Finally, some statements concerning the relative case.

Theorem 2.2.6.

$$H^*(W_n, \mathfrak{gl}(n, \mathbb{K})) = H^*(\mathrm{sk}_{2n}\mathbb{C}G(\infty, n); \mathbb{K});$$
$$H^*(\mathbb{R}W_n, \mathfrak{o}(n)) = H^*(X_n/\mathrm{SO}(n); \mathbb{R});$$
$$H^*(\mathbb{R}W_n, \mathrm{O}(n)) = H^*(X_n/\mathrm{O}(n); \mathbb{R}).$$

The first isomorphism was essentially constructed when we proved Theorem 2.2.4; the two others are established in the same way as Theorem 2.2.5 by using the relative Serre—Hochschild spectral sequences (see the end of 1.5.1). The details are left to the reader.

4. Cohomology with coefficients in formal tensor fields.
A. The case of formal differential forms. Denote by Ω_n^q the space of formal exterior differential q-forms in $\mathbb{K}^n$; this space possesses a natural W_n-module structure. In particular, Ω_n^0 is the ring of formal power series in $\mathbb{K}^n$.

Theorem 2.2.7 (first proved in [30]). The bigraded algebra $H^*(W_n; \Omega_n^*)$ is generated by the elements $\lambda_i \in H^{2i-1}(W_n; \Omega_n^0)$ $[i = 1, \ldots, n]$ and $\mu_j \in H^j(W_n; \Omega_n^j)$ $[j = 1, \ldots, n]$; these generators satisfy the relations $\lambda_i \lambda_j = -\lambda_j \lambda_i$, $\lambda_i \mu_j = \mu_j \lambda_i$, $\mu_i \mu_j = \mu_j \mu_i$, and $\mu_{j_1} \cdots \mu_{j_s} = 0$ for $j_1 + \ldots + j_s > n$.

Remarks. 1°. In particular, $H^*(W_n; \Omega_n^0) = H^*(\mathfrak{gl}(n, \mathbb{K}))$.

2°. We shall not describe the cocycles representing the classes λ_i and μ_j; we shall limit ourselves to two simple formulas: λ_1 is represented by the cocycle $\xi \mapsto \operatorname{div} \xi$, and μ_1 by the cocycle $\xi \mapsto d \operatorname{div} \xi$.

3°. The reader noted possibly that the ring $H^*(W_n; \Omega_n^*)$, whose structure was described in the statement of the theorem, is isomorphic up to a change of grading to the E_2 term of the Serre–Hochschild spectral sequence corresponding to the subalgebra $\mathfrak{gl}(n, \mathbb{K})$ of the algebra W_n (see Subsection 2): $H^p(W_n; \Omega_n^q) \cong E_2^{2q, p-q}$. This coincidence is not as superficial as might appear at first glance. Indeed, the exact sequence

$$0 \to \mathbb{K} \to \Omega_n^0 \xrightarrow{d} \Omega_n^1 \xrightarrow{d} \ldots \xrightarrow{d} \Omega_n^n \to 0 \to \ldots,$$

in which d denotes the exterior differential, induces a certain spectral sequence $\{'E_r^{p,q}, 'd_r^{p,q}\}$ with $'E_1^{p,q} = H^q(W_n; \Omega_n^p)$, converging to $H^*(W_n; \mathbb{K}) = H^*(W_n)$. It can be shown that the spectral sequence $\{'E_r^{p,q}, 'd_r^{p,q}\}$ differs from the Serre–Hochs-

child spectral sequence $\{E_r^{p,\,q},\quad d_r^{p,\,q}\}$ only by the grading:
$'E_r^{p,\,q} = E_{2r-1}^{2p,\,q-p} = E_{2r}^{2p,q-p}$, $'d_r^{p,\,q} = d_{2r}^{2p,\,q-p}$. The reader should return to this statement after the proof of Theorem 2.2.7 and try to prove it as an exercise.

Now, to prove Theorem 2.2.7, consider the L_0-module $\Lambda^j V'$, in which the action of algebra L_0 is induced by the canonical action of the algebra $\mathfrak{gl}\,(n,\,\mathbb{K})$ in $\Lambda^j V'$ by means of the projection $L_0 \twoheadrightarrow L_0/L_1 = \mathfrak{gl}\,(n,\,\mathbb{K})$. Since

$$\Omega_n^j = \mathrm{Coind}_{W_n}(\Lambda^j V')$$

(see 1.2.3), we have the canonical isomorphism

$$H^*\,(W_n;\,\Omega_n^j) = H^*\,(L_0;\,\Lambda^j V')$$

(see Theorem 1.5.4). Thus, Theorem 2.2.7 may be restated as a statement concerning the cohomology of the algebra L_0. Here is this restatement, with certain refinements.

__Theorem 2.2.7'__. (i) The inclusion $\mathfrak{gl}\,(n,\,\mathbb{K}) \to L_0$ induces the isomorphism $H^*\,(L_0) \cong H^*\,(\mathfrak{gl}\,(n,\,\mathbb{K}))$.

(ii) Suppose λ_i is the inverse image under the previous isomorphism of the class $\varphi_i \in H^{2i-1}\,(\mathfrak{gl}\,(n,\,\mathbb{K}))$ (see Theorem 2.1.5) and let $\mu_j' \in H^j\,(L_0;\,\Lambda^j V')$ be the cohomology class of the cocycle $\Lambda^j L_0 \to \Lambda^j V'$, defined by the formula

$$\xi_1 \wedge \cdots \wedge \xi_j \mapsto \{v_1 \wedge \cdots \wedge v_j \mapsto \Psi_j\,(v_1 + \xi_1, \ldots, v_j + \xi_j)\},$$

where $\xi_1, \ldots, \xi_j \in L_0, v_1, \ldots, v_j \in V$, Ψ_j is defined in Subsection 2 (see Lemma 1 and Remark 1 to its statement) and the notation $v_s + \xi_s$ implies the natural identification $W_n = V \oplus L_0$. It is claimed that the bigraded algebra $H^*\,(L_0;\,\Lambda^* V')$ is generated by its elements λ_i', μ_j', and these elements satisfy the relations $\lambda_i'\lambda_j' = -\lambda_j'\lambda_i'$, $\mu_i'\mu_j' = \mu_j'\lambda_i'$, $\mu_i'\mu_j' = \mu_j'\mu_i'$, and $\mu_{j_1}' \cdots \mu_{j_r}' = 0$ for $j_1 + \ldots + j_r > n$.

For the proof, it suffices to establish that the classes

$$\mu'_{j_1} \ldots \mu'_{j_r} \lambda'_{i_1} \ldots \lambda'_{i_s}$$

$$[1 \leqslant j_1 \leqslant \ldots \leqslant j_r \leqslant n, \; 1 \leqslant i_1 < \ldots < i_s \leqslant n, \; j_1 + \ldots + j_r = m]$$

(3)

constitute a basis in $H^* (L_0; \Lambda^m V')$. Using the Serre–Hochschild spectral sequence corresponding to the subalgebra $\mathfrak{gl}(n, \mathbb{K})$ of the algebra L_0, we can write

$$E_1^{p, q} = H^q (\mathfrak{gl}(n, \mathbb{K}); \Lambda^p (L_0/\mathfrak{gl}(n, \mathbb{K}))' \otimes \Lambda^m V')$$
$$= H^q (\mathfrak{gl}(n, \mathbb{K})) \otimes \mathrm{Inv}\, [\Lambda^p (L_0/\mathfrak{gl}(n, \mathbb{K}))' \otimes \Lambda^m V']$$
$$= H^q (\mathfrak{gl}(n, \mathbb{K})) \otimes \mathrm{Inv}\, [\Lambda^p (\overset{\infty}{\underset{r=2}{\oplus}} (S^r V \otimes V')) \otimes \Lambda^m V']$$
$$= H^q (\mathfrak{gl}(n, \mathbb{K})) \otimes \underset{p_2 + p_3 + \ldots = p}{\oplus} \mathrm{Inv}\, [[\overset{\infty}{\underset{r=2}{\otimes}} \Lambda^{p_r} (S^r V \otimes V')] \otimes \Lambda^m V'].$$

The rings which are the direct summands in the last expression were computed in Subsection 2 (see Lemma 1) — $\mathrm{Inv}[[\overset{\infty}{\underset{r=2}{\otimes}} \Lambda^{p_r} (S^r V \otimes V')] \otimes \Lambda^m V'] = 0$, if $(p_2, \; p_3, \ldots) \neq (m, \; 0, \ldots)$ — while a basis of the space $\mathrm{Inv}\, (\Lambda^m (S^2 V \otimes V') \otimes \Lambda^m V')$ is formed by the products $\Psi_{m_1} \ldots \Psi_{m_s}$ with $m_1 + \ldots + m_s = m$. In particular, $E_1^{p, q} = 0$ for $p \neq m$. Therefore, $E_\infty^{p, q} = E_1^{p, q}$ and

$$H^r (L_0; \Lambda^m V') = E_\infty^{m, \, m-r} = E_1^{m, \, m-r} = E_1^{m, \, 0} \otimes H^{m-r} (\mathfrak{gl}(n, \mathbb{K})).$$

We see that (3) is indeed a basis of $H^* (L_0; \Lambda^m V')$.

B. The general case. The cohomology of the algebra W_n with coefficients in the space of formal tensor fields of arbitrary form is known at the present time only for $n = 1$ (see §3). It should nevertheless be mentioned that its computation reduces to that of the cohomology of the algebra $L_1 (n) \subset W_n$, with trivial coefficients. To state the corresponding result, note that the algebra $\mathfrak{gl}(n, \mathbb{K})$ acts in L_1 by automorphisms and, therefore, $H^* (L_1)$ is a graded $\mathfrak{gl}(n, \mathbb{K})$-module.

Theorem 2.2.8 (Losik [66]). Suppose A is a tensor $\mathfrak{gl}(n, \mathbb{K})$-module and let $\mathcal{A}$ be the corresponding space of formal tensor fields, i.e., the W_n-module coinduced by the L_0-module A. Then

$$H^*(W_n; \mathcal{A}) \cong H^*(\mathfrak{gl}(n, \mathbb{K})) \otimes \mathrm{Inv}_{\mathfrak{gl}(n, \mathbb{K})}(H^*(L_1) \otimes A). \qquad (4)$$

Proof. First of all, $H^*(W_n; \mathcal{A}) = H^*(L_0; A)$ (Theorem 1.5.4). Further, consider the Serre—Hochschild spectral sequence corresponding to the subalgebra (ideal) L_1 of the algebra L_0. For it we have

$$E_2^{p,q} = H^p(L_0/L_1; H^q(L_1; A))$$
$$= H^p(\mathfrak{gl}(n, \mathbb{K}); H^q(L_1) \otimes A)$$
$$= H^p(\mathfrak{gl}(n, \mathbb{K})) \otimes \mathrm{Inv}(H^q(L_1) \otimes A)$$

(we make use of Theorem 2.1.2 and the fact that A is a trivial L_1-module). Thus it remains to show that, for our spectral sequence, $E_\infty = E_2$. To do this, note that the projection $L_0 \to \mathfrak{gl}(n, \mathbb{K})$ allows us to view cocycles $\Phi_j \in C^{2j-1}(\mathfrak{gl}(n, \mathbb{K}))$ as cocycles of the algebra L_0. The multiplication on Φ_j defines, in $C^*(L_0; A)$, a graded $H^*(\mathfrak{gl}(n, \mathbb{K}))$-module structure which is compatible with the Serre—Hochschild filtration with respect to L_1. Therefore, the entire spectral sequence is a sequence of $H^*(\mathfrak{gl}(n, \mathbb{K}))$-modules and its differentials are $H^*(\mathfrak{gl}(n, \mathbb{K}))$-homomorphisms. But, as can be seen from the previous computation, the $H^*(\mathfrak{gl}(n, \mathbb{K}))$-module E_2 is generated by elements of the space $E_2^{0,*} = \mathrm{Inv}(H^*(L_1) \otimes A)$, and it suffices to prove that these elements belong to the kernel of all the differentials. The last statement is equivalent to the epimorphicity of the natural map $H^*(L_0; A) \to \mathrm{Inv}(H^*(L_1) \otimes A)$, and this can be proved by a direct construction of the right inverse homomorphism. Consider the map $C^q(L_1; A) \to$

$C^q (L_0; A)$, sending each cochain $c \in C^q (L_1; A)$ into the co-chain $\bar{c} \in C^q (L_0; A)$, defined by the formula

$$\bar{c} (\xi_1, \ldots, \xi_q) = c (\pi (\xi_1), \ldots, \pi (\xi_q)),$$

where π is the projection $L_0 \to L_0/\mathfrak{gl} (n, \mathbb{K}) = L_1$. This map does not commute with the differential, but obviously $d\bar{c} = \overline{dc}$, if the cochain c is $\mathfrak{gl} (n, \mathbb{K})$-invariant. We obtain the homomorphism

$$\mathrm{Inv}\ C^{\cdot} (L_1;\ A) \to C^{\cdot} (L_0;\ A),$$

which induces the required map

$$H^q (\mathrm{Inv}\ C^{\cdot} (L_1;\ A)) = \mathrm{Inv}\ (H^q (L_1) \otimes A) \to H^q (L_0;\ A).$$

Remarks. 1°. Our proof contains the construction of the canonical isomorphism (4).

2°. In view of the classical Schur theorem, the dimension of the space $\mathrm{Inv}\ (H^* (L_1) \otimes A)$ in the case when the representation A is irreducible equals the multiplicity with which A' appears in the decomposition of the representation $H^* (L_1)$ into irreducible ones.

3°. In some cases, to be sure, fairly simple ones, Theorem 2.2.8 enables us to compute the cohomology explicitly. For example,

$$H^* (L_0;\ T_l^k (n)) = 0, \text{ if } k < l$$

(this is an exercise).

4°. The requirement that A be a tensor module, appearing in the statement of Theorem 2.2.8, is unnecessarily strong and can be weakened to the requirement of finite dimensionality of the module A. The only change which must then be made in the proof of the theorem is to replace the reference to

Theorem 2.1.2 by a reference to its strengthened version from 1.7.

As was already point out, for $n > 1$ practically nothing is known about $H^* (L_1)$ and the information on the cohomology of the algebra W_n with tensor coefficients is covered by Theorems 2.2.3 and 2.2.7 and the following statement, due to G. Segal and myself, which the reader can try to prove as an exercise.

Theorem 2.2.9. If $r + q < n$, then

$$H^q (L_0; S^r V') \cong H^r (\mathfrak{gl} (n, \mathbb{K})) \otimes H^{q-r} (\mathfrak{gl} (n, \mathbb{K})).$$

Moreover, the corresponding parts of the rings $H^* (L_0; S^* V')$ and $H^* (\mathfrak{gl} (n, \mathbb{K})) \otimes H^* (\mathfrak{gl} (n, \mathbb{K}))$ are multiplicatively isomorphic.

Probably, the inequality $r + q < n$ may be replaced by $r \leqslant n$ in this statement.

It is interesting to compare Theorems 2.2.7 and 2.2.9, having restated them in the following form: We have the following partial multiplicative isomorphisms:

$$H^* (L_0; \Lambda^* V') = H^* (\mathfrak{gl} (n, \mathbb{K})) \otimes \operatorname{Inv} S^* \mathfrak{gl} (n, \mathbb{K}),$$
$$H^* (L_0; S^* V') = H^* (\mathfrak{gl} (n, \mathbb{K})) \otimes \operatorname{Inv} \Lambda^* \mathfrak{gl} (n, \mathbb{K})$$

(see Theorem 2.1.5) which transform the bidegree (q, r) into $(q - r, r)$.

We also have the partial multiplicative isomorphism

$$H^* (L_0; \otimes^* V') = H^* (\mathfrak{gl} (n, \mathbb{K})) \otimes \operatorname{Inv} (\otimes^* \mathfrak{gl} (n, \mathbb{K})).$$

5. The computation of $H^* (W_n; W_n')$. This computation was carried out by Gelfand, Feigin, and myself in [24]. It is useful for the characteristic classes of foliations (§3.1),

and I have decided to present it here, although it is rather cumbersome.

Theorem 2.2.10.

$$H^q(W_n; W_n') \cong H^{2n+1}(W_n) \otimes H^{q-2n}(\mathfrak{gl}(n, \mathbb{K}));$$

the $H^*(W_n)$-module structure in $H^*(W_n; W_n')$ is trivial (i.e., if $x \in H^q(W_n; W_n')$, $y \in H^r(W_n)$, and $r > 0$, then $xy = 0$).

In the following subsection we shall construct the canonical isomorphism $H^q(W_n; W_n') \cong H^{2n+1}(W_n) \otimes H^{q-2n}(\mathfrak{gl}(n,\mathbb{K}))$.

Let us write out the Serre–Hochschild spectral sequence corresponding to the subalgebra $\mathfrak{gl}(n, \mathbb{K})$ of the algebra W_n. For this spectral sequence, we have

$$E_1 = H^*(\mathfrak{gl}(n, \mathbb{K}); (\Lambda^*(W_n/\mathfrak{gl}(n, \mathbb{K})) \otimes W_n)').$$

Obviously,

$$\Lambda^*((W_n/\mathfrak{gl}(n, \mathbb{K})) \otimes W_n)' = \bigoplus_{p_0, p_2, p_3, \ldots; j} [\Lambda^{p_0}V'$$

$$\otimes \Lambda^{p_2}(S^2V \otimes V') \otimes \Lambda^{p_3}(S^3V \otimes V') \otimes \ldots \otimes (S^jV \otimes V')],$$

and, therefore, E_1 is the tensor product of $H^*(\mathfrak{gl}(n, \mathbb{K}))$ by the sum of the invariant summands in the last sum. An element of the summand corresponding to $p_0, p_2, p_3, \ldots; j$ is a poly-linear functional of the variables

$$\alpha_1, \ldots, \alpha_{p_0};$$
$$\beta^1_{p_0+1}, \beta^2_{p_0+1}, \alpha_{p_0+1}; \ldots; \beta^1_{p_0+p_2}, \beta^2_{p_0+p_2}, \alpha_{p_0+p_2};$$
$$\beta^1_{p_0+p_2+1}, \beta^2_{p_0+p_2+1}, \beta^3_{p_0+p_2+1}, \alpha_{p_0+p_2+1}; \ldots, \alpha_{p_0+p_2+p_3};$$
$$\cdots \cdots \cdots \cdots \cdots \cdots$$
$$\beta^1_{p_0+p_2+p_3+\ldots+1}, \ldots, \beta^j_{p_0+p_2+p_3+\ldots+1}, \alpha_{p_0+p_2+p_3+\ldots+1}.$$

Here all the α belong to V and all the β belong to V'. To the conditions of (skew-) symmetry described in Subsection 2 (before Lemma 1) we must add that the functional is symmetric with respect to $\beta^1_{p_0+p_2+p_3+\ldots+1}, \ldots, \beta^j_{p_0+p_2+p_3+\ldots+1}$.

Note that the space $E_1^{*,\,0}$ does not possess any natural ring structure, but is a module over a similar space from the spectral sequence of Subsection 2. In particular, the elements from our $E_1^{*,\,0}$ may be multiplied by Ψ_r from Lemma 1 (Subsection 2), and this multiplication increases their degree by $2r$.

For $s,\,t,\,u \geqslant 0$ and $\varepsilon_1,\,\varepsilon_2,\,\varepsilon = 0$ or 1, let us define the elements $\rho_{st}^{\varepsilon_1\varepsilon_2}$, σ_u^{ε} of the spaces

$$\mathrm{Inv}\,[\Lambda^{s+t+\varepsilon_1+\varepsilon_2}V' \otimes \Lambda^{s+t+1-\varepsilon_2}(S^2V \otimes V')$$
$$\otimes \Lambda^{\varepsilon_2}(S^3V \otimes V') \otimes (S^{\varepsilon_1}V \otimes V')] \subset E_1^{2s+2t+\varepsilon_1+\varepsilon_2+1,\,0},$$
$$\mathrm{Inv}\,[\Lambda^{u+\varepsilon}V' \otimes \Lambda^u(S^2V \otimes V') \otimes (S^{1+\varepsilon}V \otimes V')] \subset E_1^{2u+\varepsilon,\,0}$$

as (skew-) symmetrizations of the functionals

$$(\alpha,\,\beta) \mapsto \left[\prod_{i=1}^{s+t} \beta^1_{s+t+\varepsilon_1+\varepsilon_2+i}(\alpha_i)\right]$$

$$\times \left[\prod_{j=2}^{s} \beta^2_{s+t+\varepsilon_1+\varepsilon_2+j}(\alpha_{s+t+\varepsilon_1+\varepsilon_2+j-1})\right]\left[\prod_{k=2}^{t} \beta_{2s+t+\ 1+\varepsilon_2+k}(\alpha_{2s+t+\varepsilon_1+\varepsilon_2+k-1})\right]$$

$$\times \beta^1_{s+2t+\varepsilon_1+\varepsilon_2+1}(\alpha_{2s+t+\varepsilon_1+\varepsilon_2+1})\,\beta^2_{2s+2t+\varepsilon_1+\varepsilon_2+1}(\alpha_{2s+2t+\varepsilon_1+\varepsilon_2})$$

$$\times \beta^2_{2s+t+\varepsilon_1+\varepsilon_2+1}(\alpha_{2s+2t+\varepsilon_1+\varepsilon_2+1})\,\beta^2_{s+t+\varepsilon_1+\varepsilon_2+1}(\alpha_{2s+2t+\varepsilon_1+\varepsilon_2+2})$$

$$\times \begin{cases} 1, & \text{if } \dot{\varepsilon}_1 = 0,\ \varepsilon_2 = 0, \\ \beta^3_{2s+2t+2}(\alpha_{s+t+1}), & \text{if } \varepsilon_1 = 1,\ \varepsilon_2 = 0, \\ \beta^1_{2s+2t+3}(\alpha_{s+t+1}), & \text{if } \varepsilon_1 = 0,\ \varepsilon_2 = 1, \\ \beta^3_{2s+2t+3}(\alpha_{s+t+1})\,\beta^1_{2s+2t+4}(\alpha_{s+t+2}), & \text{if } \varepsilon_1 = 1,\ \varepsilon_2 = 1; \end{cases}$$

$$(\alpha,\,\beta) \mapsto \left\lfloor\prod_{i=1}^{u} \beta^1_{u+\varepsilon+i}(\alpha_i)\right\rfloor\left[\prod_{j=2}^{u} \beta^2_{u+\varepsilon+j}(\alpha_{u+\varepsilon+j-1})\right]$$

$$\times \beta^1_{2u+\varepsilon+1}(\alpha_{2u+\varepsilon})\,\beta^2_{u+\varepsilon+1}(\alpha_{2u+\varepsilon+1}) \begin{cases} 1, & \text{if } \varepsilon = 0, \\ \beta^2_{2u+2}(\alpha_{u+1}), & \text{if } \varepsilon = 1. \end{cases}$$

The skew symmetrization consists in adding (to the given functional) all possible functionals obtained from it by admissible permutations of α and β and supplied with the ap-

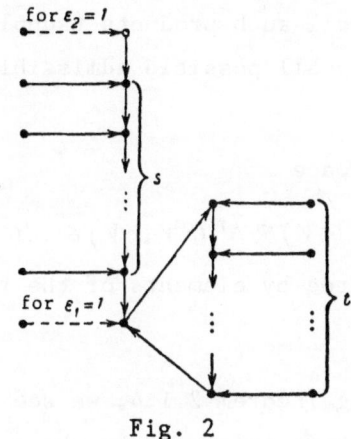

Fig. 2

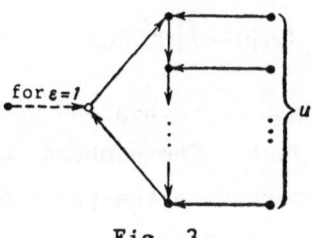

Fig. 3

propriate signs. The functionals $\rho_{st}^{\varepsilon_1\varepsilon_2}$ and σ_u^ε may be determined by pictures similar to Fig. 1 — see Figs. 2 and 3.
These pictures should be used in the same way as Fig. 1:
first one must number the points, so that the first numbers
are assigned to the black points into which no arrows enter,
then to the black points into which two arrows enter, then
to the black point into which three arrows enter (if such a
point exists), and, finally, the white point; after this, for
the arrow going from the k-th point to the l-th one, we
write $\beta_l(\alpha_k)$, adding the superscripts 1, 2, and, if necessary, 3 to β (so that no β's with identical indices are obtained), take the product of all the expressions written out

and then the sum of all such products (supplied with the appropriate signs) over all possible admissible numerations and indices.

Lemma 1. The space

$$\text{Inv}\,[\Lambda^{p_0}V' \otimes \Lambda^{p_2}(S^2V \otimes V') \otimes \Lambda^{p_3}(S^3V \otimes V') \otimes \ldots \otimes (S^jV \otimes V')]$$

is additively generated by elements of the form $\Psi_{r_1}\ldots\Psi_{r_m}\rho_{st}^{\varepsilon_1\varepsilon_2}$, $\Psi_{r_1}\ldots\Psi_{r_m}\sigma_u^{\varepsilon}$.

Proof. Applying Theorem 2.1.4, we see that the space we are concerned with is generated by skew-symmetric functionals of the form

$$(\alpha,\,\beta) \mapsto \prod_{r=1}^{q+1} \beta_{i_r^r}^j\,(\alpha_r),$$

where $q = p_0 + p_2 + p_3 + \ldots$, each β and each α appearing precisely once in the product. The numbers $i_1,\ldots,i_{p_0}$ must be pairwise distinct (compare with the proof of Lemma 1 from Subsection 2). All in all, β has $p_2 + p_3 + \ldots + 1$ pairwise distinct subscripts if $j \neq 0$, and $p_2 + p_3 + \ldots$, if $j = 0$; thus

$$p_0 \leqslant \begin{cases} p_2 + p_3 + \ldots + 1, & \text{if } j \neq 0, \\ p_2 + p_3 + \ldots, & \text{if } j = 0. \end{cases}$$

At the same time, the total number of all α's is equal to the total number of all β's, i.e., $p_0 + p_2 + p_3 + \ldots + 1 = 2p_2 + 3p_3 + \ldots + j$, or

$$p_0 = p_2 + 2p_3 + \ldots + (j-1).$$

Comparing this relation with the previous inequality, we conclude that the following five possibilities arise for p_0, p_2, p_3, $\ldots$, j:

$$(A) \ j = 0, \quad p_2 = p_0 + 1, \quad p_3 = p_4 = \ldots = 0;$$
$$(B) \ j = 0, \quad p_2 = p_0 - 1, \quad p_3 = 1, \quad p_4 = \ldots = 0;$$
$$(C) \ j = 1, \quad p_2 = p_0, \qquad\quad p_3 = p_4 = \ldots = 0;$$
$$(D) \ j = 1, \quad p_2 = p_0 - 2, \quad p_3 = 1, \ p_4 = \ldots = 0;$$
$$(F) \ j = 2, \quad p_2 = p_0 - 1, \quad p_3 = p_4 = \ldots = 0.$$

These possibilities will be studied in succession and all our arguments will be similar. We shall discuss the case (A) in detail and limit ourselves to an indication of the final result in the other cases.

(A) By an admissible permutation of the indices, we can rewrite our monomial $\prod_{r=1}^{q+1} \beta_{i_r}^{j_r}(\alpha_r)$ in the form

$$\beta_{p_0+1}^1 (\alpha_1) \ldots \beta_{2p_0}^1 (\alpha_{p_0}) \beta_{2p_0+1}^1 (\alpha_{p_0+s(0)}) \ \beta_{p_0+1}^2 (\alpha_{p_0+s(1)}) \ldots \beta_{2p_0+1}^2 (\alpha_{p_0+s(p_0+1)}),$$

where s is the bijection of the set $\{0, 1, \ldots, p_0 + 1\}$ onto $\{1, \ldots, p_0 + 1, \ p_0 + 2\}$. By an appropriate permutation of the numbers $1, \ldots, p_0 + 1$ we can rewrite this map in the form

$$s(0) = 1, \ldots, s(r_0 - 1) = r_0, \ s(r_0) = p_0 + 2,$$
$$s(r_0 + 1) = r_0 + 2, \ldots, s(r_0 + r_1 - 1) = r_0 + r_1,$$
$$s(r_0 + r_1) = r_0 + 1,$$
$$\cdot \ \cdot \ \cdot \ \cdot \ \cdot \ \cdot \ \cdot \ \cdot \ \cdot \ \cdot \ \cdot \ \cdot \ \cdot \ \cdot \ \cdot \ \cdot \ \cdot$$
$$s(r_0 + \ldots + r_m + 1) = r_0 + \ldots + r_m + 2, \ldots,$$
$$s(r_0 + \ldots + r_{m+1} - 1) = r_0 + \ldots + r_{m+1},$$
$$s(r_0 + \ldots + r_{m+1}) = r_0 + \ldots + r_m + 1,$$

where $r_0 + \ldots + r_{m+1} = p_0 + 1$. (The numbers $r_1, \ldots, r_{m+1}$ are positive, while r_0 can vanish; when it does, the upper line reduces to $s(0) = p_0 + 2$.) The symmetrization of this monomial gives us (up to a sign)

$$\Psi_{r_1} \ldots \Psi_{r_m} \rho_{st}^{00},$$

where $s = r_0$, $t = r_{m+1} - 1$; these elements are the generators of the part of the space $E_1^{*,0}$ corresponding to the case (A).

(B) The part of the space $E_1^{*,0}$ corresponding to this case is generated by elements of the form

$$\Psi_{r_1} \ldots \Psi_{r_m} \rho_{st}^{10}.$$

(C) The part of the space $E_1^{*,0}$ corresponding to this case is generated by elements of two types:

$$\Psi_{r_1} \ldots \Psi_{r_m} \rho_{st}^{01} \text{ and } \Psi_{r_1} \ldots \Psi_{r_m} \sigma_u^0.$$

(D) The corresponding part of the space $E_1^{*,0}$ is generated by the monomials

$$\Psi_{r_1} \ldots \Psi_{r_m} \rho_{st}^{11}.$$

(F) The corresponding part of the space $E_1^{*,0}$ is generated by the monomials

$$\Psi_{r_1} \ldots \Psi_{r_m} \sigma_u^1.$$

The lemma is proved.

The parts of the space $E_1^{q,0}$ generated by elements of the form $A, \ldots, F$ will be denoted, respectively, by $A_q, \ldots, F_q$. It is obvious that $E_1^{q,0}$ is the direct sum of $A_q, \ldots, F_q$ and that the spaces A_q, D_q, F_q are nontrivial only for odd q, while B_q, C_q are nontrivial only for even q. Thus,

$$E_1^{q,0} = \begin{cases} A_q \oplus D_q \oplus F_q & \text{for odd } q, \\ B_q \oplus C_q & \text{for even } q. \end{cases}$$

We have indicated generators for all the spaces A_q, $\ldots, F_q$. Now let us find the relations between them. (Note that only in the form of these relations will the dependence of $A_q, \ldots, F_q$ on n appear.)

Lemma 2. (i) The spaces A_q for $q > 2n + 1$, B_q for $q > 2n$, C_q for $q > 2n$, D_q for $q > 2n - 1$, and F_q for $q > 2n - 1$ are trivial.

(ii) In the spaces A_q for $q \leqslant 2n - 1$, B_q for $q \leqslant 2n$, C_q for $q \leqslant 2n - 2$, D_q for $q \leqslant 2n - 1$, and F_q for $q \leqslant 2n - 1$ the generators indicated previously are linearly independent.

(iii) In the space C_{2n}, a determining system of relations will be

$$\Psi_{r_1} \dots \Psi_{r_m} \sigma_u^0 = \sum_{s+t=u-1} \Psi_{r_1} \dots \Psi_{r_m} \rho_{st}^{01}$$

$$+ \sum_{i=1}^{m} (-1)^{ur_i} r_i \Psi_{r_1} \dots \widehat{\Psi}_{r_i} \dots \Psi_{r_m} \rho_{u, r_i-1}^{01}.$$

(iv) In the space A_{2n+1}, a determining system of relations will be

$$\Psi_{r_1} \dots \Psi_{r_m} \rho_{0u}^{00} = \sum_{s+t=u-1} \Psi_{r_1} \dots \Psi_{r_m} \rho_{s+1, t}^{00}$$

$$+ \sum_{i=1}^{m} (-1)^{(u+1)r_i} \Psi_{r_1} \dots \widehat{\Psi}_{r_i} \dots \Psi_{r_m} \rho_{u+1, r_i-1}^{00};$$

$$\sum_{i=1}^{m} r_i \Psi_{r_1} \dots \widehat{\Psi}_{r_i} \dots \Psi_{r_m} \rho_{0, r_i-1}^{00} = 0.$$

To prove (i), it suffices to note that $\Lambda^{p_0} V' = 0$ for $p_0 > n$. The proof of (ii) is just as simple: if the number of distinct subscripts of β (i.e., $p_2 + p_3 + \dots$ for $j = 0$ and $p_2 + p_3 + \dots + 1$ for $j > 0$) is no greater than n, then it suffices to compute the values of our basis functionals by setting

$$\beta_{p_0+i}^k = e^i \ (i = 1, 2, \dots; \ k = 1, 2, \dots), \quad \alpha_l = e_{\varphi(l)},$$

where φ is an arbitrary map $\{1, \dots, q+1\} \to \{1, \dots, n\}$. It is clear that no more than one of our basis functionals will assume a nonzero value on the family of variables indicated above and that each of these basis functionals will assume a nonzero value on this family for an appropriate choice of φ. This implies that the basis functionals are linearly independent.

These simple arguments almost solve the problem of the relations: it remains to study the relations in A_{2n+1} and C_{2n}. This study, which is the most cumbersome part of the proof of Theorem 2.2.10, will be presented in succinct form.

We begin with statement (iii). Since the space C_{2n} is contained in $\Lambda^n V' \otimes \Lambda^n (S^2 V \otimes \dot{V}') \otimes (V \otimes V')$, its elements are functionals of the variables

$$\alpha_1, \ldots, \alpha_{2n+1}; \beta^1_{n+1}, \beta^2_{n+1}, \ldots \beta^1_{2n}, \beta^2_{2n}, \beta^1_{2n+1},$$

where all the α's are in V and all the β's in V'.

We begin by noting that a cochain from C_{2n} is nontrivial if and only if for some function $\varphi: \{1, \ldots, n+1\} \to \{1, \ldots, n\}$ the value of the cochain on the following family of variables

$$\alpha_1 = e_1, \ldots, \ \alpha_n = e_n, \ \alpha_{n+1} = e_{\varphi(1)}, \ldots, \ \alpha_{2n+1} = e_{\varphi(n+1)};$$
$$\beta^1_{n+1} = \beta^2_{n+1} = e^1, \ldots, \ \beta^1_{2n} = \beta^2_{2n} = e^n, \ \beta^1_{2n+1} = e^1$$

is nontrivial. Indeed, for any cochain $P \in C_{2n}$ the function

$$(\alpha_1, \ldots, \alpha_{2n+1}) \mapsto P(\alpha_1, \ldots, \alpha_{2n+1}; e^1, e^1, \ldots, e^n, e^n, e^1)$$

is polylinear and skew symmetric with respect to $\alpha_1, \ldots, \alpha_n$. Therefore, if the cochain vanishes on the family of variables indicated above, then

$$P(\alpha_1, \ldots, \alpha_{2n+1}; e^1, e^1, \ldots, e^n, e^n, e^1) = 0$$

for all $\alpha_1, \ldots, \alpha_{2n+1}$. From the invariance of the cochain with respect to permutations of coordinates in V and its skew symmetry with respect to permutations $(\beta^1_k, \beta^2_k, \alpha_k) \leftrightarrow (\beta^1_l, \beta^2_l, \alpha_l)$ it follows that, in our case,

$$P(\alpha_1, \ldots, \alpha_{2n+1}; e^1, e^1, \ldots, e^n, e^n, e^i) = 0$$

for any i (and all $\alpha_1, \ldots, \alpha_{2n+1}$) and this, in its turn, implies (by the polylinearity of P) that

$$P(\alpha_1, \ldots, \alpha_{2n+1}; e^1, e^1, \ldots, e^n, e^n, \beta^1_{2n+1}) = 0$$

for all $\alpha_1, \ldots, \alpha_{2n+1}, \beta^1_{2n+1}$. Further, the invariance of P with
respect to linear automorphisms of the space V implies

$$P(\alpha_1, \ldots, \alpha_{2n+1}; \beta_{n+1}, \beta_{n+1}, \ldots, \beta_{2n}, \beta_{2n}, \beta^1_{2n+1}) = 0$$

for all $\alpha_1, \ldots, \alpha_{2n+1}, \beta^1_{2n+1}$ and all linearly independent β_{n+1},
$\ldots, \beta_{2n}$; this last linear independence requirement may be
ignored in view of the continuity of P. Thus,

$$P(\alpha_1, \ldots, \alpha_{2n+1}; \beta^1_{n+1}, \beta^2_{n+1}, \ldots, \beta^1_{2n}, \beta^2_{2n}, \beta^1_{2n+1}) = 0$$

under the only condition that $\beta^1_{n+1} = \beta^2_{n+1}, \ldots, \beta^1_{2n} = \beta^2_{2n}$. But
this condition may also be ignored, since P is symmetric
with respect to permutations $\beta^1_k \leftrightarrow \beta^2_k$.

Our next (obvious) remark is that the value of the cochain P on the family of variables indicated above can be
nontrivial only if among the numbers $\varphi(1), \ldots, \varphi(n+1)$ the
number 1 appears twice and each of the numbers $2, \ldots, n$ appears once.

Further, it is necessary to compute the values of our
generators of the space C_{2n} on the family of variables corresponding to the given φ.

First case: $\varphi(n+1) = 1$. Then $\varphi \mid \{1, \ldots, n\}$ is a permutation; suppose r_1 is the length of the cycle of this permutation containing 1, and $r_2, \ldots, r_m$ are the lengths of its
other cycles. It is easy to understand that on the family
of arguments corresponding to this φ, all the generators have
trivial values except three:

$$\Psi_{r_1} \dots \Psi_{r_m} \sigma_0^0, \quad \Psi_{r_2} \dots \Psi_{r_m} \sigma_{r_1}^0, \quad \Psi_{r_2} \dots \Psi_{r_m} \rho_{0,\, r_1-1}^{01};$$

the values of the latter are nonzero (and can easily be computed).

Second case: $\varphi\,(n+1) \neq 1$. Construct the sequence $l_0 = \varphi\,(n+1)$, $l_1 = \varphi\,(l_0)$, $l_2 = \varphi\,(l_1)$, ... Suppose l_u is the first term of the sequence which equals one of the previous terms, say l_s. Clearly, $0 < s < u$, $l_s = l_u = 1$, and

$$\varphi \mid \{1, \dots, n\} \setminus \{l_0, \dots, l_{u-1}\}$$

is a permutation. Suppose $r_1, \dots, r_m$ are the lengths of the cycles of this permutation. Then on the family of variables corrsponding to this φ, the values of all the generators are trivial, except three:

$$\Psi_{r_1} \dots \Psi_{r_m} \sigma_u^0, \quad \Psi_{r_1} \dots \Psi_{r_m} \Psi_{u-s} \sigma_s^0, \quad \Psi_{r_1} \dots \Psi_{r_m} \rho_{s,\, u-s-1}^{01};$$

the values of the latter are nonzero (and can easily be computed).

It follows from the above that the cochains $\Psi_{r_1} \dots \Psi_{r_m} \rho_{st}^{00}$ are linearly independent, while the cochains $\Psi_{r_1} \dots \Psi_{rm} \sigma_u^0$ can be expressed in terms of them. In order to prove the formula appearing in the statement of the lemma, it suffices to compute the nontrivial values listed above and substitute them into this formula. This is an automatic (although not very short) computation, which we shall omit.

The proof of (iv) is similar to the proof of (iii). We shall merely indicate that for the triviality of a cochain from $A_{2n+1} \mid \subset \Lambda^n V' \otimes \Lambda^{n+1} (S^2 V \otimes V') \otimes V']$ it suffices to show its triviality on the following family of variables:

$$\begin{cases} \alpha_1 = e_1, \dots, \alpha_n = e_n, \, \alpha_{n+1} = e_{\varphi(1)}, \dots, \alpha_{2n+2} = e_{\psi(n+2)}; \\ \beta_{n+1}^1 = \beta_{n+1}^2 = e^1, \dots, \beta_{2n}^1 = \beta_{2n}^2 = e^n, \, \beta_{2n+1}^1 = \beta_{2n+1}^2 = e^1; \end{cases}$$

$$\begin{cases} \alpha_1 = e_1, \ldots, \alpha_n = e_n, \ \alpha_{n+1} = e_{\varphi(1)}, \ldots, \ \alpha_{2n+2} = e_{\varphi(n+2)}; \\ \beta_{n+1}^1 = \beta_{n+1}^2 = e^1, \ldots, \ \beta_{2n}^1 = \beta_{2n}^2 = e^n, \ \beta_{2n+1}^1 = \beta_{2n+1}^2 = e^1 + e^2, \end{cases}$$

where φ is an arbitrary map $\{1, \ldots, n+2\} \to \{1, \ldots, n\}$.

The proof of Lemma 2 is complete. This lemma concludes the description of the term E_1 of our Serre–Hochschild spectral sequence and we must now compute the differential d_1. Since, obviously,

$$d_1^{p,\,q} = d_1^{p,\,0} \otimes \mathrm{id}\, H^q(\mathfrak{gl}\,(n, \mathbb{K})),$$

it suffices to compute $d_1^{p,\,0}$; this computation reduces to finding the differentials of the cochains ρ and σ, since multiplication by (cocycles) Ψ commutes with the differential. A fairly simple calculation shows that

$$\begin{aligned} d\rho_{st}^{00} &= -\rho_{st}^{10} + (-1)^{s+t-1}\rho_{st}^{01}; \\ d\rho_{st}^{10} &= -\rho_{s+1,\,t}^{00} + \rho_{st}^{11}; \\ d\rho_{st}^{01} &= (-1)^{s+t}\rho_{s+1,\,t}^{00} + (-1)^{s+t-1}\rho_{st}^{11}; \\ d\rho_{st}^{11} &= (-1)^{s+t}\rho_{s+1,\,t}^{10} - \rho_{s+1,\,t}^{01}; \\ d\sigma_u^0 &= (-1)^{u+1}\rho_{0u}^{00} + (-1)^u\sigma_u^1; \\ d\sigma_u^1 &= (-1)^{u+1}\rho_{0u}^{10} - \rho_{0u}^{01}. \end{aligned}$$

These formulas, in turn, allow us to compute the term E_2 immediately. We obtain

$$\begin{aligned} E_2^{p,\,q} &= 0 \quad \text{for } p \neq 2n, \\ E_2^{2n,\,q} &= E_2^{2n,\,0} \otimes H^q(\mathfrak{gl}\,(n, \mathbb{K})), \end{aligned}$$

$E_2^{2n,\,0}$ is generated by the homology classes $x_{r_1 \ldots r_m}$ of the elements

$$\xi_{r_1 \cdots r_m} = \sum_{i=1}^{m} (-1)^{r_i} \Psi_{r_1} \ldots \hat{\Psi}_{r_i} \ldots \Psi_{r_m} \sigma_{r_i-1}^0$$

of the space $E_1^{2n,\,0}$ for $0 < r_1 \leqslant \ldots \leqslant r_m$, $\quad r_1 + \ldots + r_m = n+1$,

and the classes $x_{r_1 \ldots r_m}$ are related by a unique linear relation (its coefficients may easily be found, but are not
needed in the sequel).

This implies $E_2 = E_\infty$; to prove the additive part of
the theorem, it suffices to note that $\dim E_2^{2n, 0} = \mathrm{p}\,(n + 1) -$
$1 = \dim H^{2n+1}\,(W_n)$.

It remains to prove that $xy = 0$ for all $x \in H^q(W_n; W_n')$
and $y \in H^r\,(W_n)$ when $r > 0$. But this is obvious: the
Serre–Hochschild filtration of the class x with respect to
the subalgebra $\mathfrak{gl}\,(n, \mathbb{K})$ equals $2n$, while the filtration of y
cannot be less than n; therefore xy has filtration $\geqslant 3n$,
implying $xy = 0$.

6. Commentary and additions to Theorem 2.2.10. A. The
homomorphism var. Suppose $\mathfrak{g}$ is an arbitrary Lie algebra.
For $q \geqslant 0$ consider the homomorphism $C^{q+1}\,(\mathfrak{g}) \rightarrow C^q(\mathfrak{g}; \mathfrak{g}')$, which
sends every cochain $c \in C^{q+1}\,(\mathfrak{g})$ into the cochain

$$(g_1, \ldots, g_q) \mapsto \{g \mapsto c\,(g_1, \ldots, g_q,\ g)\}.$$

It is easy to verify that such homomorphisms commute with
the differential. We define

$$\mathrm{var}:\ H^{q+1}\,(\mathfrak{g}) \rightarrow H^q\,(\mathfrak{g};\ \mathfrak{g}')$$

as the induced homomorphism in cohomology. (The notation
var will be explained in §3.1.) We present two key properties of the homomorphism var in the case $\mathfrak{g} = W_n$.

Theorem 2.2.11. (i) The homomorphism

$$\mathrm{var}:\ H^{2n+1}\,(W_n) \rightarrow H^{2n}\,(W_n;\ W_n')$$

is an isomorphism.

(ii) The sequence

$$\Pi^{q+1}(W_{n+1}) \to \Pi^{q+1}(W_n) \xrightarrow{\text{var}} \Pi^q(W_n; W_n'),$$

in which the first homomorphism is induced by the natural inclusion $W_n \to W_{n+1}$, is exact.

The proof consists of a direct computation and is left
to the reader.

Part (ii) of this theorem will be needed in §3.1, while
part (i) will be used immediately: it enables us to establish the canonical isomorphism between $H^q(W_n; W_n')$ and
$H^{2n+i}(W_n) \otimes H^{q-2n}(\mathfrak{gl}(n, \mathbb{K}))$ promised at the beginning of Subsection 5.

B. Results in the relative case.

Theorem 2.2.12.

$$H^q(\mathbb{R}W_n, \mathfrak{o}(n); \mathbb{R}W_n') \cong \Pi^{2n+1}(\mathbb{R}W_n) \otimes \Pi^{q-2n}(\mathfrak{gl}(n, \mathbb{R}), \mathfrak{o}(n));$$
$$H^q(\mathbb{R}W_n, O(n); \mathbb{R}W_n') \cong H^{2n+1}(\mathbb{R}W_n) \otimes H^{q-2n}(\mathfrak{gl}(n, \mathbb{R}), O(n));$$

the $\Pi^*(\mathbb{R}W_n, \mathfrak{o}(n))$-module structure in $\Pi^*(\mathbb{R}W_n, \mathfrak{o}(n); \mathbb{R}W_n')$,
as well as the $\Pi^*(\mathbb{R}W_n, O(n))$-module structure in $H^*(\mathbb{R}W_n, O(n); \mathbb{R}W_n')$, is trivial.

The proof resembles that of Theorem 2.2.10 and will be
omitted. It is easy to define and compute the homomorphisms

$$\text{var}: H^{q+1}(\mathbb{R}W_n, \mathfrak{o}(n)) \to H^q(\mathbb{R}W_n, \mathfrak{o}(n); \mathbb{R}W_n'),$$
$$\text{var}: H^{q+1}(\mathbb{R}W_n, O(n)) \to H^q(\mathbb{R}W_n, O(n); \mathbb{R}W_n').$$

I will merely mention that the sequences

$$H^{q+1}(\mathbb{R}W_{n+1}, \mathfrak{o}(n+1)) \to H^{q+1}(\mathbb{R}W_n, \mathfrak{o}(n)) \xrightarrow{\text{var}} H^q(\mathbb{R}W_n, \mathfrak{o}(n); \mathbb{R}W_n'),$$
$$H^{q+1}(\mathbb{R}W_{n+1}, O(n+1)) \to H^{q+1}(\mathbb{R}W_n, O(n)) \xrightarrow{\text{var}} H^q(\mathbb{R}W_n, O(n); \mathbb{R}W_n'),$$

in which the right-hand-side homomorphisms are induced by
natural inclusions, are exact.

7. <u>Other formal vector-field algebras</u>. I shall state
without proof certain results concerning the cohomology (with
constant coefficients) of the algebras S_n, $\hat{S}_n$, H_n, $\hat{H}_n$, and
K_n. As was already mentioned, the rings $H^*(\hat{S}_n), H^*(\hat{H}_n)$, and
$H^*(K_n)$ are finite dimensional, so that the problem of their
computation appears reasonable. However, attempts to carry
out this computation along the lines of the proof of Theorem
2.2.4 were successful only for $\hat{S}_n$.

<u>Theorem 2.2.13</u>. The ring $H^*(\mathbb{K}\hat{S}_n)$ is isomorphic to
$H^*(S^1 \times Y_n; \mathbb{K})$, where Y_n is the inverse image of the $2n$-di-
mensional skeleton of the base in the standard universal
$SU(n)$-bundle.

This theorem was proved by Rozenfeld [78] and independently by Shnider [83]; note that Rozenfeld's article also
contains a statement describing the cohomology of the algebra
$\hat{H}_n$ (this statement is repeated in an article by Gelfand,
Kalinin, and myself [33]) but the proof given in [78] is not
complete, so that the cohomology of the algebra $\hat{H}_n$ is not
known at present (the statement from [78] mentioned above
should be viewed as a likely conjecture).

The cohomology of the algebra K_n was recently computed
by Feigin [15]. His computation does not resemble the proof
of Theorem 2.2.4. He uses the Serre—Hochschild spectral sequence corresponding to the subalgebra of the algebra $\mathbb{K}K_n$,
isomorphic to $\mathfrak{sp}(n+1, \mathbb{K})$ [the inclusion of this last algebra in $\mathbb{K}K_n$ is similar to the inclusion $\mathfrak{sl}(n+1, \mathbb{K}) \to W_n$
from 1.1.3, see formula (8)]. The main distinctive trait of
this spectral sequence is that the irreducible submodules of
the $\mathfrak{sp}(n+1, \mathbb{K})$-module $\mathbb{K}K_n/\mathfrak{sp}(n+1, \mathbb{K})$ and, even more so
$\Lambda^q(\mathbb{K}K_n/\mathfrak{sp}(n+1, \mathbb{K}))$, are infinite dimensional; this infinite

dimensionality excludes, for example, the application of classical theorems from the theory of invariants. Nevertheless, the final result turns out to be similar to the statement of Theorem 2.2.4:

Theorem 2.2.14. The ring $H^*(\mathbb{K}K_n)'$ is isomorphic to the cohomology ring of the inverse image of the $(4n + 2)$-dimensional skeleton of the base in the standard universal $(S^1 \times \mathrm{Sp}\,(n))$-bundle.

The computation of the cohomology of the algebras H_n and S_n appears to be much more difficult. There is no special information about the algebras S_n (except Theorem 2.2.17 stated below). It should be mentioned that the algebra S_1 is finite dimensional (and its cohomology may be computed without difficulty), while the algebra S_2 does not differ from H_1. As to the algebras S_n for $n \geqslant 3$, the problem of computing their cohomology induces even less optimism than the same problem for H_n.

Regarding the cohomology of the algebra H_1 we can say the following. The corresponding cochain complex can be modified along the lines of Subsection 3D: the space $C^q(H_1)$ is represented as the space of polynomials of the form

$$P(x_1, \ldots, x_q; y_1, \ldots, y_q),$$

which change their sign under the simultaneous permutation of x_i with x_j and y_i with y_j and which satisfy $P(0, x_2, \ldots, x_q; 0, y_2, \ldots, y_q) = 0$. The differential is determined by the formula

$$dP(x_1, \ldots, x_{q+1}; y_1, \ldots, y_{q+1}) = \sum_{1 \leqslant s < t \leqslant q+1} (-1)^{s+t-1} \begin{vmatrix} x_s & x_t \\ y_s & y_t \end{vmatrix} P(x_s + x_t,$$

$$x_1, \ldots \hat{x}_s \ldots \hat{x}_t \ldots, x_{q+1}; y_s + y_t, y_1, \ldots \hat{y}_s \ldots \hat{y}_t \ldots, y_{q+1}).$$

By calculations, which resemble those of $H^*(W_1)$ in Subsection 3D, it can be shown that our complex is homology equivalent to its subcomplex determined by the following condition: every monomial in P has the same total degree in x and in y. This restricted complex can be decomposed into a sum $\oplus_{i=-1}^{\infty} C^{\cdot}(i)$ of finite-dimensional subcomplexes: the space $C^q(i)$ consists of polynomials of the form indicated above which are homogeneous in x and in y and have the degree $q+i$ in x and in y. The cohomology of the complex $C^{\cdot}(i)$ is denoted by $H^q(i)$. It is quite easy to show that

$$\dim H^q(-1) = \begin{cases} 1 & \text{for } q=2, 5, \\ 0 & \text{otherwise}; \end{cases}$$

$$\dim H^q(0) = \begin{cases} 1 & \text{for } q=0, 7, \\ 0 & \text{otherwise}. \end{cases}$$

Our first expectations for finding the cohomology of the algebra H_1 were related to the conjecture $H^*(i) = 0$ for $i > 0$. This conjecture, however, was refuted by a computer calculation, which showed that $H^*(i) = 0$ for $i = 1, 2, 3, 5$ and

$$\dim H^q(4) = \begin{cases} 1 & \text{for } q=7, 10, \\ 0 & \text{otherwise}. \end{cases}$$

This theorem was proved by Gelfand, Kalinin, and myself in [33], which explicitly indicates the cocycles representing the "exotic" cohomology classes of dimensions 7 and 10 of the algebra H_1. Thus:

Theorem 2.2.15. $\dim H^2(H_1) \geqslant 1$, $\dim H^5(H_1) \geqslant 1$, $\dim H^7(H_1) \geqslant 2$, $\dim H^{10}(H_1) \geqslant 1$.

The fact that there are actually very many (probably infinitely many) such exotic classes was shown later by Perchik [71]. His result is the following. Denote by $C_0^q(i)$ the subspace of the space $C^q(i)$, determined by two conditions: (i) the sum of degrees of each monomial of the polynomial P in x_s, y_s for each s differs from 2; (ii) dP possesses the same property. The spaces $C_0^q(i)$ constitute a subcomplex of the complex $C^{\cdot}(i)$; this subcomplex is denoted by $C_0^{\cdot}(i)$, its cohomology, by $H_0^q(i)$. It is easy to show that if $i \neq 0$, then $H^*(i) = H^*(S^3; \mathbb{K}) \otimes H_0^*(i)$.

Theorem 2.2.16 (Perchik). The Euler characteristic of the complex $C_0^{\cdot}(i)$ is equal to half the coefficient of x^{2i} $(= t^0 x^{2i})$ in the product

$$\prod_{\substack{-1 \leqslant \alpha \leqslant \infty, \\ \beta \equiv \alpha \bmod 2, \\ -\alpha-2 \leqslant \beta \leqslant \alpha+2, \\ (\alpha, \beta) \neq (0, 0)}} (1 - t^\beta x^\alpha).$$

Perchik calculated (also using a computer) the Euler characteristics of the complexes $C_0^{\cdot}(i)$ when $i \leqslant 28$. For small i they often vanish, but when i increases, they become different from zero as a rule and even assume fairly large values. The sum of absolute values of all the Euler characteristics found by Perchik equals 57, and this shows that the space $\oplus_i H_0^*(i)$ is of dimension $\geqslant 57$. There are no reasons which would indicate that the sequence of coefficients indicated in Theorem 2.2.16 is finite, so that we have the very probable conjecture that the space $\oplus_i H_0^*(i)$, and also the space $H^*(H_1)$, are infinite dimensional. In any case, it follows from Perchik's computations that $\dim H^*(H_1) \geqslant 112$.

Then complex $C^{\cdot}(H_n)$ is also homology equivalent to a certain subcomplex which decomposes into a sum of finite-dimensional complexes (with numbers varying from $-n$ to ∞). In Perchik's paper, a method for computing the Euler characteristics of complexes similar to $C_0^*(i)$ is provided for $n > 1$ as well, but this method is less effective.

To the above we should add that the computation of the cohomology of the algebras $S_n, H_n, \widehat{S}_n, \widehat{H}_n, K_n$ (and, of course, W_n) will be considerably simplified if we limit ourselves to not very high ("stable") dimensions. The technical reason for this simplification is that the relationship between invariants, provided by the main theorems of invariant theory, appears only in tensor spaces of sufficiently high ranks. In view of this fact, the proof of the theorem stated below necessitates an analog of only the first (easy) part of Lemma 1 from Subsection 2.

Theorem 2.2.17 (Guillemin and Shnider [45]).

$$H^q(S_n) = 0 \quad \text{for} \quad 0 < q < n;$$

$$\dim H^q(H_n) = \begin{cases} 0 & \text{for odd } q \leqslant n, \\ 1 & \text{for even } q \leqslant n; \end{cases}$$

$$H^q(\widehat{H}_n) = 0 \quad \text{for} \quad 0 < q \leqslant n;$$

$$\dim H^q(\widehat{S}_n) = \begin{cases} 1 & \text{for } q = 0, 1, \\ 0 & \text{for } 2 \leqslant q \leqslant n; \end{cases}$$

$$H^q(K_n) = 0 \quad \text{for} \quad 0 < q \leqslant n.$$

Note that a similar theorem for W_n is Lemma 2 from Subsection 2, i.e., the relation $H^q(W_n) = 0$ for $0 < q \leqslant n$; but actually $H^q(W_n) = 0$ for $0 < q \leqslant 2n$ (see Subsection 2). Theorems 2.2.13 and 2.2.14 enable us to improve the statements of Theorem 2.2.17 concerning $\widehat{S}_n$ and K_n just as considerably. One has the impression that all the statements of Theorem 2.2.17 will remain correct if the inequalities $<n$, $\leqslant n$ in it are replaced everywhere by the inequality $<2n$, $\leqslant 2n$.

At present this has been done only for the algebra $\widehat{H}_n$ (except for the algebras whose cohomology has been entirely computed — see Shnider's thesis [83]).

§3. COMPUTATIONS FOR LIE ALGEBRAS
     OF FORMAL VECTOR FIELDS ON THE LINE

As we already mentioned, for $n = 1$ the computation of the cohomology of the algebra W_n with various coefficients may be carried through much further than in the general case. Apparently the reason for this is not only the natural simplification of the situation for smaller n, but the fairly intimate analogy between the algebra W_1 and the current algebra, whose cohomology is computed in the next section. With the increase of n this analogy disappears completely.

Section 3.3 relates to the present section; in it, the zero-dimensional cohomology of the algebra W_1 with certain coefficients is in fact computed. I have presented this calculation in Chapter 3 because of its interest from the point of view of the applications.

The present section is written in the style of a survey article: many proofs are omitted or merely sketched.

1. Computation of $H^*(L_k(1))$. A. Goncharova's theorem. The ring $H^*(L_k(1))$ was computed by Goncharova [38, 39] in 1972. We begin by stating her main result.

Theorem 2.3.1. (i) $\dim H^q(L_k(1)) = \binom{q+k-1}{k-1} + \binom{q+k-2}{k-1}$;

in particular,

$$\dim H^q(L_1(1)) = 2 \text{ for } q \geqslant 1,$$
$$\dim H^q(L_2(1)) = 2q + 1,$$
$$\dim H^q(L_3(1)) = (q + 1)^2.$$

(ii) The dimension of the space $H^q_{(m)}(L_k(1))$ is equal to the number of ways in which the integer m may be represented in the form of the sum $m_1 + \ldots + m_q$, where the m_i are integers satisfying the conditions $m_1 \geqslant k, m_2 \geqslant m_1 + 3, \ldots, m_q \geqslant m_{q-1} + 3$ and

$$m_q \leqslant \begin{cases} 3q + 2k - 4 & \text{for } m_1 = k, \\ 3q + 2k - 3 & \text{for } m_1 > k; \end{cases}$$

in particular,

$$\dim H^q_{(m)}(L_1(1)) = \begin{cases} 1, & \text{if } m = \frac{3q^2 \pm q}{2}, \\ 0 & \text{otherwise}; \end{cases}$$

$$\dim H^q_{(m)}(L_2(1)) = \begin{cases} 1, & \text{if } \frac{3q^2 + q}{2} \leqslant m < \frac{3(q+1)^2 - (q+1)}{2}, \\ 0 & \text{otherwise}. \end{cases}$$

Here the subscript m in parentheses corresponds to the grading of the algebras $L_k(1)$, and is determined by the formula $\deg e_i = i$ (see 1.3.7).

Recalling that $1 \in \mathfrak{gl}(1) = \mathsf{K}$ determines a multiplication by m in $H^q_{(m)}(L_k(1))$ and applying Theorem 2.2.8, we get the following:

Corollary.

$$\dim H^q(W_1; F_\lambda) = \begin{cases} 1, & \text{if } \lambda = -\frac{3q^2 \pm q}{2} \text{ или } -\frac{3(q-1)^2 \pm (q-1)}{2}, \\ 0 & \text{otherwise}; \end{cases}$$

$$\dim H_q(W_1; F'_\lambda) = \begin{cases} 1, & \text{if } \lambda = -\frac{3q^2 \pm q}{2} \text{ или } -\frac{3(q-1)^2 \pm (q-1)}{2}, \\ 0 & \text{otherwise}. \end{cases}$$

Theorem 2.3.1 still remains one of the most difficult theorems in the cohomology theory of infinite-dimensional Lie algebras. Three proofs of it are known. The proof contained in the original works of Goncharova essentially consists of a direct calculation of the kernels and images of the differentials of the cochain complex and looks very cumbersome. The second, due to Gelfand, Feigin, and myself [35], concerns the case $k = 1$ only and is based on the calculation of the spectrum of the Laplace operator in the chain complex of the algebra $L_1(1)$ with metric $\langle e_i, e_j \rangle = \delta_{ij}$; below we provide some details of this proof. The third proof is contained in the recent paper of Retakh and Feigin [77], who showed that the cohomology of the algebra $L_k(1)$ is unchanged under a certain deformation which reduces it to an algebra whose co-homology can be computed by methods devised by Bott and Segal for computing the cohomology of algebras of smooth vector fields (see 4.3).

B. Computation of $H^*(L_1(1))$ with the help of the Laplace operator. This computation was made in 1978 by Gelfand, Feigin, and myself. As F. V. Vainshtein recently discovered, our paper contained a defect: the eigenvalues of the Laplace operator were found correctly in it, but their multiplicities were understated. The formulation of Theorem 2.3.2 given below is Vainshtein's corrected version of the basic theorem of our paper; the proof of this theorem found by Vainshtein appears in [95].

As an abbreviation, instead of $L_1(1)$ we shall write L_1.

We recall that in view of the existence of a metric in L_1 we can identify chains of this algebra with cochains: $C^q_{(m)}(L_1) = C_q^{(m)}(L_1)$.   Correspondingly, we assume that the differentials d and ∂ act on the cochain spaces:

$$d: C^q_{(m)}(L_1) \to C^{q+1}_{(m)}(L_1), \quad \partial: C^{q+1}_{(m)}(L_1) \to C^q_{(m)}(L_1).$$

The Laplace operator Δ is defined by the formula $\Delta = d \circ \partial + \partial \circ d$.   According to Theorem 1.5.3, there is a canonical isomorphism

$$H^q_{(m)}(L_1) = \mathrm{Ker}\,(\Delta \mid C^q_{(m)}(L_1)).$$

Following Goncharova, we call the collection $(i_1, \ldots, i_q)$ of natural numbers __principal__ if each of the differences $i_2 - i_1, \ldots, i_q - i_{q-1}$ is greater than or equal to 3. For a principal family $(i_1, \ldots, i_q)$ let

$$E(i_1, \ldots, i_q) = \sum_{k=1}^{q} \binom{i_k}{3} - \sum_{1 \leqslant l < m \leqslant q} i_l i_m.$$

__Theorem 2.3.2a.__ The set of eigenvalues of the Laplace operator $\Delta: C^*_{(m)}(L_1) \to C^*_{(m)}(L_1)$ coincides with the set of numbers of the form $E(i_1, \ldots, i_q)$, where $(i_1, \ldots, i_q)$ is a principal family with $i_1 + \ldots + i_q = m$, $q = 1, 2, \ldots$

For example, for $m = 7$, the Laplace operator has the eigenvalues $E(7) = 35$, $E(1,6) = 14$, $E(2,5) = 0$. For any m, the Laplace operator possesses only integer eigenvalues (this phenomenon does not have a satisfactory explanation as of yet).

The reader will verify without difficulty that $E(i_1, \ldots, i_q) \geqslant 0$ for any principal family $(i_1, \ldots, i_q)$ and that $E(i_1, \ldots, i_q) = 0$ in two cases (for a fixed q, when m varies): namely, if $(i_1, \ldots, i_q) = (1, 4, 7, \ldots, 3q-2)$ or $(2, 5, 8, \ldots, 3q - 1)$.

In order to indicate the multiplicities of these eigenvalues of the Laplace operator, let us introduce the following notation. For a principal family $i = (i_1, \ldots, i_q)$ let

$$\alpha_1(i) = \begin{cases} 0, & \text{if } i_1 < 3, \\ 1, & \text{if } i_1 \geqslant 3; \end{cases}$$

$$a_r(i) = \begin{cases} 0, & \text{if } i_r - i_{r-1} = 3, \ r > 1, \\ 1, & \text{if } i_r - i_{r-1} > 3, \ r > 1; \end{cases}$$

$$\alpha(i) = \alpha_1(i) + \ldots + \alpha_q(i).$$

__Theorem 2.3.2b.__ Suppose $i = (i_1, \ldots, i_q)$ is a principal family satisfying $i_1 + \ldots + i_q = m$. The multiplicity of the

eigenvalue $E(i)$ equals one if $\alpha(i) = 0$ and equals $2^{\alpha(i)}$ if
$\alpha(i) > 0$ [of course accidental coincidences $E(i) = E(i')$ are
possible; in this case the multiplicities must be added]. We
have $\alpha(i) = 0$ if and only if $E(i) = 0$; in this case the corresponding eigenvector is contained in $C^q_{(m)}(L_1)$.   If $\alpha(i) > 0$,
the dimensionality of intersection of the eigenspace, corresponding to the eigenvalue $E(i)$, with $C^{q+r}_{(m)}(L_1)$, is equal to
$\binom{\alpha(i)}{r}$ (if the coincidence $E(i) = E(i')$, the dimensionality is
summed).

Goncharova's "stable cycles" are eigenvectors of the
operator Δ. Their definition is as follows: Consider the
operator $\sigma: C^q_{(m)}(L_1) \to C^q_{(m+q)}(L_1)$ determined by the formula

$$\sigma(\varepsilon_{i_1} \wedge \ldots \wedge \varepsilon_{i_q}) = \varepsilon_{i_1+1} \wedge \ldots \wedge \varepsilon_{i_q+1},$$

where ε_i is a cochain in $C^1(L_1)$, acting according to the
formula $\varepsilon_i(e_j) = \delta_{ij}$. An element γ of the space $C^q_{(m)}(L_1)$ is
called a __stable cycle__ if $\partial(\sigma^r \gamma) = 0$ for $r = 0, 1, \ldots$.

Theorem 2.3.2c.   The space of stable cycles is invariant
with respect to Δ. Each of the numbers $E(i)$ is an eigenvalue of multiplicity 1 of the restriction of the operator Δ
to this subspace [for accidental coincidences $E(i) = E(i')$
the multiplicity must be increased in an appropriate way].
The stable cycle which is an eigenvector of the operator Δ
with eigenvalue $E(i_1, \ldots, i_q)$ may be written in the form

$$\varepsilon_{i_1} \wedge \ldots \wedge \varepsilon_{i_q} + \sum \alpha_{j_1 \ldots j_q} \varepsilon_{j_1} \wedge \ldots \wedge \varepsilon_{j_q},$$

where the sum is taken over all the families $(j_1, \ldots, j_q) \neq (i_1,$
$\ldots, i_q)$ with $j_1 \geqslant i_1$, $j_1 + j_2 \geqslant i_1 + i_2$, $\ldots, j_1 + \ldots + j_{q-1} \geqslant$
$i_1 + \ldots + i_{q-1}$ $[j_1 + \ldots + j_q = i_1 + \ldots + i_q = m]$.

The proofs of Theorems 2.3.2a-c consist in the explicit listing of the eigenvectors of the operator Δ and the computation of the corresponding eigenvalues.

In conclusion, let us describe a convenient method for constructing stable cycles. Elements of the space $C^q_{(m)}(L_1)$ will be identified with skew-symmetric polynomials of degree m in q variables: to the monomial $\varepsilon_{i_1} \wedge \cdots \wedge \varepsilon_{i_q}$ we assign the polynomials

$$\sum_{\alpha \in \mathrm{Symm}\,(q)} \mathrm{sgn}\,(\alpha)\, z_1^{i_{\alpha(1)}} \ldots z_q^{i_{\alpha(q)}}$$

(this identification differs from a similar identification which was used in 2.3D). Obviously, all the polynomials corresponding to cochains of the algebra L_1, are divisible by $z_1 \ldots z_q$ and

$$\Pi_q = \prod_{1 \leqslant i < j \leqslant q} (z_j - z_i).$$

<u>Lemma.</u> A cochain in L_1 is a stable cycle if and only if the corresponding polynomial is divisible by Π_q^3.

<u>Proof.</u> If $\gamma \in C^q_{(m)}(L_1)$ is represented by the polynomial

$$\Pi_q\, P\,(z_1, \ldots, z_q)$$

(where P is a symmetric polynomial), then $\partial\gamma$, as can be easily checked, is represented by the polynomial

$$\Pi_{q-1}\, Q\,(z_1, \ldots, z_{q-1}),$$

where

$$Q\,(z_1, \ldots, z_{q-1}) = \sum_{i=1}^{q-1} z_i P\,(z_1, \ldots, z_i, z_i, \ldots, z_{q-1})(z_i - z_1) \cdots (z_i - z_{q-1})$$

(the zero factor $z_i - z_i$, is, of course, omitted). In order that the cochain γ be a stable cycle, it is necessary that

the sum on the right-hand side of the last relation vanish
after the polynomial P is replaced by $z_1^r \ldots z_q^r P$ for arbitrarily large r. But after such a substitution, the i-th
summand of the sum will be multiplied by $z_1^r \ldots z_{i-1}^r z_i^{2r} z_{i+1}^r \ldots z_q^r$,
and this shows that our requirement means all the summands
of the sum vanish, i.e, the polynomial

$$P\,(z_1,\, \ldots,\, z_i,\, z_i,\, \ldots,\, z_{q-1})$$

vanishes for each i. But this means that the polynomial P
is divisible by Π_q and, therefore, by Π_q^2, since this polynomial is symmetric. Thus, γ is a stable cycle if and only
if the corresponding polynomial is divisible by Π_q^3.

Comparing this lemma with Theorems 2.3.2a-c, we see that
it is possible to give an explicit description of the cocycles
whose cohomology classes generate $H^*(L_1)$.

<u>Theorem 2.3.2</u>. If $q \geqslant 1$, then $\dot{H}^q(L_1)$ is two-dimension-
al and is generated by the cohomology classes of the cocycles

$$z_1 \ldots z_q \Pi_q^3 (z_1, \ldots, z_q) \in C_{\left(\frac{3q^2-q}{2}\right)}^q (L_1),$$

$$z_1^2 \ldots z_q^2 \Pi_q^3 (z_1, \ldots, z_q) \in C_{\left(\frac{3q^2+q}{2}\right)}^q (L_1).$$

<u>2. The cohomology of the algebra W_1 with coefficients
in $F_\lambda \otimes F_\mu$</u>. As was already mentioned in Subsection 1,
Theorem 2.3.1 allows us to find the cohomology of the algebra
W_1 with coefficients in F_λ. The next problem in order of
difficulty — the calculation of the cohomology of W_1 with
coefficients of the form $F_\lambda \otimes F_\mu$ — was solved by Feigin and
myself in [17], which also contains some other computations.
Here I shall present the results of [17], omitting the fairly
cumbersome proof of the main lemma. For technical reasons,
it is more convenient to deal with the homology of the al-

gebra W_1^{pol}, instead of the cohomology of the algebra W_1 to which it is isomorphic. To simplify notations, we shall omit the superscript pol everywhere in this subsection, understanding W_1 as the algebra of polynomial vector fields on the line. The algebras and modules L_0, L_1, F_λ, etc., should be understood in the same sense.

A. Auxiliary modules. Recall that the additive basis in the algebra $\mathbb{C}W_1$ is constituted by the fields $e_i = x^{i+1}d/dx$ $(i = -1, 0, 1, \ldots)$, and the commutator in it acts according to the formula $[e_i, e_j] = (j - i)\, e_{i+j}$. The basis in the space F_λ is constituted by its elements $f_j = z^j dz^{-\lambda}$; the action of $\mathbb{C}W_1$ in F_λ is described by the formula $e_i f_j = (j - (i + 1)\lambda)\, f_{i+j}$. By $\mathcal{F}_\lambda$ we denote the $\mathbb{C}W_1$-module whose definition differs from the previous description of the module F_λ only in that the index j can assume all possible integer values; thus $F_\lambda \subset \mathcal{F}_\lambda$.

The adjoint modules F_λ', $\mathcal{F}_\lambda'$ are understood as modules of linear functionals $F_\lambda \to \mathbb{C}$, $\mathcal{F}_\lambda \to \mathbb{C}$, which are finite in the sense that they can assume nonzero values only on a finite number of f_j. Thus F_λ' and $\mathcal{F}_\lambda'$ are generated by elements f_j' (respectively, $j \geqslant 0$ and $j \in \mathbb{Z}$), while $\mathbb{C}W_1$ acts on them according to the formula

$$e_i f_j' = \begin{cases} (-(j-i)+(i+1)\lambda)f_{j-i}', & \text{if} \quad f_j \in \mathcal{F}_\lambda' \quad \text{or} \quad j \geqslant i, \\ 0, & \text{if} \quad f_j \in F_\lambda' \quad \text{and} \quad j < i. \end{cases}$$

Respectively, $f_j' \leftrightarrow f_{-1-j}$ establishes (for any λ) the isomorphism $\mathcal{F}_\lambda' = \mathcal{F}_{-1-\lambda}$, and here we have $F_{-1-\lambda} = \operatorname{ann} F_\lambda$, so that $F_\lambda' = \mathcal{F}_{-1-\lambda}/F_{-1-\lambda}$.

As we said in 1.2.2, the W_1-modules F_λ for $\lambda \neq 0$ are irreducible. If we consider them as L_0-modules, they become reducible: in order to obtain a L_0-submodule of the module F_λ, it suffices to takes its subspace generated by f_j for $j \geqslant \mu$, where μ is some positive integer. The L_0-module which thus arises is denoted by $F_{\lambda, \mu}$. It is convenient to define it directly as the space generated, just as F_λ, by the elements f_j for $j = 0, 1, \ldots$, in which L_0 acts according to the formula

$$e_i f_j = (j + \mu - (i + 1) \lambda) f_{i+j}.$$

In this definition, μ may be viewed as an arbitrary complex number; for integers $\mu > 0$ the inclusion $F_{\lambda, \mu} \to F_\lambda$ is determined by the formula $f_j \mapsto f_{j+\mu}$. If, in the definition of $F_{\lambda, \mu}$, we no longer require j to be positive, we get the definition of the modules $\mathcal{F}_{\lambda, \mu}$ (now over W_1). Obviously, $F_{\lambda, 0} = F_\lambda$, $\mathcal{F}_{\lambda, 0} = \mathcal{F}_\lambda$, and, similarly to the above, $F_{\lambda, \mu} \subset \mathcal{F}_{\lambda, \mu}$, $\mathcal{F}'_{\lambda, \mu} = \mathcal{F}_{-1-\lambda, -\mu}$, $F'_{\lambda, \mu} = \mathcal{F}_{-1-\lambda, -\mu}/F_{-1-\lambda, -\mu}$.

All the modules listed above are graded; the grading is determined everywhere by the formulas $\deg f_j = j$, $\deg f'_j = -j$.

B. **Relations between the homology of the algebras W_1, L_0 and L_1.** In the following propositions 2.3.4a-d we assume given a graded L_0-module $A = \oplus A_{(m)}$ and a number $\lambda \in \mathbb{C}$ possessing the following property: $e_0 f = (m - \lambda) f$ for $f \in A_{(m)}$ (for example, F_λ, $\mathcal{F}_\lambda$, $F_{\lambda+\nu, \nu}$, $\mathcal{F}_{\lambda+\nu, \nu}$).

Proposition 2.3.4a.

$$H_*^{(m)} (L_0; A) = 0, \text{ if } m \neq \lambda,$$
$$H_q^{(\lambda)} (L_0; A) = H_q^{(\lambda)} (L_1; A) \oplus H_{q-1}^{(\lambda)} (L_1; A).$$

Proof. Any chain in $C_q^{(m)} (L_0; A)$ can be represented in the form $e_0 \wedge c_1 + c_2$, where $c_1 \in C_{q-1}^{(m)} (L_1; A)$, $c_2 \in C_q^{(m)} (L_1; A)$. Obviously,

$$\partial_q (e_0 \wedge c_1 + c_2) = (m - \lambda) c_1 - e_0 \wedge \partial_{q-1} c_1 + \partial_q c_2.$$

If $\partial (e_0 \wedge c_1 + c_2) = 0$, then $\partial_q c_2 = -(m - \lambda) c_1$, and therefore for $m \neq \lambda$ we have

$$e_0 \wedge c_1 + c_2 = \partial_{q+1} (e_0 \wedge c_2/(m - \lambda));$$

The first equality is proved. (Actually, it follows from the homology version of Theorem 1.5.2a.) To prove the second equality, it suffices to note that the decomposition

$$C_*^{(\lambda)} (L_0; A) = e_0 \wedge C_{*-1}^{(\lambda)} (L_1; A) \oplus C_*^{(\lambda)} (L_1; A)$$

is compatible with the differential.

Proposition 2.3.4b. Suppose E_μ is a one-dimensional L_0-module in which e_0 defines the multiplication by μ, while e_i for $i > 0$ acts trivially. Then

$$H_*^{(m)} (L_0; A \otimes E_\mu) = 0 \text{ for } m \neq \lambda - \mu,$$
$$H_*^{(\lambda-\mu)} (L_0; A \otimes E_\mu) = H_q^{(\lambda-\mu)} (L_1; A) \oplus H_{q-1}^{(\lambda-\mu)} (L_1; A).$$

This is a corollary of the previous proposition.

Now suppose A is actually a W_1-module. There are two methods for processing the information concerning the homology of the algebra L_1 into information about the homology of the algebra W_1. The first consists in representing chains from $C_q^{(m)} (W_1; A)$ in the form

$$e_{-1} \wedge e_0 \wedge c_1 + e_{-1} \wedge c_2 + e_0 \wedge c_3 + c_4,$$

where c_i are chains of the algebra L_1, and then applying an argument similar to the proof of Proposition 2.3.4a. We then obtain

Proposition 2.3.4c.
$$H_*^{(m)} (W_1; A) = 0, \qquad m \neq \lambda,$$
$$\dim H_q^{(\lambda)} (W_1; A) \leqslant \dim [H_q^{(\lambda)} (L_1; A) \oplus H_{q-1}^{(\lambda)} (L_1; A)$$
$$\oplus H_{q-1}^{(\lambda+1)} (L_1; A) \oplus H_{q-2}^{(\lambda+1)} (L_1; A)].$$

[The sum whose dimension is the right-hand side of the last inequality is the initial term of the spectral sequence

converging to the homology of the algebra W_1 with coefficients in A: this is the Serre–Hochschild spectral sequence corresponding to the pair $(W_1, \mathbb{C}e_{-1})$. The calculation
of the unique differential of the spectral sequence is hardly ever difficult.]

The second method consists in applying Theorem 1.5.4.
Obviously, the W_1-module induced by the L_0-module $A \otimes E_\mu$,
is $A \otimes F_\mu'$ (it is important that A is a W_1-module), so
that Proposition 2.3.4b gives

Proposition 2.3.4d.

$$H_*^{(m)}(W_1;\ A \otimes F_\mu') = 0, \quad \text{if} \quad m \neq \lambda - \mu,$$
$$H_q^{(\lambda-\mu)}(W_1;\ A \otimes F_\mu') = H_q^{(\lambda-\mu)}(L_1;\ A) \oplus H_{q-1}^{(\lambda-\mu)}(L_1;\ A).$$

C. The spectral sequence. Our next aim is to compute
the homology of the algebra L_1 with coefficients in $\mathcal{F}_\lambda$, F_λ,
and F_λ'.

The space $C_q^{(m)}(L_1;\ \mathcal{F}_\lambda)$ is generated by the chains

$$f_j \otimes e_{i_1} \wedge \ldots \wedge e_{i_q} \quad \text{with} \quad j + i_1 + \ldots + i_q = m;$$

denote by $G_p C_q^{(m)}(L_1;\ \mathcal{F}_\lambda)$ the subspace of the space $C_q^{(m)}(L_1;\ \mathcal{F}_\lambda)$ generated by monomials of the type indicated above
with $i_1 + \ldots + i_q \leqslant p$. Obviously $\{G_p C_q^{(m)}(L_1;\ \mathcal{F}_\lambda)\}_p$ is a de-
creasing filtration in $C_{\cdot}^{(m)}(L_1;\ \mathcal{F}_\lambda)$. The spectral sequence
corresponding to this filtration is denoted by $\mathscr{E}(\lambda, m)$. In
it, we have $E_{p,q}^0 = C_{p+q}^{(p)}(L_1;\ \mathbb{C})$, $d_{p,q}^0$ is the differential ∂_{p+q}:
$C_{p+q}^{(p)}(L_1;\ \mathbb{C}) \to C_{p+q-1}^{(p)}(L_1;\ \mathbb{C})$, and $E_{p,q}^1 = H_{p+q}^{(p)}(L_1;\ \mathbb{C})$. Thus,
$E_{p,q}^1 = \mathbb{C}$ for $p = (3r^2 \pm r)/2$, $p + q = r$ and $E_{p,q}^1 = 0$ for the
other p, q (see Theorem 2.3.1). Set $E_p^r = \bigoplus_q E_{p,q}^r$, $d_p^r = \bigoplus_q d_{p,q}^r$; obviously,
$$H_q^{(m)}(L_1;\ \mathbb{C}) \cong E_{(3q^2-q)/2}^\infty \oplus E_{(3q^2+q)/2}^\infty.$$

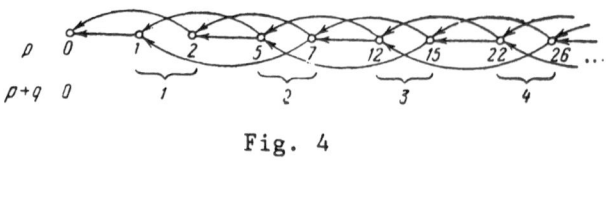

Fig. 4

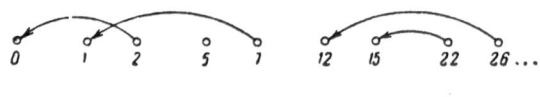

Fig. 5

The initial term of the spectral sequence is shown schemati-
cally in Fig. 4. Points on this picture represent nontrivial
spaces $E_p^1 (= \mathbb{C})$, i.e., $E_0^1, E_1^1, E_2^1, E_5^1, E_7^1, \ldots$. The numbers at
the other points denote the corresponding p, those at the
points of the lower row denote the complete degree, i.e.,
$p + q$. The arrows in Fig. 5 indicate all possible differentials: the horizontal arrows are the differentials $d_{(3r^2-r)/2}^{2r-1}$
[these differentials will be called <u>first</u>, and be denoted by
$d_1 (r)$)], the upper arrows denote the differentials $d_{(3r^2-r)/2}^{3r-2}$
for $r > 1$ and $d_{(3r^2+r)/2}^{3r-1}$ [we shall call them <u>second</u> and denote
them by $d_2 (r, 1)$ and $d_2 (r, 2)$], and the lower arrows — the differentials $d_{(3r^2+r)/2}^{4r-2}$ for $r > 1$ [we shall call them <u>third</u> and
denote them by $d_3 (r)$]. Each of these differentials may be
trivial, nontrivial, or undefined. Leaving only the arrows
corresponding to nontrivial differentials in Fig. 4, we obtain a diagram (one of its <u>a priori</u> versions is shown in Fig.
5) which entirely describes the spectral sequence. The arrows of this diagram are disjoint; in the term E^∞ only the
points which are neither initial points nor extremities of
arrows remain; for example, if for some λ and m the spectral
sequence $\mathscr{E} (\lambda, m)$ would be the one shown in Fig. 5, then we

would have $H_2^{(m)}(L_1; \mathscr{F}_\lambda) = \mathbb{C}$ and $H_q^{(m)}(L_1; \mathscr{F}_\lambda) = 0$ for $q = 0, 1, 3, 4$.

If the coefficients are taken in F_λ, instead of $\mathscr{F}_\lambda$, then the definition of the filtration remains valid. The new spectral sequence will be naturally mapped into the old one. On E_p^1, for $p \leqslant m$, this map is an isomorphism, while for $p > m$ the new spectral sequence has $E_p^1 = 0$. Thus, the diagram describing the new spectral sequence is obtained from the diagram describing the old one if we cut off part of it, leaving only the points and arrows which are entirely contained in the domain $p \leqslant m$. Thus, if Fig. 5 represents the spectral sequence $\mathscr{E}(\lambda, m)$ for $m = 20$, then $H_2^{(20)}(L_1; F_\lambda) = \mathbb{C}$, $H_3^{(20)}(L_1; F_\lambda) = \mathbb{C} \oplus \mathbb{C}$, $H_q^{(20)}(L_1; F_\lambda) = 0$ for $q \neq 2, 3$.

If we cut off the diagram of the spectral sequence $\mathscr{E}(\lambda, m)$ from the other side, leaving in it the part corresponding to $p \geqslant m$, then obviously we obtain the diagram for the spectral sequence converging to $H_*^{(m)}(L_1; F_{-1-\lambda}')$ (recall that $F_{-1-\lambda}' = \mathscr{F}_\lambda / F_\lambda$).

The definition of the spectral sequence $\mathscr{E}(\lambda, m)$ remains valid if, instead of the module $\mathscr{F}_\lambda$, we take the module generated by f_j for arbitrary $j \in \mathbb{C}$ (the action of the algebra W_1 is defined by the old formulas).

If $m \in \mathbb{Z}$, then this extension of the coefficient module will not influence either $C_q^{(m)}(L_1; \mathscr{F}_\lambda)$ or $\mathscr{E}(\lambda, m)$; however, it allows us to give a meaning to the spectral sequence $\mathscr{E}(\lambda, m)$ for any $\lambda, m \in \mathbb{C}$.

Note also that the previous constructions withstand the replacement of the modules $\mathscr{F}_\lambda$, F_λ, $F_{-1-\lambda}'$ by the modules $\mathscr{F}_{\lambda,\mu}$, $F_{\lambda,\mu}$, $F_{-1-\lambda,-\mu}'$. When we replace the module $\mathscr{F}_\lambda$ by $\mathscr{F}_{\lambda,\mu}$

the spectral sequence $\mathcal{E}(\lambda, m)$ becomes isomorphic to $\mathcal{E}(\lambda, m - \mu)$.

D. Statement of the main lemma. Set $e(t) = (3t^2 + t)/2$ (the Euler polynomial) and define the k-th parabola as the curve in the (complex) plane defined by the parametric equation

$$\lambda = e(t) - 1,$$
$$m = e(t) + e(t + k) - 1. \tag{1}$$

Obviously the "0-th parabola" is the straight line $2\lambda - m + 1 = 0$, while the other parabolas are actually real parabolas. If in Eq. (1) we take k to be a negative integer, then we obtain another parametric equation of the $|k|$-th parabola.

For $k_1, k_2 \in \mathbb{Z}$, set

$$P(k_1, k_2) = (e(k_1) - 1, e(k_1) + e(k_2) - 1).$$

The points $P(k_1, k_2)$ are two-by-two disjoint. Set $\mathbb{P} = \{P(k_1, k_2) \mid k_1, k_2 \in \mathbb{Z}\}$, and for $P = P(k_1, k_2)$ set $k(P) = |k_1 - k_2|$, $K(P) = |k_1| + |k_2|$. Clearly, if $P \in \mathbb{P}$, then $K(P) \geqslant k(P)$, $K(P) \equiv k(P) \bmod 2$ and P is on the $k(P)$-th parabola. For $k \neq 0$, all the points of the k-th parabola with integer coordinates are contained in $\mathbb{P}$. On the 0-th parabola, there is one point from $\mathbb{P}$ with $K = 0$ and two points each $K = 2$, $4, 6, \ldots$. For $k \neq 0$, the k-th parabola contains $2k + 2$ points from $\mathbb{P}$ with $K = k$ and four points each $K = k + 2$, $k + 4, k + 6, \ldots$

Figure 6 shows the first few parabolas (their numbers are indicated in parentheses), the lines $\lambda = e(r) - 1$ and $m - \lambda = e(r)$ for integer r, as well as the points from $\mathbb{P}$ which fit into the picture. The λ-axis on this picture is horizontal, the m-axis forms an acute angle with it, while the vertical direction is assumed by the lines $2\lambda - m = \text{const}$.

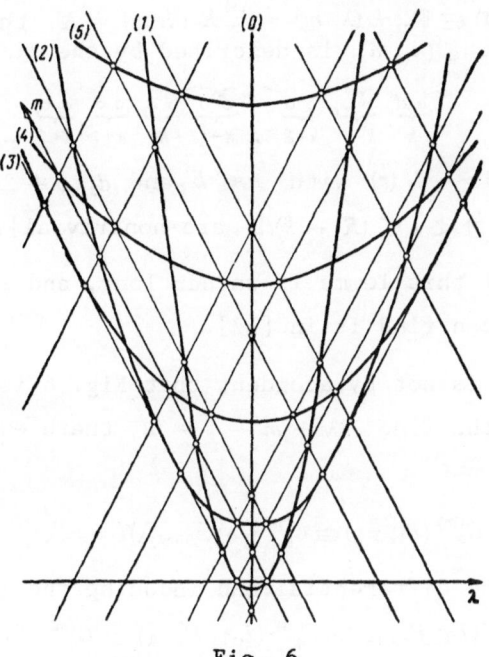

Fig. 6

<u>Main lemma</u>. (i) If the point (λ, m) is not on any pa-
rabola, then all the first differentials in the spectral sequence $\mathscr{E}(\lambda, m)$ are nonzero:

(ii) If the point (λ, m) is on the k-th parabola, but
neither on any parabola with a smaller number nor in $\mathbb{P}$, then
the spectral sequence $\mathscr{E}(\lambda, m)$ can be described by the diagram

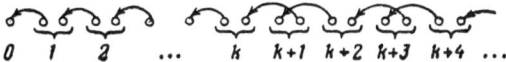

[the differentials $d_1(r)$ with $r \leqslant k$ and $d_2(k + 2s, 1), d_2(k +
2s - 1, 2)$ with $s > 0$ are nontrivial].

(iii) If $(\lambda, m) \in \mathbb{P}$, $k(\lambda, m) = k$, $K(\lambda, m) = K$, then the spectral sequence $\mathscr{E}(\lambda, m)$ is described by the diagram

$$0 \quad 1 \quad \cdots \quad k \quad k+1 \quad k+2 \quad \cdots \quad K-1 \quad K \quad K+1 \quad K+2 \quad \cdots$$

[the differentials $d_1(r)$ with $r \leqslant k$ and $d_2(k + 2s, 1)$, $d_2(k + 2s - 1, 2)$ with $0 < s \leqslant (K - k)/2$ are nontrivial].

The proof of this lemma is rather long, and I shall omit it. The reader can find it in [17].

Remark. It is not by accident that Fig. 6 is symmetric with respect to the line $2\lambda + m - 1 = 0$: there exists an isomorphism

$$C_q^{(m)}(L_1; \mathscr{F}_\lambda) \cong C_q^{(m)}(L_1; \mathscr{F}_{m-1-\lambda}),$$

commuting with the differential and inducing the isomorphisms $C_q^{(m)}(L_1; F_\lambda) \cong C_q^{(m)}(L_1; F_{m-1-\lambda})$, $C_q^{(m)}(L_1; F'_{-1-\lambda}) \cong C_q^{(m)}(L_1; F'_{m+\lambda})$, as well as numerous isomorphisms between homology and the spectral sequences. This isomorphism acts according to the formula

$$f_j \otimes e_{i_1} \wedge \ldots \wedge e_{i_q} \mapsto \sum_{u_1=0}^{i_1-1} \cdots \sum_{u_q=0}^{i_q-1} (-1)^{\sum (i_s-u_s)} \left(\prod_{r=1}^{q} \binom{i_r+1}{u_r} \right) f_{j+\sum u_s}$$
$$\otimes e_{i_1-u_1} \wedge \ldots \wedge e_{i_q-u_q}.$$

It is not really clear to us where it comes from.

E. Main results. Theorem 2.3.5a. (i) If $(\lambda, m) \notin \mathbb{P}$, then $H_*^{(m)}(L_1; \mathscr{F}_\lambda) = 0$.

(ii) If $(\lambda, m) \in \mathbb{P}$ and $K(\lambda, m) = K$, then

$$\dim H_q^{(m)}(L_1; \mathscr{F}_\lambda) = \begin{cases} 0 & \text{for } q < K, \\ 1 & \text{for } q = K, \\ 2 & \text{for } q > K. \end{cases}$$

Theorem 2.3.5b. (i) if the point (λ, m) is not on any of the parabolas, then

$$\dim H_q^{(m)}(L_1; F_\lambda) = \begin{cases} 1, & \text{if } \dfrac{3q^2 + q}{2} \leqslant m < \dfrac{3(q+1)^2 - (q+1)}{2}, \\ 0 & \text{otherwise,} \end{cases}$$

[Note that the right-hand side of this relation coincides with $\dim H_q^{(m)}(L_2)$ — see Theorem 2.3.1.]

(ii) If the point (λ, m) is on the k-th parabola and not on any parabola with smaller number, then

$$\dim H_q^{(m)}(L_1; F_\lambda) = \dim H_q^{(m)}(L_2) + \begin{cases} 1, & \text{if we have} \\ & \quad\quad \text{condition (*) below,} \\ 0 & \text{otherwise,} \end{cases}$$

$(*)\, q \geqslant k$ and $u(3r^2 - r)/2 \leqslant m < (3r^2 + r)/2$, where

$$r = \begin{cases} q, & \text{if } q \equiv k \bmod 2, \\ q+1, & \text{if } q \not\equiv k \bmod 2. \end{cases}$$

Theorem 2.3.5c. For any λ, m

$$H_q^{(m)}(L_1; F'_{-1-\lambda}) = H_{q-1}^{(m)}(L_1; F_\lambda) \oplus H_q^{(m)}(L_1; \mathcal{F}_\lambda).$$

These theorems follow from the main lemma. Note that the difference between the spectral sequences from statements (i) and (ii) of this lemma does not make itself felt in the computation of homology with coefficients in $\mathcal{F}_\lambda$. But when we compute homology with coefficients in F_λ, the spectral sequence has to be "truncated" (see Subsection C) and then not only the origin and the extremity of the arrows become important, but their lengths as well. This explains the complexity of the formulation of statement (ii) of Theorem 2.3.5b. Finally, it is easier to deduce Theorem 2.3.5c not directly from the main lemma, but from Theorems 2.3.5a,b by using the short exact sequence

$$0 \to F_\lambda \to \mathcal{F}_\lambda \to F'_{-1-\lambda} \to 0$$

(see Subsection A). The homomorphism $H_q(L_1; F_\lambda) \to H_q(L_1; \mathcal{F}_\lambda)$, as can be seen from Theorems 2.3.5a, b, is always trivial (from considerations of dimension).

The homology of the algebra L_1 with coefficients in $\mathcal{F}_{\lambda,\mu}$ and $F_{\lambda,\mu}$ can also be easily extracted from the main lemma. The complete statement is too cumbersome, and I will limit myself to the following generalizations of Theorems 2.3.5a (i) and 2.3.5b (i) needed in the sequel.

Theorem 2.3.5d. If $(\lambda, m - \mu) \not\in \mathbb{P}$, then

$$H_*^{(m)}(L_1; \mathcal{F}_{\lambda,\mu}) = 0;$$

if the point $(\lambda, m - \mu)$ is not on any of the parabolas, then

$$H_q^{(m)}(L_1; F_{\lambda,\mu}) = H_q^{(m)}(L_2).$$

We now consider the homology of the algebra W_1.

Theorem 2.3.5e.

$$\dim H_q(W_1; \mathcal{F}_{-1+\frac{3r^2\pm r}{2}}) = \begin{cases} 2 & \text{for } q = r+1, \\ 1 & \text{for } q = r, r+2; \\ 0 & \text{for } q \neq r, r+1, r+2; \end{cases}$$

$$\dim H_q(W_1; F_{-1+\frac{3r^2\pm r}{2}}) = \begin{cases} 1 & \text{for } q = r+1, r+2, \\ 0 & \text{for } q \neq r+1, r+2; \end{cases}$$

if λ does not equal $-1 + (3r^2 \pm r)/2$ for $r = 0, 1, 2, \ldots$, then $H_*(W_1; \mathcal{F}_\lambda) = 0, H_*(W_1; F_\lambda) = 0$.

This follows from the previous theorems (see Proposition 2.3.5c and the remark which follows it).

Finally, let us state a consequence of Proposition 2.3.4d.

Theorem 2.3.5f.

$$H_q(W_1; \mathcal{F}_\lambda \otimes \mathcal{F}'_\mu) = H_q^{(\lambda-\mu)}(L_1; \mathcal{F}_\lambda) \oplus H_{q-1}^{(\lambda-\mu)}(L_1; \mathcal{F}_\lambda);$$

$$H_q(W_1; F_\lambda \otimes F'_\mu) = H_q^{(\lambda-\mu)}(L_1; F_\lambda) \oplus H_{q-1}^{(\lambda-\mu)}(L_1; F_\lambda);$$
$$H_q(W_1; F'_\lambda \otimes F'_\mu) = H_q^{(\lambda+\mu+1)}(L_1; F'_\lambda) \oplus H_{q-1}^{(\lambda+\mu+1)}(L_1; F'_\lambda).$$

[Recall that $H_*^{(\lambda+\mu+1)}(L_1; F'_\lambda) = H_*^{(\lambda+\mu+1)}(L_1; F'_\mu)$ — see the remark at the end of Subsection D.]

F. First application: extensions. Recall that according to 1.4.5, short exact sequences of the form

$$0 \to F_\mu \to E \to F_\lambda \to 0 \tag{2}$$

correspond to elements of the space $H^1(W_1; \text{Hom}(F_\lambda, F_\mu)) = H_1(W_1; F_\lambda \otimes F'_\mu)$. According to Theorems 2.3.5b, f,

$$H_1(W_1; F_\lambda \otimes F'_\mu) = H_1^{(\lambda-\mu)}(L_1; F_\lambda) \ominus H_0^{(\lambda-\mu)}(L_1; F_\lambda)$$

$$= \begin{cases} \mathbb{C}, & \text{if } \lambda-\mu = 0, 2, 3, 4, \\ \mathbb{C} \oplus \mathbb{C}, & \text{if } \lambda-\mu = 1 \text{ and the point } (\lambda, \lambda-\mu) \\ & \qquad \text{is on the 0-th parabola,} \\ \mathbb{C}, & \text{if } \lambda-\mu = 5, 6 \text{ and the point } (\lambda, \lambda-\mu) \cdot \\ & \qquad \text{is on the first parabola,} \\ 0 & \text{otherwise.} \end{cases}$$

Thus, we know for what λ and μ we have the extensions (2); these λ and μ are indicated in the second column of Table 1. The third column lists cocycles characterizing these extensions; let us explain the notations used in this column. The space E is generated by elements of the form $hdz^{-\mu}$ and the inverse images $(gdz^{-\lambda})^\sim$ of elements $gdz^{-\lambda}$ of the module F_λ. The vector field fd/dz sends $(gdz^{-\lambda})^\sim$ into

$$[(fd/dz)(gdz^{-\lambda})]^\sim + C(f, g)\, dz^{-\mu}$$

(compare with 1.4.5). It is the expressions $C(f, g)$ which are indicated in the third column of the table. The first column contains the notations for the modules obtained as the result of extensions.

TABLE 1

I_λ	$\mu = \lambda$	$f'g$
II'_0	$\mu = -1, \lambda = 0$	$(f'g)'$
II''_0	$\mu = -1, \lambda = 0$	$f''g$
III_λ	$\mu = \lambda - 2$	$f'''g + 2f''g'$
IV_λ	$\mu = \lambda - 3$	$f'''g' + f''g''$
V_λ	$\mu = \lambda - 4$	$\lambda f^V g + f^{IV} g' - 6f'''g'' - 4f''g'''$
VI_0	$\mu = -5, \lambda = 0$	$2f^V g - 5f^{IV}g' + 10f'''g'' + 5f''g'''$
VI_4	$\mu = -1, \lambda = 4$	$12f^{VI}g + 22f^V g' + 5f^{IV}g'' - {} - 10f'''g''' - 5f''g^{IV}$
$VII_\pm$	$\mu = \dfrac{-7 \pm \sqrt{19}}{2}$ $\lambda = \dfrac{5 \pm \sqrt{19}}{2}$	$\dfrac{22 \pm 5\sqrt{19}}{4} f^{VII}g - \dfrac{31 \pm 7\sqrt{19}}{2} f^{VI}g' - {} - \dfrac{25 \pm 7\sqrt{19}}{2} f^V g'' - 5f^{IV}?''' + {} + 5f'''g^{IV} + 2f''g^V$

Some of these extensions are quite transparent. Thus, the W_1-module l_λ is induced by the two-dimensional L_0-module with basis u, v and an action of the algebra L_0, defined by the formulas $e_0 u = \lambda u$, $e_0 v = \lambda v + u$. The module F_0 (functions) has the one-dimensional submodule $\mathbb{C}$ (constants) and this submodule is preserved in I_0; obviously, $I_0/\mathbb{C} = II'_0$. In a similar way, $VI_4 = V_4/\mathbb{C}$. The compositional series of the module $\Lambda^2 F_\lambda$ is

$$F_{2\lambda-1}, F_{2\lambda-3}, \ F_{2\lambda-5}, \ \dots,$$

i.e., in $\Lambda^2 F_\lambda$ there is a filtration $\Lambda^2 F_\lambda = \Phi_0 \supset \Phi_1 \supset \Phi_2 \supset \dots$ with $\Phi_i/\Phi_{i+1} = F_{2\lambda-2i+1}$. The quotient module $\Lambda^2 F_\lambda/\Phi_2$ is isomorphic to $III_{2\lambda-1}$. Finally, the extensions VI_0 and VI_4 are related by the symmetry described in the concluding remark of Subsection D and a similar fact is true for the extensions VII_+ and VII_-.

G. Second application: deformations. (The results
of this subsection are due to A. Fialowski.) According to
1.4.7, to elements of the space $H^2(L_1; L_1)$ correspond deformations of the algebra L_1. As an L_1-module, L_1 is $F_{1,1}$; thus,

$$H^2_{(-m)}(L_1; L_1) = H^{(m)}_2(L_1; L_1') = H^{(m)}_2(L_1; F_{1,1}') = H^{(m)}_2(L_1; \mathcal{F}_{-2,-1}/F_{-2,-1})$$

(see Subsection A). The line $\lambda = -2$ does not intersect any
of the parabolas of Subsection D. Therefore,

$$H_*(L_1; \mathcal{F}_{-2,-1}) = 0,$$

$$H^{(m)}_q(L_1; \mathcal{F}_{-2,-1}/F_{-2,-1}) = H^{(m)}_{q-1}(L_1; F_{-2,-1}) = H^{(m)}_{q-1}(L_2)$$

(see Theorem 2.3.5d). We conclude that

$$\dim H^2_{(-m)}(L_1; L_1) = \begin{cases} 1 & \text{for } m = 2, 3, 4, \\ 0 & \text{for } m \neq 2, 3, 4. \end{cases}$$

It is not difficult to find cocycles whose cohomology classes
generate $H^2(L_1; L_1)$; such are, for example, the cocycles $\alpha \in$
$C^2_{(-2)}(L_1; L_1)$, $\beta \in C^2_{(-3)}(L_1; L_1)$, $\gamma \in C^2_{(-4)}(L_1; L_1)$, defined by the formulas

$$\alpha(e_2, e_3) = 4e_3,$$

$$\alpha(e_i, e_j) = \begin{cases} (-1)^i (j - i + 2) e_{i+j-2}, & \text{if } i = 2, 3, j \geqslant 4, \\ 0, & \text{if } i \neq 2, 3, j > i; \end{cases}$$

$$\beta(e_2, e_3) = 8e_2, \quad \beta(e_2, e_4) = 4e_3, \quad \beta(e_3, e_4) = -10e_4,$$

$$\beta(e_i, e_j) = \begin{cases} (-1)^i \binom{i-2}{2}(j - i + 3) e_{i+j-3}, & \text{if } i = 2, 3, 4, i \geqslant 5, \\ 0, & \text{if } i \neq 2, 3, 4, j > i; \end{cases}$$

$$\gamma(e_2, e_3) = 15e_1, \quad \gamma(e_2, e_4) = 0, \quad \gamma(e_2, e_5) = 8e_3,$$

$$\gamma(e_3, e_4) = -24e_3, \quad \gamma(e_3, e_5) = -16e_4, \quad \gamma(e_4, e_5) = 18e_5,$$

$$\gamma(e_i, e_j) = \begin{cases} (-1)^i \binom{i-2}{3}(j - i + 4) e_{i+j-4}, & \text{if } i = 2, 3, 4, 5, j \geqslant 6, \\ 0, & \text{if } i \neq 2, 3, 4, 5, j > i. \end{cases}$$

[The fact that α, β, γ are cocycles can be established by
a direct calculation, while the fact that the cocycles $\alpha, \beta,$

γ are not cohomology trivial is implied by the following statement (which can easily be checked): Any element of the space $H^2_{(-m)}(L_1; L_1)$ with $m \geqslant 2$ is represented by a unique cocycle which vanishes after the substitution of e_1.]

The infinitesimal deformations β and γ cannot be extended, however, to full-fledged deformations of the algebra L_1: the square $[c, c]$ of the cohomology class c of the cocycle γ, as well as the Massey cube of the cohomology class of the cocycle β, is nonzero (the verification of these statements, especially the second one, necessitates a rather tedious calculation). On the other hand, to the infinitesimal deformation α corresponds a real deformation of the algebra L_1. Here is the description of the deformation $h : L_1 \times L_1 \times \mathbb{R} \to L_1$ of the algebra L_1, for which the corresponding infinitesimal deformation actually does not coincide with α, but is only cohomologous to $-\alpha/3$. Denote by $L_1(t)$ the subalgebra of the algebra W_1, consisting of vector fields of the form $(x^2 + t)\, \varphi(x)\, d/dx$, and define the (linear) isomorphism $\varepsilon_t :$ $L_1 \to L_1(t)$ by the formula

$$\varepsilon_t(e_i) = (x^2 + t)\, x^{i-1} \frac{d}{dx}\, .$$

The deformation h is then defined by the formula

$$h(\xi_1, \xi_2, t) = \varepsilon_t^{-1}\,[\varepsilon_t(\xi_1),\, \varepsilon_t(\xi_2)].$$

§4.   COMPUTATIONS FOR LIE ALGEBRAS
      OF SMOOTH VECTOR FIELDS

At the present time, we can say that the problem of computing the cohomology of Lie algebras of smooth vector fields of a manifold with coefficients in a trivial representation and in spaces of smooth tensor fields has been solved com-

pletely. In the case of trivial coefficients, the calculations were first carried out by Haefliger [47, 48], and then
by a different method by Bott and Segal [9]; an exposition
of the work of Bott and Segal is presented below (in Subsection 2). In the case of nontrivial coefficients, the final
result is due to Tsujishita [91]. His calculation follows
along the lines indicated by Haefliger, but there are no
doubts that in the case of nontrivial coefficients the Bott—
Segal method can also be applied. The results of Tsujishita
are given without detailed proofs in Subsection 3.

These works were preceded by a large number of more particular results, among which the finite dimensionality of
the space H^q (Vect M) for any q and any compact manifold M,
established in 1969 by Gelfand and myself in [28], should
be mentioned. Some of these earlier results have not lost
their interest today: the definition and computation of "diagonal" cohomology of the algebra Vect M and the explicit
computation of the ring H^* (Vect S^1) , for instance. These
results are presented in Subsection 1.

Until the end of this section, M denotes a closed ori-
ented n-dimensional smooth ($\mathscr{C}^\infty$-) manifold. (Note that although we always limit ourselves, for the sake of simplicity,
to the closed oriented case, the main results of this section,
in particular Theorems 2.4.3, 2.4.4, 2.4.9, and 2.4.10, remain
valid without any changes for arbitrary smooth manifolds.)

1. Diagonal cohomology. A. The diagonal filtration.
We shall say that a family of vector fields $\xi_1, \ldots, \xi_q \in$ Vect
M possesses the property (Δ_k) if for any set of k points
$\Gamma \subset M$ at least one of the fields ξ_i is identically zero in
a neighborhood of Γ. For example, the property (Δ_1)   is

equivalent to the intersection $\bigcap_{i=1}^{q} \operatorname{Supp} \xi_i$ being void, while the property (Δ_k) for $k > q$ means that at least one of the fields ξ_i is identically zero.

We shall say further that the cochain $\alpha \in C^q (\operatorname{Vect} M)$ has (underline: diagonal) underline: filtration $\leqslant k$ if $\alpha (\xi_1, \ldots, \xi_q) = 0$ for any family of vector fields $\xi_1, \ldots, \xi_q$ possessing the property (Δ_k). The subspace of the space $C^q (\operatorname{Vect} M)$, consisting of cochains of diagonal filtration $\leqslant k$, is denoted by $\Delta_k C^q (\operatorname{Vect} M)$. Obviously, we have

$$C^q (\operatorname{Vect} M) = \Delta_q C^q (\operatorname{Vect} M) \supset \Delta_{q-1} C^q (\operatorname{Vect} M) \supset \ldots$$
$$\supset \Delta_1 C^q (\operatorname{Vect} M) \supset \Delta_0 C^q (\operatorname{Vect} M) = 0,$$
$$d (\Delta_k C^q (\operatorname{Vect} M)) \subset \Delta_k C^{q+1} (\operatorname{Vect} M),$$
$$\Delta_k C^q (\operatorname{Vect} M) \Delta_l C^r (\operatorname{Vect} M) \subset \Delta_{k+l} C^{q+r} (\operatorname{Vect} M).$$

Thus, $\{\Delta_k C^q (\operatorname{Vect} M)\}$ is an increasing multiplicative filtration in the complex $C^{\cdot} (\operatorname{Vect} M)$.

The complex $\Delta_1 C^{\cdot} (\operatorname{Vect} M) \subset C^{\cdot} (\operatorname{Vect} M)$ is of interest in itself and is called the underline: diagonal cochain complex of the algebra $\operatorname{Vect} M$. It consists of cochains which are annihilated by families of vector fields whose supports have an empty intersection. We shall use the shorter notation $C_\Delta^q (\operatorname{Vect} M)$, $H_\Delta^q (\operatorname{Vect} M)$ instead of $\Delta_1 C^q (\operatorname{Vect} M)$, $H^q (\Delta_1 C^{\cdot} (\operatorname{Vect} M))$.

A convenient description of the diagonal filtration, which also explains where its name comes from, may be given by using the language of "generalized sections." A underline: generalized section of a smooth vector bundle whose base is a compact manifold is, by definition, any continuous functional on the space of smooth sections of this bundle (this generally accepted term has an obvious defect: ordinary sections of a bundle do not belong to the class of its generalized sections). Cochains from $C^q (\operatorname{Vect} M)$ may be identified with

generalized sections of the bundle $\otimes^q \operatorname{tang} M = \otimes_{i=1}^q \operatorname{pr}_i^* \operatorname{tang} M$
over $M^q = M \times \ldots \times M$, where pr_i denotes the projec-
tion of M^q on the i-th factor and $\operatorname{tang} M$ is the tangent
bundle of the manifold M satisfying the condition of skew
symmetry: the section is multiplied by -1 under the action
of the automorphism interchanging two factors of the bundle
(such an identification is possible in view of the well-known
Schwartz theorem on the kernel — see [25]). A cochain be-
longs to the space $\Delta_k C^q (\operatorname{Vect} M)$ if and only if the corresponding generalized section is concentrated on the subset M_k^q
of the product M^q consisting of such $(x_1, \ldots, x_q)$ that among
the points $x_1, \ldots, x_q \in M$ there are no more than k different
ones. In particular, cochains of the diagonal complex are
cochains concentrated on the diagonal

$$\Delta = \{(x_1, \ldots, x_q) \in M^q \mid x_1 = \ldots = x_q\}.$$

B. __Spectral sequences.__ Here we do not discuss the
spectral sequence associated with the diagonal filtration,
but consider certain spectral sequences used to compute its
initial term, i.e., the cohomology of the quotient complexes
$\Delta_k C^{\cdot} (\operatorname{Vect} M)/\Delta_{k-1} C^{\cdot} (\operatorname{Vect} M)$. These spectral sequences were constructed by Gelfand and myself in [28]. Here we shall main-
ly be concerned with the case $k = 1$, i.e., with diagonal cohomology.

We say that the generalized section of a vector bundle
concentrated on the subset S of its base is of order $\leqslant l$
on S if it vanishes on any section which has a zero of order
$>l$ at every point of the set S (or, as one says, a trivi-
al l-jet). Denote by $F^m C_\Delta^q (\operatorname{Vect} M)$ the subspace of the space
$C_\Delta^q (\operatorname{Vect} M)$, consisting of cochains which are of order $\leqslant q -$
m on Δ (as generalized sections of the bundle $\otimes^q \operatorname{tang} M$).
Obviously,

$$\ldots \supset F^{q-1}C_\Delta^q \left(\text{Vect } M\right) \supset F^q C_\Delta^q \left(\text{Vect } M\right)$$
$$\supset F^{q+1}C_\Delta^q \left(\text{Vect } M\right) = 0,$$
$$\bigcup_m F^m C_\Delta^q \left(\text{Vect } M\right) = C_\Delta^q \left(\text{Vect } M\right),$$
$$d\left(F^m C_\Delta^q \left(\text{Vect } M\right)\right) \subset F^m C_\Delta^{q+1} \left(\text{Vect } M\right),$$

i.e., $\{F^m C_\Delta^q \left(\text{Vect } M\right)\}$ is a decreasing filtration of the complex $C_\Delta^{\cdot}\left(\text{Vect } M\right)$. Our next aim is to compute the initial term of the spectral sequence $\{E_r^{p,q}, d_r^{p,q}\}$ associated with this filtration.

Theorem 2.4.1a. For $p+q > 0$, we have

$$E_2^{p,q} = H_{-p}\left(M; \mathbb{R}\right) \otimes H^q(\mathbb{R}W_n).$$

Proof. We begin with the computation of $E_0^{p,q}$. By definition, $E_0^{p,q}$ is the quotient space of the space of cochains from $C_\Delta^{p+q}\left(\text{Vect } M\right)$, which are of order $\leqslant q$ with respect to Δ , factored by cochains of order $< q$. In other words, an element of the space $E_0^{p,q}$ is a generalized section of the bundle

$$\hat{\mathcal{E}}_0^{p,q} = \text{Hom}\left(S^q \text{norm}_{M^{p+q}}\Delta, \left(\bigotimes^{p+q} \text{tang } M\right)|\Delta\right)$$

over M, where $\text{norm}_{M^{p+q}}\Delta$ is the normal bundle of the diagonal Δ in M^{p+q}. These generalized sections must satisfy the skew-symmetry conditions, which reduces (in view of the triviality of the action of the symmetry group in Δ) to the requirement that these sections are sections of the subbundle

$$\mathcal{E}_0^{p,q} = \text{Alt}\left(S^q \text{norm}_{M^{p+q}}\Delta, \left(\bigotimes^{p+q} \text{tang } M\right)|\Delta\right)$$

of the bundle $\hat{\mathcal{E}}_0^{p,q}$, consisting of the homomorphisms which are multiplied by -1 when we interchange any two factors of the product M^{p+q}. The bundles $\hat{\mathcal{E}}_0^{p,q}$ and $\mathcal{E}_0^{p,q}$ are associated with the tangent bundle $\text{tang } M$ and have the following fibers. If V is a fiber of the bundle $\text{tang } M$ over the point $x \in M$, then the fiber of the bundle $\hat{\mathcal{E}}_0^{p,q}$ over x is

$$\operatorname{Hom}(S^q(\underbrace{V\oplus\ldots\oplus V}_{p+q})/V_\Delta),\ \underbrace{V\otimes\ldots\otimes V}_{p+q}),$$

where V_Δ is the image of the diagonal inclusion $V\to V\oplus\ldots$ $\oplus V$, while a fiber of the bundle $\mathscr{E}_0^{p,q}$ is the part of the previous bundle consisting of homomorphisms which are multiplied by -1 under the simultaneous interchange of any two summands of the sum $V\oplus\ldots\oplus V$ and of the factors with the same numbers in the product $V\otimes\ldots\otimes V$. Writing briefly

$$\underbrace{V\oplus\ldots\oplus V}_{p+q}=\oplus^{p+q}V\quad\text{and}\quad\underbrace{V\otimes\ldots\otimes V}_{p+q}=\otimes^{p+q}V,$$

consider the exact sequence

$$0\leftarrow S^q((\oplus^{p+q}V)/V_\Delta)\xleftarrow{\text{pr}}S^q(\oplus^{p+q}V)\xleftarrow{\sigma_1}S^{q-1}(\oplus^{p+q}V)\otimes V$$

$$\leftarrow\ldots\xleftarrow{\sigma_{i-1}}S^{q-i+1}(\oplus^{p+q}V)\otimes\Lambda^{i-1}V\xleftarrow{\sigma_i}\ldots,\qquad(1)$$

in which the homomorphism σ_i is defined by the formula

$$\sigma_i(w\otimes(v_1\wedge\ldots\wedge v_i))=\sum_{s=1}^{i}(-1)^{s-1}w\underbrace{(v_s,\ldots,v_s)}_{p+q}\otimes(v_1\wedge\ldots\hat{v}_s\ldots\wedge v_i)$$

where $w\in S^{q-i}(\oplus^{p+q}V)$ and $v_1,\ldots,v_i\in V$. We stress that the sequence (1) is finite. Obviously,

$$\operatorname{Alt}(S^*(\oplus^r V),\ \otimes^r V)=\Lambda^r(S^*V'\otimes V)=C^r(W_n(V'))$$

[where $W_n(V')$ is the algebra of formal vector fields in V']; to be more precise,

$$\operatorname{Alt}(S^t(\oplus^r V),\ \otimes^r V)=C^r_{(t-r)}(W_n(V'))=C^r_{(t-r)}(W_n(V))'$$

(the index in parentheses, as usual, relates to the natural grading of the algebra W_n). Thus,

$$\operatorname{Alt}(S^{q-i}(\oplus^{p+q}V)\otimes\Lambda^i V,\ \otimes^{p+q}V)=\Lambda^i V'\otimes\operatorname{Alt}(S^{q-i}(\oplus^{p+q}V),\ \otimes^{p+q}V)$$

$$=\Lambda^i V'\otimes C^{p+q}_{(-p-i)}(W_n(V))'.$$

Denote the bundle with standard fiber $C_{(-p-i)}^{p+q}(W_n(V))'$ associated with $\tan g\,M$ by $\gamma_i^{p,q}$ and the tensor product $\Lambda^i(\tan g\,M)'\otimes$ $\gamma_i^{p,q}$ by $\xi_i^{p,q}$. Applying the functor $\mathrm{Alt}\,(\ ,\otimes^{p+q}V)$ to (1) and passing from the sequence of fibers to the sequence of bundles, we obtain an exact sequence

$$0 \to \mathcal{E}_0^{p,q} \to \xi_0^{p,q} \to \xi_1^{p,q} \to \dots$$

and the corresponding exact sequence

$$0 \leftarrow E_0^{p,q} \leftarrow \mathrm{Sec}'\,\xi_0^{p,q} \leftarrow \mathrm{Sec}'\,\xi_1^{p,q} \leftarrow \dots \qquad (2)$$

(Sec' denotes the space of generalized sections.) Moreover, the sequences (2) with different q can be used to construct the commutative diagram

$$
\begin{array}{ccccccc}
\vdots & & \vdots & & \vdots & & \\
\uparrow & & \uparrow & & \uparrow & & \\
0 \leftarrow E_0^{p,\,q+1} & \leftarrow & \mathrm{Sec}'\xi_0^{p,\,q+1} & \leftarrow & \mathrm{Sec}'\xi_1^{p,\,q+1} & \leftarrow & \dots \\
\uparrow d_0^{p,\,q+1} & & \uparrow & & \uparrow & & \\
0 \leftarrow E_0^{p,\,q} & \leftarrow & \mathrm{Sec}'\xi_0^{p,\,q} & \leftarrow & \mathrm{Sec}'\xi_1^{p,\,q} & \leftarrow & \dots \\
\uparrow & & \uparrow & & \uparrow & & \\
\vdots & & \vdots & & \vdots & &
\end{array}
$$

whose vertical arrows, except the ones in the left column, are induced by the homomorphisms

$$\mathrm{id}\otimes d'_{p+q}\colon \Lambda^i V'\otimes C_{(-p-i)}^{p+q+1}(W_n(V))' \to \Lambda^i V'\otimes C_{(-p-i)}^{p+q}(W_n(V))'.$$

In other words, for every p we obtain a finite exact sequence of complexes

$$0 \leftarrow E_0^{p,\,\cdot} \leftarrow \mathrm{Sec}'\xi_0^{p,\,\cdot} \leftarrow \mathrm{Sec}'\xi_1^{p,\,\cdot} \leftarrow \dots \qquad (3)$$

Lemma. Except for $E_0^{p,\cdot}$ and $\mathrm{Sec}'\,\xi_{-p}^{p,\cdot}$, the complexes from the sequence (3) are acyclic.

This follows from the acyclicity of the complexes $C_{(i)}^{\cdot}(W_n)$ for $i \neq 0$ (see Theorem 1.5.2).

Together with the exactness of sequence (3), this lemma shows that the homologies of the complexes $E_0^{p, \cdot}$ and $\mathrm{Sec}' \xi_{-p}^{p, \cdot}$ coincide after an appropriate shift of dimensions:

$$E_1^{p, q} = H^q(E_0^{p, \cdot}) = H^{q-p}(\mathrm{Sec}' \xi_{-p}^{p, \cdot}) = \mathrm{Sec}'(\Lambda^{-p}(\mathrm{tang}\, M)') \otimes H_{(0)}^q(\mathbb{R}W_n)$$

$$= \Omega^{-p}(M)' \otimes H^q(\mathbb{R}W_n).$$

Finally, a direct calculation shows that the differential $d_1^{p, q}: E_1^{p, q} \to E_1^{p+1, q}$ coincides with

$$d' \otimes \mathrm{id}: \Omega^{-p}(M)' \otimes H^q(\mathbb{R}W_n) \to \Omega^{-p-1}(M)' \otimes H^q(\mathbb{R}W_n),$$

which implies $E_2^{p, q} = H^{-p}(M)' \otimes H^q(\mathbb{R}W_n) = H_{-p}(M) \otimes H^q(\mathbb{R}W_n)$. The theorem is proved.

The homology of the complexes $\Delta_k C^\cdot(\mathrm{Vect}\, M)/\Delta_{k-1}C^\cdot(\mathrm{Vect}\, M)$ for $k > 1$ can be calculated by using the spectral sequence associated with the filtration defined by the order of generalized sections with respect to M_k^q. We denote this spectral sequence by $\{{}^{(k)}E_r^{p, q}, {}^{(k)}d_r^{p, q}\}$ and state the result of our calculation of its second term.

<u>Theorem 2.4.1b</u>. The space ${}^{(k)}E_2^{p, q}$ coincides with the subspace of the space

$$H_{-p}(M^k, M_{k-1}^k) \otimes [\bigoplus_{q_1 + \ldots + q_k = q} H^{q_1}(\mathbb{R}W_n) \otimes \ldots \otimes H^{q_k}(\mathbb{R}W_n)],$$

consisting of elements which are multiplied by $(-1)^{q_i q_j}$ under the action of the transformation induced by the simultaneous interchange of the i-th and j-th factor in M^k and in $H^{q_1}(\mathbb{R}W_n) \otimes \ldots \otimes H^{q_k}(\mathbb{R}W_n)$.

The proof is similar to that of Theorem 2.4.1a.

The spectral sequences of this subsection, in principle, may be used to compute the cohomology of the algebra $\mathrm{Vect}\, M$: first one must calculate (by using the spectral sequences of Theorem 2.4.1a, b) the cohomology of the complexes

$\Delta_k C^{\cdot}/\Delta_{k-1}C^{\cdot}$ (Vect M) and then apply the spectral sequence associated with the diagonal filtration. This program was
finally carried out by Haefliger [47], but he had to overcome considerable difficulties. Since a presentation of
Haefliger's work does not enter in our plans, we shall
mention three earlier results based on Theorems 2.4.1a, b.

C. Finite dimensionality. Theorems 2.4.1a, b, together
with statements 1° and 2° from Subsection 2.3A, show that

$$\dim H^* \left(\Delta_k C^{\cdot}/\Delta_{k-1}C^{\cdot} \text{ (Vect } M)\right) < \infty,$$
$$H^q \left(\Delta_k C^{\cdot}/\Delta_{k-1}C^{\cdot} \text{ (Vect } M)\right) = 0 \text{ for } \begin{cases} k=1, \ 0 < q < n+1, \\ k>1, \ 0 \leqslant q < k\,(n+1). \end{cases}$$

Applying the spectral sequence associated with the diagonal
filtration, we conclude that the spaces H^q (Vect $\tilde{M}$) are finite dimensional for all q.

D. The case $M = S^1$.· Theorems 2.4.1a, b in this case
give us

$$\dim E_2^{p, q} = \begin{cases} 1, & \text{if } (p, q) = (0, 0), (-1, 3), (0, 3), \\ 0 & \text{otherwise,} \end{cases}$$

and for $k > 1$

$$\dim^{(k)} E_2^{p, q} = \begin{cases} 1, & \text{if } (p, q) = (-k, 3k), (-k+1, 3k), \\ 0 & \text{otherwise.} \end{cases}$$

Thus, for $q > 0$

$$\dim H^q \left(\Delta_k C^{\cdot}/\Delta_{k-1}C^{\cdot} \text{ (Vect } S^1)\right) = \begin{cases} 1, & \text{if } q = 2k, \ 2k+1, \\ 0 & \text{otherwise.} \end{cases}$$

Moreover, if we look into our spectral sequences in more detail, we can indicate explicitly the cocycles which represent
the cohomologies indicated above. In order to state the re-

sult, find the cochain $\alpha \in C^2$ (Vect S^1), $\beta \in C^3$ (Vect S^1) by the formulas

$$\alpha \left(f_1 (\varphi) \frac{d}{d\varphi}, f_2 (\varphi) \frac{d}{d\varphi} \right) = \int_{S^1} \begin{vmatrix} f_1' (\varphi) & f_2' (\varphi) \\ f_1'' (\varphi) & f_2'' (\varphi) \end{vmatrix} d\varphi,$$

$$\beta \left(f_1 (\varphi) \frac{d}{d\varphi}, f_2 (\varphi) \frac{d}{d\varphi}, f_3 (\varphi) \frac{d}{d\varphi} \right) = \int_{S^1} \begin{vmatrix} f_1 (\varphi) & f_2 (\varphi) & f_3 (\varphi) \\ f_1' (\varphi) & f_2' (\varphi) & f_3' (\varphi) \\ f_1'' (\varphi) & f_2'' (\varphi) & f_3'' (\varphi) \end{vmatrix} d\varphi$$

(where φ is the angular parameter on the circle). An obvious verification shows that α, β are cocycles and that the cohomology classes of the cocycles α^k, $\alpha^{k-1}\beta$ are the generators of the spaces

$$H^{2k} (\Delta_k C^{\cdot}/\Delta_{k-1} C^{\cdot} (\text{Vect } S^1)),$$
$$H^{2k+1} (\Delta_k C^{\cdot}/\Delta_{k-1} C^{\cdot} (\text{Vect } S^1)).$$

Now, when we see that these generators are represented by cocycles of the complex $C^{\cdot}$ (Vect S^1), we can conclude that the spectral sequence associated with the diagonal filtration is trivial and thus state the final result.

Theorem 2.4.2. The ring H^* (Vect S^1) is isomorphic to the tensor product of the polynomial ring with one two-dimensional generator and the exterior algebra with one three-dimensional generator; these generators are represented by the cocycles α, β.

This theorem was proved by Gelfand and myself in 1968 (see [27]), before the spectral sequences of Subsection B were constructed.

Note that the cocycle β is cohomologous to the cocycle

$$\left(f_1 (\varphi) \frac{d}{d\varphi}, f_2 (\varphi) \frac{d}{d\varphi}, f_3 (\varphi) \frac{d}{d\varphi} \right) \rightarrow \begin{vmatrix} f_1 (\varphi_0) & f_2 (\varphi_0) & f_3 (\varphi_0) \\ f_1' (\varphi_0) & f_2' (\varphi_0) & f_3' (\varphi_0) \\ f_1'' (\varphi_0) & f_2'' (\varphi_0) & f_3'' (\varphi_0) \end{vmatrix},$$

where φ_0 is any fixed value of the variable φ; thus, the co-homology class of this cocycle is the image of the generator of the space $H^3(\mathbb{R}W_1)$ under the homomorphism induced by the projection Vect $S^1 \to \mathbb{R}W_1$, which sends each vector field into its ∞-jet at an arbitrarily chosen point.

E.   The Losik–Guillemin theorem.   The differentials of the spectral sequence of Theorem 2.4.1a were independently computed by Losik [64, 65] and Guillemin [43].   In order to state the result, denote by $X(M)$ the total space of the bundle $x(M)$ over M with standard fiber X_n, associated with the complexification of the tangent bundle tang M.

Theorem 2.4.3.   When $q > 0$ we have

$$H_\Delta^q (\text{Vect } M) \cong H^{q+n}(X(M); \mathbb{R}).$$

We shall now give the proof of this theorem, although it is not by any means a consequence of the results concerning the ring $H^*(\text{Vect } M)$ developed below.   The relationship between Theorem 2.4.3 and these results will be discussed in Sub-subsection 2A.

2.   Calculation of the ring $H^*(\text{Vect } M)$.   A.   Statement of the main theorem.   As in Sub-subsection 1E, we denote by $x(M)$ the bundle with base M, structural group $U(n)$, and standard fiber X_n, associated with the complexification of the tangent bundle tang M; by $X(M)$ we denote the total space of the bundle $x(M)$.

Theorem 2.4.4.   We have the multiplicative isomorphism

$$H^*(\text{Vect } M) \cong H^*(\text{Sec } x(M); \mathbb{R}),$$

where   Sec   denotes the space of sections.

For example, $x(S^1)$) is the trivial bundle $S^1 \times S^3 \to S^1$, Sec $(S^1, S^3) = $ Map $(S^1, S^3) = S^3 \times \Omega S^3$, and H^* (Sec $x(S^1)$; $\mathbb{R}$) is the free anticommutative algebra with two generators of dimension 2 and 3; thus Theorem 2.4.4 agrees with Theorem 2.4.2.

Theorem 2.4.4 was stated as a conjecture by Bott and myself as early as 1972. As we already mentioned, its first proof was given in 1974 by Haefliger; a year later, Bott and Segal published a new proof of this theorem [9], which considerably clarified the matter. Later in this subsection we give the Bott–Segal proof of Theorem 2.4.4.

The diagonal filtration plays an episodic role in this proof, and the relationship between Theorems 2.4.3 and 2.4.4 is not clear from it. Nevertheless, the reader might try to extract the following statement from the proof.

<u>Theorem 2.4.5</u>. We have the commutative diagram

$$\begin{array}{ccc} H^q_\Delta (\text{Vect}\, M) & \to & H^{q+n} (X(M); \mathbb{R}) \\ \downarrow & & \downarrow \\ H^q (\text{Vect}\, M) & \to & H^q (\text{Sec}\, x(M); \mathbb{R}), \end{array}$$

in which the horizontal arrows denote the isomorphisms from Theorems 2.4.3, 2.4.4, the left vertical arrow is induced by the inclusion of the diagonal complex in the complete cochain complex of the algebra Vect M, and the right vertical arrow is the composition

$$H^{q+n} (X(M); \mathbb{R}) \to H^{q+n} (\text{Sec}\, x(M) \times M; \mathbb{R}) \to H^q (\text{Sec}\, x(M); \mathbb{R})$$

of the homomorphism induced by the map Sec $x(M) \times M \to X(M)$, which acts according to the formula $(s, y) \mapsto s(y)$, and integration over M.

The assumptions on the compactness and orientability of
M made at the beginning of the section hold good in this
subsection; however, they are in fact not used anywhere in
the proof of Theorem 2.4.4, except for one reference to the
previous subsection. This reference may have been avoided,
since the proof developed there works almost without any
changes in the general case. Also, in the noncompact case,
we have the analog of Theorem 2.4.4 for the algebra $\text{Vect}_c(M)$
of vector fields with compact supports; in this analog, the
space of all sections is replaced by the space of finite sec-
tions. [For example, $H^*(\text{Vect}_c \mathbb{R}^i) \cong H^*(\Omega S^3; \mathbb{R})$ is the polynomial ring in one two-dimensional generator.] The proof
is similar to that of Theorem 2.4.4.

B. The case of Euclidean space. Lemma 1. The homo-
morphism $\text{Vect } \mathbb{R}^n \to \mathbb{R}W_n$ sending each vector field into its
∞-jet at 0 induces the isomorphism

$$H^*(\mathbb{R}W_n) \overset{\cong}{\to} H^*(\text{Vect } \mathbb{R}^n).$$

Proof. Obviously, we have the decomposition

$$C^{\cdot}(\text{Vect } \mathbb{R}^n) = C^{\cdot}(W_n) \oplus B^{\cdot},$$

where $B^{\cdot}$ is the subcomplex of the complex $C^{\cdot}(\text{Vect } \mathbb{R}^n)$, consisting of cochains which become zero when any polynomial
vector field is substituted into them. In other words, a
cochain $\gamma \in C^q(\text{Vect } \mathbb{R}^n)$ belongs to B^q if and only if

$$\lim_{t \to 0} T_t \alpha = 0, \tag{4}$$

where T_t is the homothety of the space $\mathbb{R}^n$ of center 0 and
coefficient t. Suppose $\theta(\xi)\colon C^q(\text{Vect } \mathbb{R}^n) \to C^q(\text{Vect } \mathbb{R}^n)$ and
$i(\xi)\colon C^q(\text{Vect } \mathbb{R}^n) \to C^{q-1}(\text{Vect } \mathbb{R}^n)$ are the Lie derivation along
the vector field $\xi \in \text{Vect } \mathbb{R}^n$ and the substitution of the
vector field ξ; by definition of Lie derivation,

$$\theta\,(\xi) = i\,(\xi) \circ d + d \circ i\,(\xi).$$

Therefore, for any cochain $\alpha \in C^q\,(\mathrm{Vect}\,\mathbb{R}^n)$,

$$\frac{d}{dt}T_t\alpha = H_t d\alpha + dH_t\alpha,$$

where $H_t = t^{-1}T_t i\,(\sum x_i d/dx_i) = i\,(\sum x_i d/dx_i)\,t^{-1}T_t$; if we have condition (4), then

$$\alpha = T_1\alpha - \lim T_t\alpha = Kd\alpha + dK\alpha,$$

where $K\,(\cdot) = i\,\left(\sum x_i d/dx_i\right)\int_0^1 t^{-1}T_t\,(\cdot)\,dt.$ Thus, K is the homotopy which joins the identical map of the complex $B^{\cdot}$ with the trivial map, so that the complex $B^{\cdot}$ is acyclic.

C. Lemma on cochain G-algebras and Weyl algebras. Suppose G is a Lie group, $\mathfrak{g}$ the corresponding Lie algebra, and $A^{\cdot}$ a multiplicative cochain (this word simply means that the differential is of degree +1) complex of vector spaces over $\mathbb{C}$ with unit. We shall say that $A^{\cdot}$ is a <u>topological cochain G-algebra</u> if it is supplied with an action of the group G in $A^{\cdot}$ by means of automorphisms and with G-homomorphisms

$$i = i_q\colon \mathfrak{g} \to \mathrm{Hom}\,(A^q, A^{q-1}) \qquad (q \in \mathbb{Z}),$$

such that for every $\xi \in \mathfrak{g}$ the homomorphism

$$\theta\,(\xi) = di\,(\xi) + i\,(\xi)\,d\colon A^q \to A^q$$

is the derivative of the action of the group G along the tangent vector ξ; the condition that i is a G-homomorphism means that $i\,(g\xi) = g \circ i\,(\xi) \circ g^{-1}$, where $g\xi$ is understood in the sense of the adjoint action of G in $\mathfrak{g}$. In an obvious way, one defines G-homomorphisms between topological cochain G-algebras (it is required that a G-homomorphism send the unit into the unit).

For example, if G acts smoothly on the manifold M, then $\Omega^{\cdot}M$ and $C^{\cdot}(\text{Vect}\, M)$ are topological cochain G-algebras in a natural sense. Another important example of a topological cochain G-algebra is given by the Weyl algebra $W(\mathfrak{g})$ of the algebra $\mathfrak{g}$ [the action of G in $W(\mathfrak{g})$ is induced by the adjoint representation of group G, while $i(\xi)$ acts on $S^*\mathfrak{g}'$ trivially and on $\Lambda^*\mathfrak{g}'$ by substitutions.]

The element α of the topological cochain G-algebra $A^{\cdot}$ is said to be <u>horizontal</u> if $i(\xi)\alpha = 0$ for any $\xi \in \mathfrak{g}$. The space of horizontal elements is denoted by A_{horiz}; in general, it is not closed with respect to the differential. For example, $W(\mathfrak{g})_{\text{horiz}} = S^*\mathfrak{g}'$.

By a <u>connection</u> in the topological cochain G-algebra $A^{\cdot}$ we mean an arbitrary G-homomorphism $W(\mathfrak{g}) \to A$. Any connection induces the homomorphism $S^*\mathfrak{g}' = W(\mathfrak{g})_{\text{horiz}} \to A_{\text{horiz}}$ and, by its means, the action of the commutative algebra $S^*\mathfrak{g}'$ in A_{horiz} as well as the homomorphism

$$W(\mathfrak{g}) \otimes_{S^*\mathfrak{g}'} A_{\text{horiz}} \to A. \qquad (5)$$

A connection is said to be <u>standard</u> if (5) is an isomorphism. (The fundamental example which explains the terms is: If M is the space of a principal smooth G-bundle, then the choice of a connection in $\Omega^{\cdot}M$ is equivalent to the choice of a connection in this bundle; this connection is always standard.)

A horizontal element of the topological cochain G-algebra is said to be <u>basic</u> if it is G-invariant. The space of basic elements is denoted by A_{basic}; unlike A_{horiz} the space A_{basic} is closed with respect to the differential. For example, $W(\mathfrak{g})_{\text{basic}} = \text{Inv}_{\mathfrak{g}}\, S^*\mathfrak{g}'$; we denote this space by $I(\mathfrak{g})$.

By Theorem 2.1.4,

$$\mathrm{Inv}_{\mathfrak{gl}(n,\,\mathbb{C})}\, S^*\,(\mathfrak{gl}\,(n,\,\mathbb{C})')$$

is the ring of polynomials in n variables contained in $S^1\,(\mathfrak{gl}\,(n,\;\mathbb{C})'),\,\ldots,\,S^n\,(\mathfrak{gl}\,(n,\,\mathbb{C})')$. Thus $I\,(\mathfrak{gl}\,(n,\,\mathbb{C}))$ is the ring of polynomials with generators of degrees $2, 4, \ldots, 2n$.

The connection $W\,(\mathfrak{g}) \to A^{\cdot}$ induces the G-homomorphism

$$W\,(\mathfrak{g}) \otimes_{I\,(\mathfrak{g})} A_{\text{basic}} \to A^{\cdot}. \tag{6}$$

Lemma 2. Suppose $G = \mathrm{GL}\,(n,\,\mathbb{C})$ and A is the topological cochain G-algebra with standard connection. If A is represented as a G-module in the form of a (possibly infinite) product of tensor modules, then the homomorphism (6) induces such an isomorphism in cohomology. Moreover, if A' is another algebra, a G-homomorphism $A \to A'$ induces an isomorphism in cohomology whenever the corresponding homomorphism $A_{\text{basic}} \to A'_{\text{basic}}$ induces an isomorphism in cohomology.

Proof. Using the isomorphism $A \cong W\,(\mathfrak{g}) \otimes_{S^*\mathfrak{g}'} A_{\text{horiz}} \cong \Lambda^*\mathfrak{g}' \otimes A_{\text{horiz}}$, define a filtration $A = A_0 \supset A_1 \supset A_2 \supset \cdots$ in A by the formula $A_p = \oplus_{q \geqslant p} \Lambda^*\mathfrak{g}' \otimes A_{\text{hor.z}}^q$; obviously this filtration is compatible with the differential. The corresponding spectral sequence has the initial term

$$E_2 = H^*\,(\mathfrak{g};\,A_{\text{horiz}}).$$

On the other hand, $(W\,(\mathfrak{g}) \otimes_{I\,(\mathfrak{g})} A_{\text{basic}})_{\text{horiz}} = S^*\mathfrak{g}' \otimes_{I\,(\mathfrak{g})} A_{\text{basic}}$, so that a similar spectral sequence for $W\,(\mathfrak{g}) \otimes_{I\,(\mathfrak{g})} A_{\text{basic}}$ begins with

$$E_2 = H^*\,(\mathfrak{g};\,S^*\mathfrak{g}' \otimes_{I\,(\mathfrak{g})} A_{\text{basic}}),$$

and it remains to show that the natural map $S^*\mathfrak{g}' \otimes_{I\,(\mathfrak{g})} A_{\text{basic}} \to A_{\text{horiz}}$ induces the isomorphism

$$H^*\,(\mathfrak{g};\,S^*\mathfrak{g}' \otimes_{I\,(\mathfrak{g})} A_{\text{basic}}) \cong H^*\,(\mathfrak{g};\,A_{\text{horiz}}).$$

But the latter follows from Theorem 2.1.2, since

$$\mathrm{Inv}_{\mathfrak{g}}\,(S^*\mathfrak{g}'\otimes_{I(\mathfrak{g})}A_{\mathrm{basic}})=I\,(\mathfrak{g})\otimes_{I(\mathfrak{g})}A_{\mathrm{basic}}=A_{\mathrm{basic}}=\mathrm{Inv}_{\mathfrak{g}}\,A_{\mathrm{horiz}}.$$

D. The fundamental map. Suppose M and N are smooth manifolds and P is a smooth family of immersions $M\to N$. In other words, we are given a smooth map $f\colon P\times M\to N$, such that $f_p=f\,|\,p\times M$ is an immersion for every point $p\in P$. We shall construct a homomorphism $\hat{f}$ of the cochain algebra $C^{\cdot}(\mathrm{Vect}\,M)$ into the bigraded cochain algebra $\Omega^{\cdot}(P;C^{\cdot}(\mathrm{Vect}\,N))$ of differential forms of the manifold P with coefficients in $C^{\cdot}(\mathrm{Vect}\,N)$. To do this, note that f defines a map F of the manifold of tangent vectors to the manifold P into $\mathrm{Vect}\,M$: for each vector v, applied to the point $p\in P$, we put $[F\,(v)]\,(x)=(df_p)^{-1}df\,(v)$. Now, for $c\in C^q\,(\mathrm{Vect}\,M)$, set

$$\hat{f}\,(c)=\sum_{r=0}^{q}\omega^r,$$

where the form $\omega^r\in\Omega^r\,(P;C^{q-r}\,(\mathrm{Vect}\,N))$ is defined by the formula

$$[\omega^r\,(v_1,\ldots,v_r)]\,(\eta_1,\ldots,\eta_{q-r})$$
$$=c\,(F\,(v_1),\ldots,F\,(v_r),f_p^*\eta_1,\ldots,f_p^*\eta_{q-r}),$$

in which $v_1,\ldots,v_r$ are tangent vectors to P at the point p and $\eta_1,\ldots,\eta_{q-r}\in\mathrm{Vect}\,N$. The fact that the map $\hat{f}$ commutes with the differential is checked directly. It is just as easy to establish that, if the manifolds P and N are supplied with the action of some group G and the map f is equivariant, then $\hat{f}\,(C^*\,(\mathrm{Vect}\,M))$ is contained in $\Omega^*\,(P;C^*\,(\mathrm{Vect}\,N))_{\mathrm{basic}}$ [in the case considered, $\Omega^{\cdot}\,(P;C^{\cdot}\,(\mathrm{Vect}\,N))$ is a cochain G-algebra in an obvious sense].

E. Topological lemmas. Our next goal is to describe a convenient cochain algebra for the computation of the cohomology of the space of maps of one topological space into an-

other, or, in a more general way, of the space of sections
of a bundle.

   We begin with the case of the space of maps. Suppose
X is a topological space with open covering $\mathfrak{u} = \{U_\alpha\}_{\alpha \in S}$,
all finite nonempty intersections of its elements being con-
tractible; such coverings will be referred to as contractible
for brevity. For $\sigma = \{\alpha_0, \ldots, \alpha_q\} \subset S$ we put $U_\sigma = U_{\alpha_0} \cap \ldots$
$\cap U_{\alpha_q}$; for $\sigma, \tau \subset S$ we shall write $\sigma \leqslant \tau$, if $\sigma \supset \tau$ (and
thus $U_\sigma \subset U_\tau$). By the <u>nerve</u> of the covering $\mathfrak{u}$ we mean
the simplicial set Σ, the set Σ_p of whose p-simplices is
the set of chains $\sigma_0 \leqslant \ldots \leqslant \sigma_p$ of finite subsets of S with
nonempty U_{σ_0}. (This is the barycentric subdivision of what
is usually called the nerve.) Our goal is to show that under
favorable circumstances the cohomology of the function space
Y^X coincides with the complete cohomology of the double complex

$$C\,(Y^\Sigma) = \{C\,(Y^{\Sigma_0}) \leftarrow C\,(Y^{\Sigma_1}) \leftarrow C\,(Y^{\Sigma_2}) \leftarrow \ldots\},$$

where Y^{Σ_p} denotes the product of a collection of copies
of the space Y, indexed by elements of the set Σ_p. The horizontal differentials in $C\,(Y^\Sigma)$ come from the boundary operators in the nerve; C denotes some reasonable space of cochains. In order to clarify the formulation of the problem,
we split it into two parts. The assignment $[p] \to Y^{\Sigma_p}$ defines a covariant functor from the category of finite ordered
sets to the category of spaces, i.e., a <u>cosimplicial space</u>.
We recall that the <u>realization</u> $|Z|$ of a cosimplicial space
$[p] \mapsto Z_p$ is defined to be the subspace of the product $\Pi_{p \geqslant 0} Z_p^{\Delta_p}$
(where Δ_p is the standard p-simplex), consisting of those
sequences $\{\alpha_p \colon \Delta_p \to Z_p\}$, such that the diagram

$$\begin{array}{ccc} \Delta^p & \xrightarrow{\alpha_p} & Z_p \\ \downarrow{\scriptstyle\theta_*} & & \downarrow{\scriptstyle\theta_*} \\ \Delta^q & \xrightarrow{\alpha_q} & Z_q \end{array}$$

is commutative for any monotone mapping θ. The following assertion is obvious.

The realization of a cosimplicial space $[p] \to Y^{\Sigma_p}$ coincides with the space of maps of the realization of the simpli-
cial set $[p] \to \Sigma_p$ into Y (in short: $|Y^\Sigma| = Y^{|\Sigma|}$).   In
particular, if the covering $\mathfrak{u}$ is contractible, then the space
$|Y^\Sigma|$ is homotopy equivalent to Y^X.

If Z is a simplicial space, i.e., a contravariant functor $[p] \to Z_p$, it is then well known that the cohomology of
its realization coincides with the cohomology of the double
complex $\{C^q(Z_p)\}$.   We need a similar statement for cosimplicial spaces, but it is true only under considerable restrictions.   First of all, let us note that cochains of a cosimplicial space constitute a cosimplicial cochain complex $[p] \to$
$C(Z_p)$,   i.e., the diagram

$$\begin{array}{c} \cdots \quad \cdots \quad \cdots \\ \uparrow \quad \uparrow \quad \uparrow \\ C_0^1 \xleftarrow{} C_1^1 \overset{\leftarrow}{\leftarrow} C_2^1 \overset{\leftarrow}{\underset{\leftarrow}{\leftarrow}} \cdots \\ \uparrow \quad \uparrow \quad \uparrow \\ C_0^0 \xleftarrow{} C_1^0 \overset{\leftarrow}{\leftarrow} C_2^0 \overset{\leftarrow}{\underset{\leftarrow}{\leftarrow}} \cdots, \end{array}$$

in which $C_p^q = C^q(Z_p)$.   The total complex C_p^q is of degree
$q - p$.   It shall be convenient to replace the simplicial co-
chain complex $C = \{C_0 \rightleftarrows C_1 \overset{\leftarrow}{\rightleftarrows} \ldots\}$ by the total complex of
the "normalized bicomplex" $\{C_0 \leftarrow \bar{C}_1 \leftarrow \bar{C}_2 \leftarrow \ldots\}$, where $\bar{C}_p$ is
the quotient complex of the complex C_p by the sum of images
of homomorphisms $C_{p-1} \to C_p$, induced by the degeneracy operators.   This bicomplex will be called the <u>realization</u> of the

simplicial complex C and denoted by $|C|$. The part of degree p of the total complex of this bicomplex is by definition the <u>sum</u> (not the product) of all the $\bar{C}_p^q$ with $q - p = r$.

We shall assume that our cochains from C constitute a commutative cochain algebra. Then the realization $|C|$ is also supplied with a commutative cochain algebra structure: the product $x_1 \in C_{p_1}^{q_1}$ and $x_2 \in C_{p_2}^{q_2}$ is defined as

$$\sum_\theta \text{sign} (\theta) (\theta_1^* x_1) (\theta_2^* x_2) \in C_{p_1+p_2}^{q_1+q_2},$$

where θ ranges over pairs of maps $\theta_1 \colon [p_1 + p_2] \to [p_1]$, $\theta_2 \colon [p_1 + p_2] \to [p_2]$ with $\theta_1 (0) \leqslant \ldots \leqslant \theta_1 (p_1 + p_2)$, $\theta_2 (0) \leqslant \ldots \leqslant \theta_2 (p_1 + p_2)$, $\theta_1 (m) + \theta_2 (m) = m$ for $m = 0, \ldots, p_1 + p_2$. We shall also assume that the integration $\int_{\Delta^p} \colon C^{p+q}(\Delta^p \times X) \to C^q (X)$ is defined for our cochains, that it satisfies the relation $(-1)^p d\int_{\Delta^p} \alpha = \int_{\Delta^p} d\alpha - \int_{\partial \Delta^p} \alpha$, and that it vanishes on the image of the map $C^{r+q} (\Delta^{p-1} \times X) \to C^{p+q} (\Delta^p \times X)$, induced by the degeneracy operator. (All of this holds if the Z_p are manifolds and C is the de Rham theory.)

Since for a cosimplicial space $[p] \mapsto Z_p$ we have the natural map $\Delta^p \times |Z| \to Z_p$, integration defines homomorphisms $C^{p+q}(Z_p) \to C^q (|Z|)$, which constitute the cochain map $|C (Z)| \to C (|Z|)$. It is easy to check that this is a multiplicative homomorphism of one cochain algebra into another.

We can limit our considerations to the special class of cosimplicial spaces. Namely, we shall assume that we are given an inverse system of spaces $\{P_\alpha\}_{\alpha \in A}$, i.e., that A is a partially ordered set and for $\alpha \leqslant \beta$ we are given the map

$P_\beta \to P_\alpha$.   Our cosimplicial space $[p] \to Z_p$ is defined by
the formula

$$Z_p = \prod_{\alpha_o \leqslant \ldots \leqslant \alpha_p} P_{\alpha_\bullet}.$$

<u>Lemma 3</u>. If in the previous situation the set A is
finite and each P_α is n-connected, where n is the dimension
of the nerve A, then the natural map $|C(Z)| \to C(|Z|)$ is
a cohomology equivalence.

<u>Corollary</u>. If $\mathfrak{u}$ is an open covering of the space X
with the nerve Σ and $\pi_i(Y) = 0$ for $i \leqslant \dim \Sigma$, then
$|C(Y^\Sigma)| \to C(Y^{|\Sigma|})$ is a cohomology equivalence.

I shall not give the proof of Lemma 3 here, since it
would lead us far away from the subject. This proof is contained in Anderson's article [2] and in the Bott–Segal article mentioned previously. The reader with sufficient topological sophistication, in particular one who knows the
Eilenberg–Moore spectral sequence, will certainly be able to
recover this proof.

The passage from the space of maps to the space of sections of a bundle can be carried through without much difficulty. First of all we associate, to the open covering $\mathfrak{u} =$
$\{U_\alpha\}_{\alpha \in S}$ of the space X, its "thickened nerve" — the simplicial space X_Σ, defined by the formula

$$(X_\Sigma)_p = \bigsqcup_{\alpha_o \leqslant \ldots \leqslant \alpha_p} U_{\alpha_\bullet};$$

there is a natural map $\pi\colon |X_\Sigma| \to X$, which is a homotopy equiv-
alence for paracompact X. Suppose now $p\colon E \to X$ is a continuous mapping; denote by Γ_α the space of sections of the
map p over U_α. Then $\{\Gamma_\alpha\}$ is an inverse system of spaces to

which we associate some cosimplicial space Γ_Σ . The following proposition is obvious.

The realization $|\Gamma_\Sigma|$ of the cosimplicial space Γ_Σ is
the space of sections of the maps $\pi^* E \to |X_\Sigma|$, induced by
the map p by means of π . In particular, if p is a Hure-
wiecz fibration and X is paracompact, then $|\Gamma_\Sigma|$ is homo-
topy equivalent to the space $\operatorname{Sec} p$ of sections of the fibration p.

Further, we assume that p is a Hurewiecz fibration. If
the covering $\mathfrak{u}$ is contractible, then all the spaces Γ_α which
constitute Γ_Σ are homotopy equivalent to the fiber of the
fibration p and, under the condition that the covering is
finite and the fibers are sufficiently highly connected, Lemma 3 yields a cochain complex for $\Gamma(E)$.

It shall be convenient to replace Γ_Σ by an equivalent
but smaller cosimplicial space which is constructed in the
following way. For subsets σ of the set S with a nonempty
U_σ denote by U^σ the union $\bigcup_{\alpha \in \sigma} U_\alpha$ (the "star" of the set
U_σ). If the covering $\mathfrak{u}$ is contractible, then the set U^σ
will also be contractible. Let, further, $E^\sigma = p^{-1}(U^\sigma)$. Then
$\{E^\sigma\}$ is an inverse system of spaces. The cosimplicial space
E^Σ which arises in this way is comparable in its size with
the cosimplicial space Y^Σ (which was used as a model of the
space of maps). The following lemma shows that E^Σ is an
equivalent replacement of the space Γ_Σ.

<u>Lemma 4</u>. If X is paracompact and possesses a finite
contractible covering whose nerve is of dimension $\leqslant n$, and if
$p: E \to X$ is a Hurewiecz fibration with n-connected fiber,
then we have the homotopy equivalence $|C(E^\Sigma)| \to C(\operatorname{Sec} p)$.

Proof. Let $D = \bigcup_{\alpha \in S} (U_\alpha \times U_\alpha) \subset X \times X$. A slice function for the fibration p will be, by definition, a family of
homotopy equivalences $T_{xy}\colon p^{-1}(x) \to p^{-1}(y)$, defined for $(x, y) \in$
D and continuously depending on x, y. Obviously, any Hure-
wiecz fibration possesses a slice function with respect to
any contractible covering. By means of such a function, we
can define a family of maps $E^\sigma \to \Gamma_\sigma$ by sending each point
$e \in p^{-1}(x)$ into the section whose value at the point y is
$T_{xy}(e)$. Obviously, these maps are homotopy equivalences and
constitute a map of the cosimplicial space E^Σ into the cosimplicial space Γ_Σ. Since, as we already know, the realization of the space Γ_Σ is $\operatorname{Sec} p$, it remains to use the following standard statement, which shall also be useful later on.

Lemma 5. If $C \to C'$ is a morphism of one simplicial
chain complex into another such that $C_p \to C'_p$ is a cohomol-
ogy equivalence for every p, then $|C| \to |C'|$ is also a
homology equivalence.

For the proof it suffices to establish that $\bar{C}_p \to \bar{C}'_p$
is a homology equivalence for each p, and this can be done
by means of the 5-lemma and the exact sequence

$$0 \leftarrow \bar{C}_n \leftarrow C_n \leftarrow \bigoplus_i C_{n-1}^{(i)} \leftarrow \bigoplus_{i<j} C_{n-2}^{(ij)} \leftarrow \ldots \leftarrow C_0^{(12\ldots n)} \leftarrow 0,$$

in which $C_{n-k}^{(i_1\ldots i_k)}$ is a copy of C_{n-k}.

F. The simplicial cochain complex associated with
$C^\cdot(\operatorname{Vect} M)$. Our next goal is to represent the cohomology
of the algebra $\operatorname{Vect} M$, as we did with the cohomology of the
space $\operatorname{Sec} x(M)$, in the form of the cohomology of a certain
simplicial cochain complex. Suppose we are given a finite
contractible covering $\mathfrak{u}$ of the manifold M; moreover, assume
that M is supplied with a Riemann metric and the elements

of the covering 𝔲 are geodesically convex. Since $C^{\cdot}$ (Vect □)
is a covariant functor with respect to dimension-preserving
immersions, the "thickened nerve" M_{Σ} of the covering 𝔲 determines a simplicial cochain algebra $C^{\cdot}$ (Vect M_{Σ}); and the
inclusion homomorphism $C^{\cdot}$ (Vect U_{α}) → $C^{\cdot}$ (Vect M) constitutes
an "augmentation" $C^{\cdot}$ (Vect M_{Σ}) → $C^{\cdot}$ (Vect M).

Lemma 6. The augmentation determines a cohomology equivalence

$$| C^{\cdot} \text{ (Vect } M_{\Sigma}) | \to C^{\cdot} \text{ (Vect } M).$$

Proof. If

$$
\begin{array}{ccccccc}
\vdots & & \vdots & & \vdots & & \vdots \\
\uparrow & & \uparrow & & \uparrow & & \uparrow \\
A^2 & \leftarrow & C_0^2 & \leftarrow & C_1^2 & \leftarrow & C_2^2 \leftarrow \ldots \\
\uparrow & & \uparrow & & \uparrow & & \uparrow \\
A^1 & \leftarrow & C_0^1 & \leftarrow & C_1^1 & \leftarrow & C_2^1 \leftarrow \ldots \\
\uparrow & & \uparrow & & \uparrow & & \uparrow \\
A^0 & \leftarrow & C_0^0 & \leftarrow & C_1^0 & \leftarrow & C_2^0 \leftarrow \ldots
\end{array}
$$

is a double complex with augmentation whose lines are exact,
then it has the some cohomology as the "infinite total com-
plex" whose cochain space in dimension r is $\prod_{q \geqslant 0} C_q^{q+r}$. However, this is not the complex which we use, since we are us-
ing direct sums and not direct products. To take care of
the "convergence problem," we shall use the filtration introduced in Subsection 1:

$$
0 = \Delta_0 C^q \text{ (Vect } M) \subset \Delta_1 C^q \text{ (Vect } M) \subset
$$
$$
\ldots \subset \Delta_q C^q \text{ (Vect } M) = C^q \text{ (Vect } M).
$$

For every k construct the augmented double complex by re-
placing the function $C^{\cdot}$ in $| C^{\cdot}$ (Vect M_{Σ}) $|$ by the functor
$\Delta_k C^{\cdot}$. For this augmented double complex we shall prove two
statements:

(i) each line is trivial, beginning with the term of number dk, where $d = \dim \Sigma$;

(ii) each line is exact.

It follows from these two statements that $\mid \Delta_k C^{\cdot}\,(\mathrm{Vect}\ M_\Sigma)\mid$ $\to \Delta_k C^{\cdot}\,(\mathrm{Vect}\ M)$ is a cohomology equivalence, since the normalized double complex associated with $\Delta_k C^{\cdot}\,(\mathrm{Vect}\ M_\Sigma)$, has no more than $kd + 1$ nontrivial columns. Since $\Delta_k C^q\,(\mathrm{Vect}\ M_\Sigma) = C^q\,(\mathrm{Vect}\ M_\Sigma)$ for $k \geqslant q$, we conclude that $\mid C^{\cdot}\,(\mathrm{Vect}\ M_\Sigma)\mid \to C^{\cdot}\,(\mathrm{Vect}\ M)$ is also a cohomology equivalence. Thus it remains to prove the statements (i) and (ii).

The proof of statement (i) is based on the following obvious property of the functor $F = \Delta_k C^{\cdot}\,(\mathrm{Vect}\ \square)$: for any manifolds $X_1, \ldots, X_m$ with $m \geqslant k$, the natural map

$$\bigoplus_{i_1 < \ldots < i_k} F\,(X_{i_1} \sqcup \ldots \sqcup X_{i_k}) \to F\,(\bigsqcup_{1 \leqslant i \leqslant m} X_i)$$

is epimorphic. Suppose $q > kd$. In dimension q the simplicial vector space $F\,(M_\Sigma)$ is $F\,(\bigsqcup_c U_c)$, where c ranges over chains $\sigma_0 \leqslant \ldots \leqslant \sigma_q$ of elements of the set Σ and U_c denotes U_{σ_0}. Each element of the space $F\,(\bigsqcup_c U_c)$ is a sum of elements which come from $F\,(U_{c_1} \sqcup \ldots \sqcup U_{c_k})$ for some chains $c_1, \ldots, c_k$. But the relation $\dim \Sigma = d$ means that in each chain $c_l = \{\sigma_{i0} \leqslant \ldots \leqslant \sigma_{iq}\}$ of q symbols $\leqslant$ no more than d may be replaced by the symbol $<$. Therefore, there exists a j such that $\sigma_{i,\,j-1} = \sigma_{ij}$ for all i, so that the entire space $F\,(U_{c_1} \sqcup \ldots \sqcup U_{c_k})$ is covered by the image of the j-th degeneracy operation. Statement (i) is proved.

To prove statement (ii), we shall construct a null-homotopy compatible with filtration in the simplicial vector space $C^p\,(\mathrm{Vect}\ M_\Sigma)$. The simplicial space M_Σ comes from the

open covering $\{U_\alpha\}_{\alpha \in S}$ of the manifold M. If we are given
a refinement $\{U'_\alpha\}_{\alpha \in S}$ of the covering $\{U_\alpha\}$ (i.e., $\bar{U}'_\alpha \subset U_\alpha$),
then a subspace M'_Σ of the simplicial space arises. Since
cochains from C^p (Vect M_Σ) have compact supports, C^p (Vect M_Σ)
is the sum of all the C^p (Vect M'_Σ), constructed over all possible refinements of the covering $\{U_\alpha\}$. Therefore, it suf-
ices to construct a null-homotopy preserving filtration of
the inclusion C^p (Vect M'_Σ) $\rightarrow C^p$ (Vect M_Σ). But for any manifold M the space C^p (Vect M) is the space of Symm (p)-
invariants of the larger space $\tilde{C}^p$ (Vect M), obtained by elim-
inating the skew-symmetry condition. Since the passage to
Symm (p)-invariants is an exact functor, it suffices to
construct a null-homotopy compatible with filtration of the
inclusion $\tilde{C}^p$ (Vect M'_Σ) $\rightarrow \tilde{C}^p$ (Vect M_Σ).

For the family $\pi = (\sigma_1, \ldots, \sigma_p)$ of p subsets of the set
S, denote by U_π the product $U_{\sigma_1} \times \ldots \times U_{\sigma_p}$. The set Π
of such π possesses a natural partial order such that $\pi \leqslant \pi'$
implies $U_\pi \subset U_{\pi'}$. Furthermore, the q-th term of the simplicial space $(M_\Sigma)^p$ is

$$\bigsqcup_{\pi_\bullet \leqslant \ldots \leqslant \pi_q} U_{\pi_\bullet},$$

while the action of the group Symm (p) in $(M_\Sigma)^p$ is due to
its action in Π. Similarly, using $\{U'_\alpha\}$ instead of $\{U_\alpha\}$, we
can define U'_π. For $\pi \in \Pi$ we put $W_\pi = U_\pi \setminus \bigcup_{\rho > \pi} \bar{U}'_\rho$. Obviously, $\{W_\pi\}$ is an open covering of the product M^p; sup-
pose $\{\varphi_\pi\}$ is a partition of unity subordinated to this covering.

The vector space $\tilde{C}^p$ (Vect $M_{\Sigma,q}$) is generated by elements
which can be denoted by the symbols $[\pi_0, \ldots, \pi_q; f]$, where
$\pi_0 \leqslant \ldots \leqslant \pi_q$ is a chain of elements of the set Π and f is

a generalized section of the bundle $(\operatorname{tang} M)^p$ with support contained in $U'_{\pi_\bullet}$. The differential

$$\partial\colon\ C^p\,(\operatorname{Vect} M_{\Sigma,\,q})\to C^p\,(\operatorname{Vect} M_{\Sigma,\,q-1})$$

in these notations is described by the formula

$$\partial\,[\pi_0,\,\ldots,\pi_q;\,f]=\sum\,(-1)^i\,[\pi_0,\,\ldots\,\hat\pi_i\,\ldots,\pi_q;\,f];$$

similar formulas may be written for $C^p\,(\operatorname{Vect} M_\Sigma)$. The required homotopy

$$h\colon\ C^p\,(\operatorname{Vect} M'_{\Sigma,\,q})\dashrightarrow C^p\,(\operatorname{Vect} M_{\Sigma,\,q+1})$$

is defined by the formula

$$h\,[\pi_0,\,\ldots,\pi_q;\,f]=\sum_\pi\,[\pi,\,\pi_0,\,\ldots,\pi_q;\,\varphi_\pi f];$$

this formula is meaningful since for $\pi_0>\pi$ the sets $U'_{\pi_\bullet}$ and W_π do not intersect and $\varphi_\pi f=0$. The fact that h is indeed a null-homotopy of the inclusion $C^p\,(\operatorname{Vect} M'_{\Sigma,q})\to C^p\,(\operatorname{Vect} M_{\Sigma,q})$ is checked automatically; it is more delicate to verify the compatibility of h with filtration. We must show that if $[\pi_0,\,\ldots,\pi_q;\,f]$ is a filtration $\leqslant k$, then $[\pi,\pi_0,\,\cdots,\pi_q;\varphi_\pi f]$ is a filtration $\leqslant k$; to do this, one must impose supplementary conditions on $\{\varphi_\pi\}$. Namely, we must require that, for every $x\in\operatorname{Supp}\varphi_\pi\subset M^p$ and every $g\in\operatorname{Symm}(p)$ with $gx=x$, we have the relation $g\pi=\pi$. In other words, we must require that the support of the function φ_π be contained in

$$W'_\pi=W_\pi\diagdown\bigcup_{g\pi=\pi}\{x\in M\,|\,gx=x\}.$$

But it is easy to see that the sets W'_π also constitute a covering of the manifold M^p, so that the partition of unity with the property indicated above exists. The lemma is proved.

G. Conclusion of the proof of Theorem 2.2.4. It remains to construct the cohomology equivalence between the
simplicial cochain algebras $|C^{\cdot}(\text{Vect}\, M_{\Sigma})|$ and $|C^{\cdot}(X(M))^{\Sigma}|$.
Before we begin to do this, we need three more specifications.
First of all, we shall replace our cochain algebras by their
complexification: from the point of view of cohomology this
does not change anything. Secondly, we shall improve the
bundle $x(M)$ by making it smooth; to do this, instead of
the space X_n, as the fiber we shall takes its "thickening"
$\tilde{X}_n$ – the inverse image of an appropriate $\mathbb{C}G(2n, n)$-neigh-
borhood B of the skeleton $\text{sk}_{2n}\mathbb{C}G(2n, n)$ in the space of the
universal $U(n)$-bundle. Finally, our cohomology equivalence
will not be a map directed in one direction or the other,
but a chain of maps pointing in different directions.

In view of Subsection B, $\Omega^{\cdot}(\tilde{X}_n)$ is a cochain $U(n)$-
algebra. Hence $\Omega^{\cdot}(B) = \Omega^{\cdot}(\tilde{X}_n)_{\text{basic}}$ is a $I(\mathfrak{u}(n))$-module and
the same is true of $H^{\cdot}(B; \mathbb{R})$. Furthermore, the space B is
a formal space in the sense of Sullivan [86]; this means
that there exists a cochain algebra $L^{\cdot}$ with cohomology
equivalences

$$H^{\cdot}(B; \mathbb{R}) \leftarrow L^{\cdot} \leftarrow \Omega^{\cdot}(B).$$

We may assume that $L^{\cdot}$ is also a $I(\mathfrak{u}(n))$-module and that
the equivalences indicated above are $I(\mathfrak{u}(n))$-homomorphisms.

Consider the homormophism (6)

$$W(\mathfrak{u}(n)) \otimes_{I(\mathfrak{u}(n))} \Omega^{\cdot}(B) \to \Omega^{\cdot}(\tilde{X}_n).$$

To its complexification

$$W(\mathfrak{gl}(n, \mathbb{C})) \otimes_{I(\mathfrak{gl}(n, \mathbb{C}))} \Omega^{\cdot}(B; \mathbb{C}) \to \Omega^{\cdot}(\tilde{X}_n; \mathbb{C}) \qquad (7)$$

we can apply Lemma 2 [the fact that $\Omega^{\cdot}(\tilde{X}_n; \mathbb{C})$ can be decomposed into a product of tensor modules follows from the compactness of the group $U(n)$], so that (7) is a cohomology equivalence. We put

$$A^{\cdot} = W(\mathfrak{gl}(n, \mathbb{C})) \otimes_{I(\mathfrak{gl}(n,\mathbb{C}))} (L^{\cdot} \otimes \mathbb{C})$$

and obtain the cohomology equivalences

$$A^{\cdot} \rightarrow W(\mathfrak{gl}(n, \mathbb{C})) \otimes_{I(\mathfrak{gl}(n,\mathbb{C}))} \Omega^{\cdot}(B; \mathbb{C}) \rightarrow \Omega^{\cdot}(\tilde{X}_n; \mathbb{C}), \qquad (8)$$

$$A^{\cdot} \rightarrow W(\mathfrak{gl}(n, \mathbb{C})) \otimes H^{\cdot}(B; \mathbb{C}) \rightarrow C^{\cdot}(\mathbb{C}W_n)$$
$$= C^{\cdot}(\mathbb{R}W_n) \otimes \mathbb{C} \rightarrow C^{\cdot}(\text{Vect } \mathbb{R}^n) \otimes \mathbb{C}. \qquad (9)$$

In the composition (9), the second arrow is induced by the map $W(\mathfrak{gl}(n, \mathbb{C})) \rightarrow C^{\cdot}(\mathbb{C}W_n)$ from 2.3C and by the map $H^{\cdot}(B; \mathbb{C}) \rightarrow C^{\cdot}(\mathbb{C}W_n)$, which sends c_i into ψ_i (see Subsection 2.2); the fact that this is a cohomology equivalence obviously follows from Theorem 2.2.4. The last arrow in the composition (9) is the complexification of the cohomology equivalence from Lemma 1.

We shall now assume that $\mathfrak{u}$ is a finite covering constituted by geodesically convex open sets. Denote by P the manifold of orthonormed tangent n-frames on M and by P_α the inverse image of the set U_α under the projection $P \rightarrow M$; further, for $\sigma \in \Sigma$, let $P^\sigma = \bigcup_{\alpha \equiv \sigma} P_\alpha$. In an obvious sense P^σ can be viewed as a set of embeddings $U_\sigma \rightarrow \mathbb{R}^n$, and the map $P^\sigma \times U_\sigma \rightarrow \mathbb{R}^n$ which arises is equivariant with respect to the natural action of the group $O(n)$ in P^σ and $\mathbb{R}^n$. Consider the chain of maps

$$C^{\cdot}(\text{Vect } U_\sigma) \otimes \mathbb{C} \rightarrow \Omega^{\cdot}(P^\sigma; C^{\cdot}(\text{Vect } \mathbb{R}^n) \otimes \mathbb{C})_{\text{basic}} \leftarrow$$
$$\Omega^{\cdot}(P^\sigma; A^{\cdot})_{\text{basic}} \rightarrow \Omega^{\cdot}(P^\sigma; \Omega^{\cdot}(\tilde{X}_n; \mathbb{C}))_{\text{basic}} = \Omega^{\cdot}(X(M)^\sigma; \mathbb{C}),$$

where the first arrow is a map from Subsection D, while the two others are induced by the maps (8) and (9). All of these

maps are cohomology equivalences. Indeed, the O (n)-equiv-
alence $P^\sigma = O(n) \times U^\sigma$ allows us to rewrite the middle part of
this sequence in the form

$$\Omega^{\boldsymbol{\cdot}}(U^\sigma; C^{\boldsymbol{\cdot}}(Vect \, \mathbb{R}^n) \otimes \mathbb{C}) \leftarrow \Omega^{\boldsymbol{\cdot}}(U^\sigma; A^{\boldsymbol{\cdot}}) \rightarrow \Omega^{\boldsymbol{\cdot}}(U^\sigma; \Omega^{\boldsymbol{\cdot}}(\bar{X}_n; \mathbb{C})),$$

after which the required statement will be a corollary of
the contractibility of U^σ.

We have constructed the natural cohomology equivalence
between $C^{\boldsymbol{\cdot}}(Vect \, U_\sigma) \otimes \mathbb{C}$ and $\Omega^{\boldsymbol{\cdot}}(X \, (M)^\sigma; \mathbb{C})$. Exactly in the
same way, one can construct the cohomology equivalence be-
tween $C^{\boldsymbol{\cdot}}(Vect \, U_\sigma) \otimes \mathbb{C}$ and $\Omega^{\boldsymbol{\cdot}}(\prod_\sigma X \, (M)^\sigma; \mathbb{C})$, which constitutes the equivalence between $C^{\boldsymbol{\cdot}}(Vect \, M_\Sigma) \otimes \mathbb{C}$ and $\Omega^{\boldsymbol{\cdot}}(X \, (M)^\Sigma;$
$\mathbb{C})$.

The theorem is proved.

3. Cohomology of the algebra Vect M with coefficients in tensor fields.

In this subsection, we assume given a ten-
sor $\mathfrak{gl} \, (n, \mathbb{R})$-module A. By $a = a \, (M)$ we denote the bundle
with standard fiber A, associated with tang M, and by $\mathcal{A}$ the
space of smooth sections of the bundle a. The space $\mathcal{A}$ possesses a natural Vect M-module structure. For example, if
A is the trivial one-dimensional module, then $\mathcal{A} = \mathscr{C}^\infty (M)$,
while if $A = \Lambda^r V'$ [V, as usual denotes the standard n-dimen-
sional representation of the algebra $\mathfrak{gl} \, (n, \mathbb{R})$], then $\mathcal{A} =$
$\Omega^r (M)$.   The object of this subsection is to study the co-
homology of the algebra Vect M with coefficients in $\mathcal{A}$.

A. Diagonal cohomology. We shall say that the cochain
$\alpha \in C^q (Vect \, M; \, \mathcal{A})$ has (diagonal) filtration $\leqslant k$, if for
all vector fields $\xi_1, \ldots, \xi_q \in Vect \, M$ such that the section
$\alpha \, (\xi_1, \ldots, \xi_q)$ of the bundle a is not zero at some point $x \in M$,
there exist points $x_1, \ldots, x_k \in M$ such that each of the

fields $\xi_1, \ldots, \xi_q$ is not identically zero in any neighborhood (as small as we wish) of the set $\{x, x_1, \ldots, x_k\}$. [Compare with the definition of diagonal filtration in $C^{\cdot}(\text{Vect} M)$, given in Subsection 1A.] The subspace of the space $C^q(\text{Vect } M; \mathcal{A})$, consisting of cochains of diagonal filtration $\leqslant k$, is denoted by $\Delta_k C^q(\text{Vect } M; \mathcal{A})$. Obviously,

$$C^q(\text{Vect} M; \mathcal{A}) = \Delta_q C^q(\text{Vect} M; \mathcal{A})$$

$$\supset \Delta_{q-1} C^q(\text{Vect } M; \mathcal{A}) \supset \ldots \supset \Delta_0 C^q(\text{Vect} M; \mathcal{A}),$$

$$d(\Delta_k C^q(\text{Vect} M; \mathcal{A})) \subset \Delta_k C^{q+1}(\text{Vect} M; \mathcal{A})$$

and in the multiplicative situation [i.e., when A is a $\mathbb{R}$-algebra in which $\mathfrak{gl}(n, \mathbb{R})$ acts by means of derivations; in this case $\mathcal{A}$ is also a $\mathbb{R}$-algebra and $\text{Vect} M$ acts in $\mathcal{A}$ by means of derivations] we have

$$\Delta_k C^q(\text{Vect} M; \mathcal{A}) \Delta_l C^r(\text{Vect} M; \mathcal{A}) \subset \Delta_{k+l} C^{q+r}(\text{Vect} M; \mathcal{A}).$$

The subcomplex $\Delta_0 C^{\cdot}(\text{Vect} M; \mathcal{A})$ of the complex $C^{\cdot}(\text{Vect} M; \mathcal{A})$ is called <u>diagonal</u>. For its cochains and cohomology, we use the notation $C_\Delta^q(\text{Vect} M; \mathcal{A})$ and $H_\Delta^q(\text{Vect} M; \mathcal{A})$. Let us stress that this diagonal complex is not completely analogous to the diagonal complex from Subsection 1A; in particular, it is multiplicative in the multiplicative case. Its cochains α are best described by the condition: the value of the section $\alpha(\xi_1, \ldots, \xi_q)$ at the point $x \in M$ depends only on the germs of the fields $\xi_1, \ldots, \xi_q$ at the point x.

We begin by constructing a spectral sequence which converges to the cohomology of the diagonal complex. Denote by $F^m C_\Delta^q(\text{Vect} M; \mathcal{A})$ the subspace of the space $C_\Delta^q(\text{Vect} M; \mathcal{A})$, consisting of cochains α, such that if $q + 1 - m$ of the vector fields $\xi_1, \ldots, \xi_q$ vanish at the point $x \in M$, then the section $\alpha(\xi_1, \ldots, \xi_q)$ of the bundle a also vanishes at the point x. Obviously,

$$C_\Delta^q \left(\text{Vect}\, M;\, \mathcal{A} \right) = F^0 C_\Delta^q \left(\text{Vect}\, M;\, \mathcal{A} \right) \supset \cdots$$

$$\supset F^q C_\Delta^q \left(\text{Vect}\, M;\, \mathcal{A} \right) \supset F^{q+1} C_\Delta^q \left(\text{Vect}\, M;\, \mathcal{A} \right) = 0.$$

The inclusion $d\, (F^m C_\Delta^q \left(\text{Vect}\, M;\, \mathcal{A} \right)) \subset F^m C_\Delta^{q+1} \left(\text{Vect}\, M;\, \mathcal{A} \right)$ is veri-
fied by direct calculation, using the following property of
the filtration F, which is also easy to check: If $\alpha \in$
$F^m C_\Delta^q (\text{Vect}\, M;\, \mathcal{A})$ and $\xi \in \text{Vect}\, M$, then $\xi \alpha \in F^{m-1} C_\Delta^q$ ($\text{Vect}\, M;$
$\mathcal{A}$). Finally, in the multiplicative case, the filtration F
is multiplicative.

Theorem 2.4.6 [30]. The spectral sequence associated
with the filtration F begins with

$$E_2^{p,q} = H^p \left(M;\, \mathbb{R} \right) \otimes H^q \left(L_0;\, A \right).$$

Proof. First of all, it is clear that the cochain $\alpha \in$
$C_\Delta^{p+q} (\text{Vect}\, M;\, \mathcal{A})$, up to a summand from $F^{p+1} C_\Delta^{p+q} (\text{Vect}\, M;\, \mathcal{A})$, is
determined by its values

$$\alpha\, (\xi_1, \ldots, \xi_{p+q})\, (x),$$

where $\xi_1, \ldots, \xi_{p+q}$ are germs of vector fields at the point x,
such that $\xi_1 (x) = \ldots = \xi_q (x) = 0$. If we have $\alpha \in F^p C_\Delta^{p+q} \times$
$(\text{Vect}\, M;\, \mathcal{A})$, then $\alpha\, (\xi_1, \ldots, \xi_{p+q})(x)$ depends only on $\xi_1, \ldots,$
ξ_q and $\xi_{q+1} (x), \ldots, \xi_{p+q} (x)$. Thus,

$$E_0^{p,q} = F^{p+1} C_\Delta^{p+q} / F^p C_\Delta^{p+q} \left(\text{Vect}\, M;\, A \right)$$

is the space of differential forms of degree p on M with
values in sections of the bundle with fiber $C^q (L_0;\, A)$, associated with $\text{tang}\, M$. Further, it is obvious that the differential $d_0^{p,q}$ is induced by the differential

$$d\colon C^q (L_0;\, A) \to C^{q+1} (L_0;\, A),$$

so that

$$E_1^{p,q} = \Omega^p (M) \otimes H^q (L_0;\, A),$$

while $d_1^{p,q} = d \otimes \mathrm{id}$, so that

$$E_2^{p,q} = H^p(M;\ \mathbb{R}) \otimes H^q(L_0;\ A).$$

The computation of the subsequent differentials of our spectral sequence will be carried out in a special case.

<u>Theorem 2.4.7 [63, 30]</u>. For any p, r,

$$H_\Delta^p(\mathrm{Vect}\,M; \Omega^r(M)) \cong H^{p-r}(U(M), \mathbb{R}) \otimes H^{2r}(\mathrm{BU}(n); \mathbb{R}),\qquad (10)$$

where $U(M)$ is the total space of the principal $\mathrm{GL}(n, \mathbb{C})$-bundle (M), associated with the complexification of the tangent bundle $\mathrm{tang}\,M$; the isomorphisms (10) constitute the multiplicative isomorphism

$$H_\Delta^*(\mathrm{Vect}\,M; \Omega^q(M)) \cong H^*(U(M); \mathbb{R}) \otimes H^*(\mathrm{sk}_{2n}\mathrm{BU}(n); \mathbb{R}).$$

<u>Remark</u>. In 1.3.3 we constructed an inclusion of the de Rham complex $\Omega^{\cdot}(M)$ into the complex $C^{\cdot}(\mathrm{Vect}\,M; \mathscr{C}^\infty(M))$. Now we can say that the image of this inclusion is contained in $C_\Delta^{\cdot}(\mathrm{Vect}\,M; \mathscr{C}^\infty(M))$, so that this inclusion induces the homomorphism

$$H^*(M;\ \mathbb{R}) \to H_\Delta^*(\mathrm{Vect}\,M; \mathscr{C}^\infty(M)) = H^*(U(M);\ \mathbb{R}).$$

It turns out that this homomorphism coincides with the homomorphism induced by the projection $U(M) \to M$ (this is an exercise to the proof of Theorem 2.4.7 given below).

<u>Proof of Theorem 2.4.7</u>. Recall that the homomorphism

$$C^{\cdot}(\mathfrak{gl}(n, \mathbb{C})) \to \Omega^{\cdot}(\mathrm{GL}(n, \mathbb{C}); \mathbb{C})$$

sending cochains into right-invariant complex-valued differential forms induces an isomorphism in cohomology

$$H^*(\mathfrak{gl}(n, \mathbb{C})) \xrightarrow{\cong} H^*(\mathrm{GL}(n, \mathbb{C}); \mathbb{C})$$

(Theorem 2.1.1'). Suppose $\Omega_0^{\cdot}(U(M); \mathbb{C})$ is a subcomplex of the complex $\Omega^{\cdot}(U(M); \mathbb{C})$, constituted by $GL(n, \mathbb{C})$-invariant (complex-valued) forms. The standard filtration in $\Omega^{\cdot}(U(M); \mathbb{C})$, determined by the bundle $U(M) \to M$ ($F^m \Omega^q(U(M); \mathbb{C})$ consisting of forms which are annihilated by the substitution of $q + 1 - m$ vector fields tangent to the fiber), cuts out a certain filtration in $\Omega_0^{\cdot}(U(M); \mathbb{C})$. Two spectral sequences arise and converge to the cohomology of the complexes $\Omega^{\cdot}(U(M); \mathbb{C})$, $\Omega_0^{\cdot}(U(M); \mathbb{C})$, and a homomorphism from the first one to the second. The terms $E_2^{p,q}$ of the spectral sequences equal, respectively,

$$H^p(M; H^q(GL(n, \mathbb{C}); \mathbb{C})), H^p(M; H^q(\mathfrak{gl}(n, \mathbb{C}))),$$

and it follows from the above that our homomorphism establishes an isomorphism between these $E_2^{p,q}$. Thus the inclusion $\Omega_0^{\cdot}(U(M); \mathbb{C}) \to \Omega^{\cdot}(U(M); \mathbb{C})$ induces an isomorphism in cohomology.

Further, 1-jets of vector fields at the point $x \in M$ may be identified with $GL(n, \mathbb{R})$-invariant families of tangent vectors to $U(M)$ at a point of the fiber over x. The homomorphisms

$$\pi_q: \Omega_0^q(U(M)) \to C_\Delta^q(\text{Vect } M; \mathscr{C}^\infty(M)),$$

defined by the formula

$$\pi_q(\omega)(\xi_1, \ldots, \xi_q)(x) = \omega(j_x^1 \xi_1, \ldots, j_x^1 \xi_q),$$

where $\omega \in \Omega_0^q(U(M))$, $\xi_1, \ldots, \xi_q \in \text{Vect } M$, $x \in M$, and j_x^1 denotes the 1-jet at the point x, constitute a homomorphism

$$\pi: \Omega_0^{\cdot}(U(M)) \to C_\Delta^{\cdot}(\text{Vect } M; \mathscr{C}^\infty(M)).$$

This homomorphism is compatible with filtration and the same is true for its complexification. As seen from definitions, the latter induces the identical map between terms $E_2^{p,q}$ of associated spectral sequences [in both cases $E_2^{p,q} = H^p (M;$ $H^q (\mathfrak{gl} (n, \ \mathbb{C})))$] and, therefore, is a cohomology equivalence. Thus

$$H_\Delta^* (\operatorname{Vect} M; \mathscr{C}^\infty (M)) \otimes \mathbb{C} = H^* (U (M); \mathbb{C}).$$

The theorem is proved in the case $r = 0$.

In the case $r > 0$, we use the following notation for our spectral sequence $\{^{(r)}E_m^{p,\ q}, \ ^{(r)}d_m^{p,\ q}\}$. We have

$$^{(r)}E_2^{p,q} = H^p (M; \mathbb{R}) \otimes H^q (L_0; \Lambda^r V') = H^p (M; \mathbb{R})$$
$$\otimes \bigoplus_u [H^{q-u} (\mathfrak{gl} (n, \mathbb{R})) \otimes \operatorname{Inv}_{\mathfrak{gl} (n, \mathbb{R})} (H^u (L_1) \otimes \Lambda^r V')]$$

(see Theorem 2.2.8). But by Theorem 2.2.7

$$\operatorname{Inv}_{\mathfrak{gl} (n, \mathbb{R})} (H^u (L_1) \otimes \Lambda^r V') = \begin{cases} 0 & \text{for } u \ne r, \\ H^{2r} (\operatorname{BU} (n); \mathbb{R}) & \text{for } u = r. \end{cases}$$

This implies $^{(r)}E_2^{p,\ q} = {}^{(r)}E_2^{0,\ r} \otimes E_2^{p,\ q-r} = H^{2r} (\operatorname{BU} (n); \ \mathbb{R}) \otimes E_2^{p,\ q-r}$ But the entire spectral sequence $\{^{(r)}E_m^{p,q}, \ ^{(r)}d_m^{p,q}\}$ is a module over the spectral sequence $\{E_m^{p,q}, \ d_m^{p,q}\}$. We see that all the generators of this module have the same bidegree $(0, r)$. Therefore, we conclude that $^{(r)}E_m^{p,\ q} = H^{2r} (\operatorname{BU} (n); \ \mathbb{R}) \otimes E_m^{p,\ q-r}$ for all m, which in turn implies formula (10). The theorem is proved.

In the case of an arbitrary A, the differentials of the spectral sequence in Theorem 2.4.6 were calculated by Losik in [66]. Here is his result.

<u>Theorem 2.4.8.</u>

$$H_\Delta^* (\operatorname{Vect} M; \mathcal{A}) \cong H^* (U (M); \mathbb{R}) \otimes \operatorname{Inv}_{\mathfrak{gl} (n, \mathbb{R})} (H^* (L_1) \otimes A).$$

For the proof, it suffices to establish that all the differentials of the spectral sequence are trivial on the part $1 \otimes 1 \otimes \operatorname{Inv} (H^* (L_1) \otimes A)$ of the term

$$E_2 = H^*(M; \ \mathbb{R}) \otimes H^*(\mathfrak{gl}(n, \mathbb{R})) \otimes \mathrm{Inv}(H^*(L_1) \otimes A),$$

i.e., that these elements of the term E_2 are represented by cocycles from the complex $C_\Delta^{'}(\mathrm{Vect}\,M; \ \mathcal{A})$. These cocycles may actually be exhibited by explicit formulas, but these formulas are too cumbersome and I shall not present them.

B. Complete cohomology. As we already pointed out, this cohomology was computed by Tsujishita [91]. In order to state his result, note that the inclusion $U(n) \to X_n$ of the fiber of the bundle $X_n \to \mathrm{sk}_{2n} BU(n)$ into its total space can be extended to an inclusion of the bundle $u(M)$ into the bundle $x(M)$; in this connection, we view the bundle $u(M)$ as a subbundle of the bundle $x(M)$.

Theorem 2.4.9. (i) the ring $H^*(\mathrm{Vect}\,M; \ \mathscr{C}^\infty(M))$ is isomorphic to the ring of real cohomology of the space

$$Y(M) = \{(y, s) \in M \times \mathrm{Sec}\,x(M) \mid s(y) \in U(M)\}.$$

(ii) For an arbitrary tensor $\mathfrak{gl}(n, \mathbb{R})$-module A , we have

$$H^*(\mathrm{Vect}\,M; \mathcal{A}) \cong H^*(Y(M); \mathbb{R}) \otimes \mathrm{Inv}_{\mathfrak{gl}(n, \mathbb{R})}(H^*(L_1) \otimes A).$$

Theorem 2.4.10. (i) the inclusion homomorphism

$$H_\Delta^*(\mathrm{Vect}\,M; \ \mathscr{C}^\infty(M)) \to H^*(\mathrm{Vect}\,M; \ \mathscr{C}^\infty(M)) \qquad (11)$$

coincides with the homomorphism

$$H^*(U(M); \ \mathbb{R}) \to H^*(Y(M); \ \mathbb{R}),$$

induced by the map $(y, s) \mapsto s(y)$ of the space $Y(M)$ into $U(M)$.

(ii) The inclusion homomorphism

$$H_\Delta^*(\mathrm{Vect}\,M; \ \mathcal{A}) = H^*(U(M); \mathbb{R}) \otimes \mathrm{Inv}(H^*(L_1) \otimes A)$$
$$\to H^*(\mathrm{Vect}\,M; \ \mathcal{A}) = H^*(Y(M); \mathbb{R}) \otimes \mathrm{Inv}(H^*(L_1) \otimes A)$$

is the tensor product of the homomorphism (11) by $\mathrm{id}\,\mathrm{Inv}(H^*(\dot{L_1}) \otimes A)$.

Note that the map $Y(M) \to U(M)$ from statement (i) of Theorem 2.4.10 is a bundle with n-connected fiber, so that (11) is an isomorphism in dimensions $\leqslant n$. In particular,

$$H^1(\mathrm{Vect}\, M; \mathscr{C}^\infty(M)) = H^1_\Delta(\mathrm{Vect}\, M; \mathscr{C}^\infty(M)) \cong H^1(M; \mathbb{R}) \oplus \mathbb{R}.$$

It is easy to indicate the cocycles which represent these cohomologies.

Theorem 2.4.11. If $\omega_1, \ldots, \omega_k$ are closed 1-forms on M whose cohomology classes form a basis in $H^1(M; \mathbb{R})$, then the basis in $H^1(\mathrm{Vect}\, M; \mathscr{C}^\infty(M))$ is constituted by the cohomology classes of the cocycles

$$\xi \mapsto \omega_1(\xi), \ldots, \xi \mapsto \omega_k(\xi), \xi \mapsto \mathrm{div}\, \xi$$

(the divergence is taken with respect to an arbitrary volume form).

In conclusion, consider the case when $M = S^1$. The space $Y(S^1)$ is homeomorphic to $S^1 \times S^1 \times \Omega S^3$, so that the ring $H^*(Y(S^1); \mathbb{R})$ is the free skew-commutative $\mathbb{R}$-algebra with generators in dimensions 1, 1, 2. Together with Theorems 2.4.9 and 2.3.1, this enables us to give an explicit description of the cohomology ring of the algebra $\mathrm{Vect}\, S^1$ with coefficients in spaces of tensor fields of arbitrary form:

$$H^q\left(\mathrm{Vect}\, S^1; \mathscr{C}^\infty(S^1)\, d\varphi^{\frac{3r^2 \pm r}{2}}\right) = H^{q-r}(Y(S^1); \mathbb{R}),$$

$$H^q(\mathrm{Vect}\, S^1; \mathscr{C}^\infty(S^1)\, d\varphi^s) = 0,$$

$$\text{if } s \neq 0, 1, 2, 5, 7, \ldots, \frac{3r^2 \pm r}{2}, \ldots$$

(φ is the angular coordinate on the circle), and $H^*(\mathrm{Vect}\, S^1;$ $\mathscr{C}^\infty(S^1)\, d\varphi^{\frac{3r^2 \pm r}{2}})$ is a free $H^*(Y(S^1); \mathbb{R})$-module in one generator contained in $H^r(\mathrm{Vect}\, S^1; \mathscr{C}^\infty(S^1)\, d\varphi^{\frac{3r^2 \pm r}{2}})$. In the most interesting cases, this cohomology can be represented explicitly by the appropriate cocycles.

   Theorem 2.4.12.   (i) The ring H^* (Vect S^1; $\mathcal{C}^\infty$ (S^1)) is a free skew-commutative algebra whose generators are the cohomology classes of the cocycles

$$f(\varphi)\frac{d}{d\varphi} \mapsto f(\varphi), \quad f(\varphi)\frac{d}{d\varphi} \mapsto f'(\varphi),$$

$$\left(f(\varphi)\frac{d}{d\varphi}, \; g(\varphi)\frac{d}{d\varphi}\right) \mapsto \text{const}(f, g) = \int_{S^1}\begin{vmatrix} f'(\varphi) & g'(\varphi) \\ f''(\varphi) & g''(\varphi) \end{vmatrix} d\varphi.$$

   (ii) H^* (Vect S^1; Ω^1 (S^1)) is a free H^* (Vect S^1; $\mathcal{C}^\infty$ (S^1)) - module with one one-dimensional generator represented by the cocycle

$$f(\varphi)\frac{d}{d\varphi} \mapsto f'(\varphi)\, d\varphi.$$

§5.   COMPUTATIONS FOR CURRENT ALGEBRAS

   These computations were carried out by Feigin [14] and Lepowsky [62].   The calculation methods used in this section are due to Feigin; they differ from the methods of both articles cited above.

   We begin with the calculation of the homology of the subalgebra

$$N^+(\mathfrak{sl}(n, \mathbb{C})^{S^1})^{\text{pol}} = N^+(\mathfrak{sl}(n, \mathbb{C})) \oplus [\bigoplus_{r>0} \mathfrak{sl}(n, \mathbb{C})\, t^r]$$

of the Lie algebra

$$(\mathfrak{sl}(n, \mathbb{C})^{S^1})^{\text{pol}} = \mathfrak{sl}(n, \mathbb{C}) \otimes \mathbb{C}[t, t^{-1}] = \bigoplus_{r\in\mathbb{Z}} \mathfrak{sl}(n, \mathbb{C})\, t^r,$$

where $N^+(\mathfrak{sl}(n, \mathbb{C}))$ is the subalgebra of the algebra $\mathfrak{sl}(n, \mathbb{C})$, consisting of upper triangular matrices with zero diagonal. From the point of view of the applications, these homologies are much more important than the homology of the current al-

gebra themselves, since they constitute the technical basis
of the proof of the combinatorial identities of Euler–Gauss–
Jacobi–Macdonald (see §3.2). Actually, there exists a method for estimating, and in certain cases for entirely comput-
ing, the homology of the algebra $(\mathfrak{g}^{S^1})^{\mathrm{pol}}$ by means of the homology of an algebra similar to $N^+(\mathfrak{sl}(n, \mathbb{C})^{S^1})^{\mathrm{pol}}$; this method
is presented in Subsection 2. In Subsection 3 we give some
other results, without detailed proofs, concerning the homology and cohomology of current algebras.

For brevity, we set

$$(\mathfrak{sl}(n, \mathbb{C})^{S^1})^{\mathrm{pol}} = T(n), \quad N^+(\mathfrak{sl}(n, \mathbb{C})^{S^1})^{\mathrm{pol}} = T^+(n),$$
$$(\mathfrak{gl}(n, \mathbb{C})^{S^1})^{\mathrm{pol}} = \bar{T}(n), \quad N^+(\mathfrak{gl}(n, \mathbb{C})^{S^1})^{\mathrm{pol}} = \bar{T}^+(n).$$

1. Homology of the algebra $T^+(n)$.

The algebra
$T^+(n)$ is one of the few Lie algebras whose homology can be
entirely computed by means of Theorem 1.5.3, i.e., by using
the Laplace operator (compare Subsection 3.1B). This computation will be started by a description of a certain special grading in the algebra $T^+(n)$, as well as in the algebras $T(n)$, $\bar{T}(n)$, $\bar{T}^+(n)$.

A. The grading.

Denote (as in Subsection 1.2) by e^i_j
the matrix whose only nonzero element is 1 and stands at the
intersection of the i-th row and the j-th column, and assign
to the matrix e^i_j the weight $\underbrace{(0, \ldots, 0)}_{n}$, if $i = j$, and the
weight

$$(\underbrace{0, \ldots, 0}_{i-1}, \underbrace{1, \ldots, 1}_{j-i}, \underbrace{0, \ldots, 0}_{n-j+1}), \quad \text{if } i < j,$$

$$(\underbrace{0, \ldots, 0}_{j-1}, \underbrace{-1, \ldots, -1}_{i-j}, \underbrace{0, \ldots, 0}_{n-i+1}, \quad \text{if } i > j.$$

The weight of the element $e_j^i t^r$ of the algebra $T(n)$ is obtained from the weight of the matrix e_j^i by adding the number r to all of its entries. For example, the elements $e_4^2 t^3$, $e_2^4 t^2$ of the algebra $T(5)$ have weights (3, 4, 4, 3, 3), (2, 1, 1, 2, 2). Linear combinations of elements of the same weight are assigned this same weight. The space of elements of weight $(k_1, \ldots, k_n)$ of the algebra $T(n)$ is denoted by $T_{(k_1, \ldots, k_n)}(n)$. Obviously,

$$T(n) = \bigoplus_{(k_1, \ldots, k_n) \in \mathbb{Z}^n} T_{(k_1, \ldots, k_n)}(n),$$

and a direct verification shows that

$$[T_{(k_1, \ldots, k_n)}(n), T_{(l_1, \ldots, l_n)}(n)] \subset T_{(k_1+l_1, \ldots, k_n+l_n)}(n).$$

Thus, $\{T_{(k_1, \ldots, k_n)}(n)\}$ is a $\mathbb{Z}^n$-graded algebra $T(n)$. It is also clear that the spaces

$$T_{(k_1, \ldots, k_n)}^+(n) = T^+(n) \cap T_{(k_1, \ldots, k_n)}(n),$$
$$T_{(k_1, \ldots, k_n)}(n) = T(n) \cap T_{(k_1, \ldots, k_n)}(n),$$
$$T_{(k_1, \ldots, k_n)}^+(n) = T^+(n) \cap T_{(k_1, \ldots, k_n)}(n)$$

constitute $\mathbb{Z}^n$-gradings of the algebras $T^+(n)$, $T(n)$, $T^+(n)$. Notice that for $k \geqslant 1$, $\alpha + \beta + \gamma = n$, $0 < \beta < n$

$$\dim T_{(k, \ldots, k)}^+(n) = n - 1,$$
$$\dim T_{\underbrace{(k-1, \ldots, k-1)}_{\alpha}, \underbrace{k, \ldots, k}_{\beta}, \underbrace{k-1, \ldots, k-1)}_{\gamma}}^+(n) = 1,$$
$$\dim T_{\underbrace{(k, \ldots, k}_{\alpha} \underbrace{k-1, \ldots, k-1}_{\beta}, \underbrace{k, \ldots, k)}_{\gamma}}^+(n) = 1,$$

while the other spaces $T_{(k_1, \ldots, k_n)}^+(n)$ are trivial.

B. Calculation of the Laplace operator. Introduce a metric in $T(n)$ by requiring that $\{e_j^i t^r\}$ be an orthonormed basis. This metric is compatible with gradings and the same is true of the induced metric in $T^+(n)$. Along the lines of Subsection 1.5.3, we identify the spaces

$$C_q^{(k_1,\ldots,k_n)}(T^+(n)), \quad C_{(k_1,\ldots,k_n)}^q(T^+(n)),$$

thus assuming that the differentials d and ∂ act in the same spaces

$$d: C_{q-1}^{(k_1,\ldots,k_n)}(T^+(n)) \to C_q^{(k_1,\ldots,k_n)}(T^+(n)),$$
$$\partial: C_q^{(k_1,\ldots,k_n)}(T^+(n)) \to C_{q-1}^{(k_1,\ldots,k_n)}(T^+(n)),$$

and put

$$\Delta = d\circ\partial + \partial\circ d: C_q^{(k_1,\ldots,k_n)}(T^+(n)) \to C_q^{(k_1,\ldots,k_n)}(T^+(n)).$$

<u>Theorem 2.5.1.</u> On $C_q^{(k_1,\ldots,k_n)}(T^+(n))$ the Laplace operator Δ is the multiplication by

$$Q(k_1,\ldots,k_n) = -\sum_i k_i^2 + \sum_i k_i k_{i+1} + \sum_i k_i$$

(we have put $k_{n+1} = k_1$).

<u>Proof.</u> First we shall prove one general property of the Laplace operator.

<u>Lemma.</u> For any Lie algebra $\mathfrak{g}$ with metric and any $c_1 \in C_{q_1}(\mathfrak{g}), \ldots, c_r \in C_{q_r}(\mathfrak{g})$ we have

$$\Delta(c_1 \wedge \ldots \wedge c_r) = \sum_{1 \leqslant s < t \leqslant r} (-1)^{(q_s+q_t)(q_1+\ldots+q_{s-1})+q_t(q_{s+1}+\ldots+q_{t-1})} \Delta(c_s \wedge c_t)$$
$$\wedge c_1 \wedge \ldots \hat{c}_s \ldots \hat{c}_t \ldots \wedge c_r$$
$$- (r-2) \sum_{1 \leqslant s \leqslant r} (-1)^{q_s(q_1+\ldots+q_{s-1})} (\Delta c_s) \wedge c_1 \wedge \ldots \hat{c}_s \ldots \wedge c_r. \quad (1)$$

This is proved by a direct calculation based on the obvious formulas

$$\partial(c_1 \wedge \ldots \wedge c_r) = \sum_{1 \leqslant s < t \leqslant r} (-1)^{(q_s+q_t)(q_1+\ldots+q_{s-1})+q_t(q_{s+1}+\ldots+q_{t-1})} \partial(c_s \wedge c_t)$$
$$\wedge c_1 \wedge \ldots \hat{c}_s \ldots \hat{c}_t \ldots \wedge c_r$$
$$- (r-2) \sum_{1 \leqslant s \leqslant r} (-1)^{q_s(q_1+\ldots+q_{s-1})} (\partial c_s) \wedge c_1 \wedge \ldots \hat{c}_s \ldots \wedge c_r, \quad (2)$$
$$d(c_1 \wedge \ldots \wedge c_r) = \sum_{1 \leqslant s \leqslant r} (-1)^{q_s(q_1+\ldots+q_{s-1})} (dc_s) \wedge$$

$$c_1 \wedge \ldots \hat{c}_s \ldots \wedge c_r. \tag{3}$$

Formula (3) expresses the fact that d is a derivation or, better said, a superderivation of the ring $C_*(\mathfrak{g})$. We can also say that d is a (super)differential operator of the first order in $C_*(\mathfrak{g})$. The relations (1) and (2) show that ∂ and Δ are second-order operators [compare these relations with the well-known calculus formula $(f_1 \ldots f_r)'' = \sum_{s<t} (f_s f_t)'' f_1 \ldots \hat{f}_s \ldots \hat{f}_t \ldots f_r - (r-2) \sum_s f_s f_1 \ldots \hat{f}_s \ldots f_r)$. Thus, the computation which proves our lemma is simply a verification of the standard statement that the (super)commutator of first- and second-order operators is an operator of the second order.

Since the function Q, as any polynomial of degree 2, satisfies the relation

$$Q \left(\sum_{i=1}^{r} k_{i1}, \ldots, \sum_{i=1}^{r} k_{in} \right)$$

$$= \sum_{s<t} Q (k_{s1} + k_{t1}, \ldots, k_{sn} + k_{tn}) - (r-2) \sum_s Q (k_{s1}, \ldots, k_{sn}),$$

the lemma shows that the statement of the theorem need only be proved for $q = 1$ and $q = 2$.

Fix a real orthonormed basis $d_1 = \sum \alpha_{1i} e_i^i, \ldots, d_{n-1} = \sum \alpha_{n-1, i} \times e_i^i$ in the space of complex diagonal $(n \times n)$-matrices with zero trace. If we supplement the matrix $\| \alpha_{ji} \|$ with a row consisting of numbers equal to $1/\sqrt{n}$, then we obtain an orthogonal matrix. Therefore, for any s and $t \neq s$, we have

$$\sum_{i=1}^{n} \alpha_{si} = 0, \quad \sum_{i=1}^{n} \alpha_{si}^2 = 1, \quad \sum_{i=1}^{n} \alpha_{si} \alpha_{ti} = 0,$$

$$\sum_{j=1}^{n-1} \alpha_{js}^2 = \frac{n-1}{n}, \quad \sum_{j=1}^{n-1} \alpha_{js} \alpha_{jt} = -\frac{1}{n}. \tag{4}$$

An orthonormed homogeneous basis in $C_1(T^+(n))$ is constituted by chains of the form $d_p t^a$, $e_j^i t^a$, an orthonormed homogeneous basis in $C_2(T^+(n))$, by chains of the form $e_j^i t^a \wedge e_l^k t^b$, $d_p t^a \wedge e_j^i t^b$, $d_p t^a \wedge d_q t^b$. We must prove that the matrices of the Laplace operator in these bases are diagonal elements.

In order to compute the matrix of the Laplace operator $\Delta: C_q(\mathfrak{g}) \to C_q(\mathfrak{g})$, we may use the following convenient rule. Choose bases $\{b_i\}$, $\{c_i\}$, $\{d_i\}$ in $C_{q-1}(\mathfrak{g})$, $C_q(\mathfrak{g})$, $C_{q+1}(\mathfrak{g})$, write the matrices of the operators $\partial: C_q(\mathfrak{g}) \to C_{q-1}(\mathfrak{g})$ and $d: C_q(\mathfrak{g}) \to C_{q+1}(\mathfrak{g})$ in these bases and fit these two matrices together, as follows:

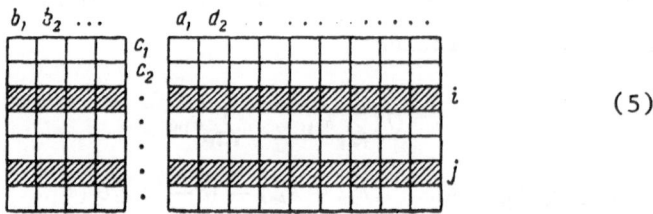

$$(5)$$

In order to get the (i, j)-th element of the matrix of the operator Δ in the basis $\{c_i\}$, it suffices to find the "scalar product" of the i-th and the j-th rows of the compound matrix (5), i.e., the sum of products of elements of these rows placed one under the other.

In order to prove that the matrix of the Laplace operators $\Delta: C_1(T^+(n)) \to C_1(T^+(n))$, $\Delta: C_2(T^+(n)) \to C_2(T^+(n))$ is indeed diagonal, it suffices to show that for an arbitrary pair of noncoinciding elements of the basis of the space $C_1(T^+(n))$ or $C_2(T^+(n))$ the scalar product of the corresponding rows of the matrix (5) vanishes. In so doing we may limit ourselves to pairs of elements of the same weight; here are all such pairs:

$1°.\ d_p t^a,\ d_q t^a;$

$2°.\ e_i^j t^a \wedge e_k^l t^b,\ e_i^l t^c \wedge e_k^j t^d;$

$3°.\ e_i^j t^a \wedge e_j^l t^b,\ e_i^k t^c \wedge e_k^l t^d;$

$4°.\ e_i^j t^a \wedge e_j^k t^b,\ d_p t^c \wedge e_i^k t^d;$

$5°.\ e_i^j t^a \wedge e_j^i t^b,\ e_k^l t^c \wedge e_l^k t^d;$

$6°.\ e_i^j t^a \wedge e_j^i t^b,\ e_i^k t^c \wedge e_k^i t^d;$

$7°.\ e_i^j t^a \wedge e_j^i t^b,\ d_p t^c \wedge d_q t^d;$

$8°.\ d_p t^a \wedge d_q t^b,\ d_p t^c \wedge d_r t^d;$

$9°.\ d_p t^a \wedge d_q t^b,\ d_r t^c \wedge d_s t^d$

(we assume everywhere that $a + b = c + d$ and that i, j, k, l are pairwise distinct, as are p, q, r, s). For each of the pairs 1°-9°, we consider the part of the matrix (5) consisting of the two rows which interest us at this point and those columns on the intersection of which with these rows there

$1°.$		$e_i^{s,a'} \wedge e_s^{t,a-a'}$	
	$\vdots$	$\vdots$	$\vdots$
$d_p t^a$	$\cdots$	$\alpha_{ps} - \alpha_{pt}$	$\cdots$
$d_q t^a$	$\cdots$	$\alpha_{qs} - \alpha_{qt}$	$\cdots$

$(s > t,\ 0 < a' \leqslant a)$

$2°.$	$e_i^j t^a \wedge e_i^l t^c$	$e_k^l t^b \wedge e_k^j t^d$	$e_i^j t^a \wedge e_i^l t^c$	$e_k^l t^b \wedge e_k^j t^a$
	$e_i^{t,d-a} \wedge e_k^{t,a-d}$		$e_j^{l,c-a} \wedge e_i^{t,a-c}$	
$e_i^j t^a \wedge e_k^l t^b$	-1	1	1	-1
$e_i^l t^c \wedge e_k^j t^d$	-1	1	-1	1

$3°.$	$e_i^{l,a+b}$	$e_j^l t^a \wedge e_k^l t^d$	$e_i^k t^c \wedge e_i^l t^b$
		$e_k^{l,d} \wedge e_j^{k,c-a}$	$e_j^{l,c} \wedge e_i^{t,a-c}$
1	$e_i^j t^a \wedge e_j^l t^b$	-1	-1
1	$e_i^k t^c \wedge e_k^l t^d$	1	1

4°.

$e_i^{l}t^{a+b}$		$e_i^{j}t^a \wedge d_p t^c \wedge e_j^{l}t^{d-a}$	$e_j^{l}t^b \wedge e_i^{j}t^c \wedge e_j^{l}t^{a-d}$	$e_j^{l}t^b \wedge d_p t^c \wedge e_i^{j}t^{a-c}$	$e_i^{j}t^a \wedge e_j^{l}t^d \wedge e_i^{j}t^{c-a}$
-1	$e_i^{j}t^a \wedge e_j^{l}t^b$	$\alpha_{pj}-\alpha_{pl}$	-1	$\alpha_{pj}-\alpha_{pi}$	-1
$\alpha_{pl}-\alpha_{pj}$	$d_p t^c \wedge e_i^{l}t^d$	-1	$\alpha_{pj}-\alpha_{pl}$	1	$\alpha_{pi}-\alpha_{pj}$

5°.

$\vdots$	$d_s^{l}t^{a+b}$	$\vdots$	
$\cdots$	$\alpha_{sj}-\alpha_{si}$	$\cdots$	$e_i^{j}t^a \wedge e_j^{i}t^b$
$\cdots$	$\alpha_{sl}-\alpha_{sk}$	$\cdots$	$e_k^{l}t^c \wedge e_l^{k}t^d$

$(1 \leqslant s \leqslant n-1)$

8°.

$\vdots$	$d_s^{l}t^{a+b}$	$\vdots$		$e_i^{j}t^a \wedge e_k^{i}t^d \wedge e_j^{k}t^{c-a}$	$e_j^{l}t^b \wedge e_i^{k}t^c \wedge e_k^{i}t^{a-c}$
$\cdots$	$\alpha_{sj}-\alpha_{si}$	$\cdots$	$e_i^{j}t^a \wedge e_j^{l}t^b$	-1	-1
$\cdots$	$\alpha_{sk}-\alpha_{si}$	$\cdots$	$e_i^{k}t^c \wedge e_k^{i}t^d$	1	1

$(1 \leqslant s \leqslant n-1)$

are nonzero numbers. Then, in the case 9°, there are no such columns, while in the case 8°, there are none if $a \neq c$. The described part of the matrix (5) for the cases 1° to 8° are shown on pages 183-185.

7°.

	$e_i^{j}t^a \wedge d_p t^c \wedge e_j^{i}t^{d-a}$	$e_j^{i}t^b \wedge d_q t^d \wedge e_i^{j}t^{a-d}$	$e_j^{i}t^a \wedge d_q t^d \wedge e_j^{i}t^{c-a}$	$e_j^{i}t^b \wedge d_p t^c \wedge e_j^{i}t^{a-c}$
$e_i^{j}t^a \wedge e_j^{i}t^b$	$\alpha_{pj}-\alpha_{pi}$	$\alpha_{qi}-\alpha_{qj}$	$\alpha_{pj}-\alpha_{pi}$	$\alpha_{qi}-\alpha_{qj}$
$d_p t^c \wedge d_q t^d$	$\alpha_{qj}-\alpha_{qi}$	$\alpha_{pi}-\alpha_{pj}$	$\alpha_{qi}-\alpha_{qj}$	$\alpha_{pj}-\alpha_{pi}$

8°.

		$d_p t^a \wedge e_s^{t}t^{b,b'} \wedge e_s^{t}t^{b-b'}$	
$d_p t^a \wedge d_q t^b$	$\cdots$	$\alpha_{qt}-\alpha_{qs}$	$\cdots$
$d_p t^a \wedge d_r t^b$	$\cdots$	$\alpha_{rt}-\alpha_{rs}$	$\cdots$

$(s>t,\,0<b'\leqslant b)$

Note that some of the three-dimensional chains written out in those diagrams contain elements which are not necessarily contained in $T^+(n)$, so that some of the columns may turn out to be superfluous. Namely, in all the cases exactly one of two columns not separated by a vertical line will actually be present.

In all the cases, it is clear that the scalar product of the lines vanishes [in the cases 1°, 5°, 6°, 8° the verification of this fact necessitates formula (4)]. The fact that the Laplace operator matrix is diagonal is proved.

In order to find the diagonal elements, we must calculate the scalar squares of the lines of matrix (5) corresponding to the basis elements of the spaces $C_1(T^+(n))$,

$C_2 (T^+ (n))$.   This calculation must be carried out separately for the basis chains

$$1^\circ.\ d_p t^a, \qquad\qquad 5^\circ.\ e_j^i t^a \wedge e_i^j t^b,$$
$$2^\circ.\ e_j^i t^a, \qquad\qquad 6^\circ.\ d_p t^a \wedge e_j^i t^b,$$
$$3^\circ.\ e_j^i t^a \wedge e_l^k t^b, \qquad 7^\circ.\ d_p t^a \wedge d_q t^b.$$
$$4^\circ.\ e_j^i t^a \wedge e_i^k t^b,$$

In all these cases, we must prove that the scalar square of the line equals $Q(k_1, \ldots, k_n)$, where $(k_1, \ldots, k_n)$ is the weight of the chain under consideration. For the cases 4°, 5°, 6° the verification may be omitted, since these chains do not belong to the kernel of the operator ∂ and the latter pre- serves the weight of chains and commutes with the Laplace operator. In the other cases the line of the matrix (5) which interests us is of the form

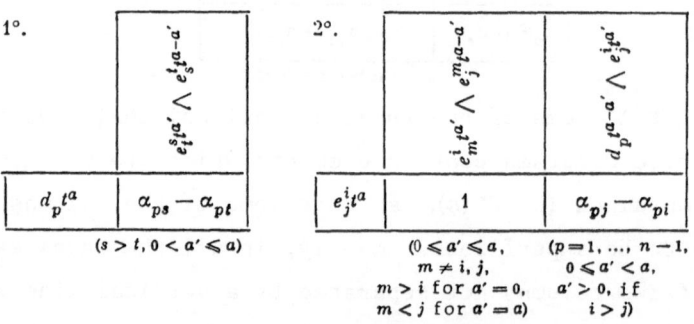

7°.

$e_j^i{}_t{}^a \wedge e_m^k{}_t{}^b \wedge e_l^m{}_t{}^{b-b'}$	$e_m^k{}_t{}^b \wedge e_i^m{}_t{}^{a'} \wedge e_j^i{}_t{}^{a-a'}$	$e_j^i{}_t{}^a \wedge e_i^k{}_t{}^{b'} \wedge d_p{}_t{}^{b-b'}$	$e_i^k{}_t{}^b \wedge e_j^i{}_t{}^{a'} \wedge d_p{}_t{}^{a-a'}$

$e_j^i{}_t{}^a \wedge e_l^k{}_t{}^b$	-1	1	$\alpha_{pl} - \alpha_{pk}$	$\alpha_{pi} - \alpha_{pj}$

$$\begin{array}{llll}
(0 \leqslant b < b', & (0 \leqslant a' < a, & (p = 1, \ldots, n-1, & \\
m \neq k, l, & m \neq i, j, & 0 \leqslant a' < a, & \\
m < k \text{ for} & m < i \text{ for} & 0 \leqslant b' < b, & \\
b' = 0, m > l & a' = 0, m > j & a' > 0, \text{ if } i > j, & \\
\text{for } b' = b) & \text{for } a' = a) & b' > 0, \text{ if } k > l) &
\end{array}$$

The calculation of the scalar square of the row in all the cases gives the required result. The theorem is proved.

C. Main theorem. As above, we put

$$Q(k_1, \ldots, k_n) = -\sum k_i^2 + \sum k_i k_{i+1} + \sum k_i$$

(where $k_{n+1} = k_1$).

Theorem 2.5.2. (i) If $Q(k_1, \ldots, k_n) \neq 0$, then

$$H_*^{(k_1, \ldots, k_n)}(T^+(n)) = 0.$$

(ii) If $Q(k_1, \ldots, k_n) = 0$, then there exists a

$$q = q(k_1, \ldots, k_n),$$

such that

$$\dim H_r^{(k_1, \ldots, k_n)}(T^+(n)) = \begin{cases} 1 & \text{for } r = q, \\ 0 & \text{for } r \neq q. \end{cases}$$

Proof. Part (i) follows immediately from Theorems 2.5.1 and 1.5.2. By the same theorems, part (ii) is equivalent to the statement: If $Q(k_1, \ldots, k_n) = 0$, then

$$\dim C_*^{(k_1,\dots,k_n)}(T^+(n)) = 1.$$

Lemma 1. If $m > 0$, then $Q(k_1 - m, \dots, k_n - m) < Q(k_1, \dots, k_n)$.

This is obvious.

Lemma 2. If $Q(k_1, \dots, k_n) = 0$, then the space $C_*^{(k_1,\dots,k_n)}(T^+(n))$ is generated by the product of chains $c_{ij}(m) = e_j^i t^\varepsilon \wedge e_j^i t^{\varepsilon+1} \wedge \dots \wedge e_j^i t^{\varepsilon+m-1}$, where $\varepsilon = 0$ if $i < j$, and $\varepsilon = 1$ if $i > j$.

Proof. Indeed, suppose $Q(k_1, \dots, k_n) = 0$, and let

$$c = d_{i_1} t^{j_1} \wedge \dots \wedge d_{i_r} t^{j_r} \wedge e_{m_1}^{l_1} t^{p_1} \wedge \dots \wedge e_{m_s}^{l_s} t^{p_s}.$$

If $r \neq 0$, then the chain $c/d_{i_1} t^{j_1}$ is of weight $(k_1 - j_1, \dots, k_n - j_1)$ and, therefore, by Theorem 2.5.1, is an eigenvector of the Laplace operator with eigenvalue $Q(k_1 - j_1, \dots, k_n - j_1)$. But this contradicts the fact that the Laplace operator is positive definite, since by Lemma 1

$$Q(k_1 - j_1, \dots, k_n - j_1) < Q(k_1, \dots, k_n) = 0.$$

Thus $r = 0$. Similarly, if for some u the factor $e_{m_u}^{l_u} t^{p_u-1}$ is contained in $T^+(n)$, but does not appear in the product $e_{m_1}^{l_1} t^{p_1} \wedge \dots \wedge e_{m_s}^{l_s} t^{p_s}$, then the chain $(c/e_{m_u}^{l_u} t^{p_u}) \wedge e_{m_u}^{l_u} t^{p_u-1}$ is of weight $(k_1 - 1, \dots, k_n - 1)$ and again is an eigenvector of Laplace operator with negative eigenvalue. The lemma is proved.

Denote the weight of the chain $c_{ij}(m)$ by $w_{ij}(m)$. We must show that if $Q(k_1, \dots, k_n) = 0$, then there exists a unique family of nonnegative integers $\{m_{ij} \mid 1 \leqslant i \leqslant n, 1 \leqslant j \leqslant n, i \neq j\}$, such that

$$(k_1, \dots, k_n) = \sum_{i \neq j} w_{ij}(m_{ij}). \tag{6}$$

Let us solve Eq. (6) in nonegative integers with respect to m_{ij} under the condition $Q(k_1, \ldots, k_n) = 0$. Let

$$\varepsilon_{ijk} = \begin{cases} 1, & \text{if } i < j < k \text{ or } j < k < i, \text{ or } k < i < j, \\ 0 & \text{otherwise.} \end{cases}$$

Obviously, ε_{ijk} does not change under a cyclic permutation of its indices. Let, further,

$$A_{ij} = \{k \mid \varepsilon_{ikj} = 1\}, \quad B_{ij} = i \cup A_{ij}$$

and denote by a_{ij} the number of elements of the set A_{ij}. Obviously,

$$\sum_{i \neq j} w_{ij}(m_{ij}) = \left(\sum_{i \neq j} \binom{m_{ij}}{2} + \sum_{1 \in B_{ij}} m_{ij}, \ldots, \sum_{i \neq j} \binom{m_{ij}}{2} + \sum_{n \in B_{ij}} m_{ij} \right).$$

Therefore,

$$k_i - k_{i-1} = \sum_s m_{is} - \sum_t m_{ti}$$

(we put $k_0 = k_n$), and since $Q(k_1, \ldots, k_n) = \sum_i k_i - \frac{1}{2} \sum_i (k_i - k_{i-1})^2$, we have

$$Q\left(\sum_{i \neq j} w_{ij}(m_{ij}) \right) = n \sum_{i \neq j} \binom{m_{ij}}{2} + \sum_{i \neq j} a_{ij} m_{ij} - \sum_{i \neq j} m_{ij}^2 + \sum_{\substack{i \neq s \\ i \neq t}} m_{is} m_{ti}.$$

The right-hand side of the last relation can be put in the form

$$\sum_{j \in A_{ik}} [(m_{ij} + m_{jk} + m_{ki} - m_{ik} - m_{kj} - m_{ji})^2$$

$$- (m_{ij} + m_{jk} + m_{ki} - m_{ik} - m_{kj} - m_{ji})] + 2n \sum_{i \neq j} m_{ij} m_{ji}.$$

Obviously, for nonnegative integers m_{ij}, all the summands of both sums in the last expression are nonnegative and therefore the relation $Q\left(\sum_{i, j} w_{ij}(m_{ij}) \right) = 0$ is equivalent to the system of relations

$$m_{ij} + m_{jk} + m_{ki} - m_{ik} - m_{kj} - m_{ji} = 0 \quad \text{or} \quad 1$$

$$\text{for} \quad \varepsilon_{ijk} = 1,$$

$$m_{ij} = 0 \quad \text{or} \quad m_{jt} = 0 \quad \text{for any} \quad i, j$$

(under the second condition, of the six summands appearing on the left-hand side of the first condition, three vanish). These conditions may be restated as follows: There exists a permutation $\sigma \in \text{Symm}(n)$, such that

$$m_{ij} = 0, \quad \text{if} \quad \sigma(i) > \sigma(j),$$

$$m_{ik} = m_{ij} + m_{jk} - (0 \text{ or } 1), \quad \text{if} \quad \sigma(i) < \sigma(j) < \sigma(k), \ j \in A_{ik},$$

$$m_{ik} = m_{ij} + m_{jk} + (0 \text{ or } 1), \quad \text{if} \quad \sigma(i) < \sigma(j) < \sigma(k), \ j \notin A_{ik}.$$

The last two conditions are equivalent to one: for $\sigma(i) < \sigma(j) < \sigma(k)$

$$m_{ik} = m_{ij} + m_{jk} - \varepsilon_{ijk} + (0 \quad \text{or} \quad 1) = m_{ij} + m_{jk} + \varepsilon_{kji} - (0 \quad \text{or} \quad 1).$$

We see that for $\sigma(i) < \sigma(j)$ we have

$$(k_i - k_{i-1}) - (k_j - k_{j-1}) = \left(\sum_s m_{is} - \sum_t m_{ti}\right) - \left(\sum_u m_{ju} - \sum_v m_{vj}\right)$$

$$= \sum_{\sigma(s) < \sigma(i) < \sigma(j)} (-m_{si} + m_{sj}) + \sum_{\sigma(i) < \sigma(s) < \sigma(j)} (m_{is} + m_{sj})$$

$$+ \sum_{\sigma(i) < \sigma(j) < \sigma(s)} (m_{is} - m_{js}) + 2m_{ij} = nm_{ij} - \sum_{s \neq i, j} \varepsilon_{sij} + \sum_{s \neq i, j} (0 \quad \text{or} \quad 1)$$

$$= nm_{ij} - a_{ji} + (0, 1, \ldots \quad \text{or} \quad n - 2).$$

Thus

$$m_{ij} = \max\left(0, \left[\frac{(k_i - k_{i-1}) - (k_j - k_{j-1}) + a_{ji}}{n}\right]\right). \tag{7}$$

We see that Eq. (6) has a unique solution. The theorem is proved.

Note that we have also found an explicit formula for $q(k_1, \ldots, k_n)$ in the statement of the theorem:

$$q(k_1, \ldots, k_n) = \sum_{i,j} m_{ij}, \tag{8}$$

where the m_{ij} are given by formula (7).

Since the numbers $k_1, \ldots, k_n$ can be recovered from the m_{ij} by means of formula (6), formula (8) yields the following important conclusion.

Theorem 2.5.3. The space $H_q(T^+(n))$ is finite dimensional for all q and n.

In conclusion, we state a corollary to Theorem 2.5.2 concerning the case $n = 2$.

Theorem 2.5.4.

$$\dim H_r^{(k_1,\, k_2)}(T^+(2)) = \begin{cases} 1 & \text{for } (k_1, k_2) = \left(\binom{r}{2}, \binom{r+1}{2}\right) \text{ or } \left(\binom{r+1}{2}, \binom{r}{2}\right), \\ 0 & \text{otherwise}; \end{cases}$$

in particular, $\dim H_r(T^+(2)) = 2$ for $r > 0$.

2. The homology of the algebra $T(n)$. A complete description of the homology and cohomology of the algebra $T(n)$ and other current algebras was presented without detailed proofs in Subsection 3C. In the present subsection, we limit ourselves to some partial results; our main goal will be to establish relationships between the homology of the algebras $T^+(n)$ and $T(n)$.

A. Application of Theorem 1.5.2b. As in Subsection 1, let us fix a basis $d_1 = \sum \alpha_{1i} e_i^i, \ldots, d_{n-1} = \sum \alpha_{n-1,\, i} e_i^i$ in the space D of complex diagonal $(n \times n)$-matrices with trivial trace. Obviously in $T_{(k_1, \ldots,\, k_n)}(n)$ commutation with d_j coincides with multiplication by the number

$$\alpha_{j1}(k_1 - k_n) + \alpha_{j2}(k_2 - k_1) + \ldots + \alpha_{jn}(k_n - k_{n-1}), \qquad (9)$$

so that, by Theorem 1.5.2b, the homology of the complex $C_{\cdot}(T(n))$ coincides with the homology of the subcomplex con-

sisting of the spaces $C_q^{(k_1, \ldots, k_n)}(T(n))$, for which the sum (9)
vanishes for each j. Since $d_1, \ldots, d_{n-1}$ is a basis in the
space of diagonal matrices with zero trace, the last condition is equivalent to the relations $k_1 = \ldots = k_n$. We come
to the following conclusion.

Proposition 2.5.5. The homology of the algebra $T(n)$
coincides with the homology of the subcomplex of the complex
$C.(T(n))$, consisting of the spaces

$$\bigoplus_{k \in \mathbf{Z}} C_q^{(k, \ldots, k)}(T(n)).$$

This subcomplex is later denoted by $C_\cdot^=(T(n))$.

B. The Goncharova spectral sequence. In 1973, Goncharova discovered a spectral sequence relating the cohomology
of the algebra $L_1(1)$ with the cohomology of the algebra of
vector fields on the circle. This sequence did not lead to
any serious applications and remained unpublished. It turns
out that a similar spectral sequence for the current algebra
is of much greater interest. Here we construct the homology
version of the spectral sequence.

Put

$$F_p C_q(T(n)) = \bigoplus_{\substack{\sum_{i,j} |k_{ij}| \leqslant p}} T_{(k_{11}, \ldots, k_{1n})}(n) \wedge \ldots \wedge T_{(k_{q1}, \ldots, k_{qn})}(n),$$

$$F_p C_q^=(T(n)) = F_p C_q(T(n)) \cap C_q^=(T(n)).$$

Obviously, $\partial (F_p C_q(T(n)) \subset F_p C_{q-1}(T(n))$, so that $\{F_p C_q(T(n))\}$,
$\{F_p C_q^=(T(n))\}$ are filtrations in the complexes $C.(T(n))$,
$C_\cdot^=(T(n))$. The spectral sequence associated with the second
one of these filtrations is called the Goncharova spectral
sequence.

Usually we shall ignore the double grading of the terms of the Goncharova spectral sequence, limiting ourselves to the complete grading (as was done for another spectral sequence in Subsection 3.2C), so that the spectral sequence is of the form $\{E_q^r, \, d_q^r : E_q^r \to E_{q-1}^r\}$.

Proposition 2.5.6. In the Goncharova spectral sequence,

$$E_q^1 \cong \bigoplus_r [\Lambda^{q-2r} D \otimes H_r (T^+ (n))]. \tag{10}$$

Proof. Consider the chain $g_1 \wedge \cdots \wedge g_q \in C_q^= (T (n))$ for $g_i \in T_{(k_{i1}, \ldots, k_{in})} (n)$. Assume that $\sum_{i,j} |k_{ij}| = p$ and

$$k_{i1} \leqslant 0, \ldots, k_{in} \leqslant 0, k_{i1} + \ldots + k_{in} < 0 \text{ for } i \leqslant q_-,$$
$$k_{i1} = \ldots = k_{in} = 0 \text{ for } q_- < i \leqslant q - q_+,$$
$$k_{i1} \geqslant 0, \ldots, k_{in} \geqslant 0, k_{i1} + \ldots + k_{in} > 0 \text{ for } q - q_+ < i.$$

If $s \leqslant q_-$, while $t > q - q_+$, then the summand

$$(-1)^{s+t-1} [g_s, g_t] \wedge g_1 \wedge \cdots \hat{g}_s \cdots \hat{g}_t \cdots \wedge g_q,$$

appearing in the expression for the differential $\partial (g_1 \wedge \cdots \wedge g_q)$, belongs to $F_{p-1} C_{q-1} (T (n))$. Furthermore, if $q_- < s \leqslant q - q_+$, then [as can be seen from the inclusion $g_1 \wedge \cdots \wedge g_q \in C_q^= (T (n))$]

$$\sum_t (-1)^{s+t-1} [g_s, g_t] \wedge g_1 \wedge \cdots \hat{g}_s \cdots \hat{g}_t \cdots \wedge g_q = 0.$$

Therefore [we identify E_q^0 with $C_q^= (T (n))$]

$$d_q^0 (g_1 \wedge \ldots \wedge g_q) = \partial (g_1 \wedge \ldots \wedge g_{q_-}) \wedge g_{q_-+1} \wedge \ldots \wedge g_q$$
$$\pm g_1 \wedge \ldots \wedge g_{q-q_+} \wedge \partial (g_{q-q_++1} \wedge \ldots \wedge g_q),$$

which implies that

$$E_q^1 = \bigoplus_{\substack{q_-+q_++q=q, \\ k_1-l_1=\ldots=k_n-l_n}} H_{q_-}^{(l_1, \ldots l_n)} (T^+ (n)) \otimes \Lambda^{q_0} D \otimes H_{q_+}^{(k_1, \ldots, k_n)} (T^+ (n)).$$

It follows from Theorem 2.5.2 that if both spaces

$$H_{q_-}^{(l_1,\ldots,l_n)}(T^+(n)), \quad H_{q_+}^{(k_1,\ldots,k_n)}(T^+(n))$$

are nontrivial and $k_1 - l_1 = \ldots = k_n - l_n$, then $k_1 = l_1, \ldots,$ $k_n = l_n$, and the spaces are one-dimensional. We obtain the relation (10).

Although the computation of the subsequent differentials of the Goncharova spectral sequence encounters difficulties, the theorem proved above already allows us to reach the following conclusion.

Theorem 2.5.7.

$$\dim H_q(T(n)) \leqslant \sum_r \binom{n-1}{r} \dim H_{q-2r}(T^+(n));$$

in particular, the spaces $H_q(T(n))$ are finite dimensional for all q and n.

For example, for the algebra $T(2)$, Theorem 2.5.7 gives

$$\dim H_q(T(2)) \leqslant \begin{cases} 1 & \text{for } q = 0, 1, \\ 2 & \text{for } q > 1. \end{cases}$$

Actually,

$$\dim H_q(T(2)) = \begin{cases} 0 & \text{for } q = 1, \\ 1 & \text{for } q > 1 \end{cases}$$

(see 3C). The reader may try to prove this fact as an exercise.

C. Homology modulo D. The Goncharova spectral sequence yields its best results if we compute its relative homology modulo the $(n-1)$-dimensional commutative subalgebra D instead of its absolute homology of the algebra $T(n)$.

Theorem 2.5.8.

$$H_q(T(n), D) \cong \begin{cases} 0 & \text{for} \quad \text{odd} \quad q, \\ H_r(T^+(n)) & \text{for} \quad q = 2r > 0. \end{cases}$$

The proof repeats the previous calculation with appropriate simplifications: the Goncharova filtration exists in the quotient complex $C.(T(n), D)$ of the complex $C^{\cdot}.(T(n))$, and the corresponding spectral sequence is such that $E_q^1 = 0$ for odd q and $E_{2r}^1 = H_r(T^+(n))$. The differentials of the spectral sequence beginning from the first one are trivial from considerations of dimension.

Theorem 2.5.8 shows that the first term of the Goncharova spectral sequence from Sub-subsection B is of the form

$$H_*(T(n), D) \otimes \Lambda^*D,$$

i.e., coincides with the first term of the Serre–Hochschild spectral sequence of the pair $T(n), D$. It is easy to understand that after appropriate changes of gradings the subsequent terms of the spectral sequences also become identical.

3. Other results. A. The Kac–Moody algebras. Suppose

$$A = \begin{Vmatrix} a_{11} & \cdots & a_{1n} \\ \cdots & \cdots & \cdots \\ a_{n1} & \cdots & a_{nn} \end{Vmatrix}$$

is an integer matrix with $a_{11} = \ldots = a_{nn} = 2$ and $a_{ij} \leqslant 0$ for $i \neq j$. Assume that the matrix A is symmetrizable, i.e., that there exist positive numbers $b_1, \ldots, b_n$ such that the matrix $\| b_i a_{ij} \|$ is symmetric. By definition, the Kac–Moody algebra $\mathfrak{g}^A$ with Cartan matrix A is the complex Lie algebra with generators

$$e_1, \ldots, e_n, \ f_1, \ldots, f_n, \ h_1, \ldots, h_n$$

and relations

$$[e_i, f_j] = \delta_{ij}h_i, \quad [h_i, h_j] = 0,$$

$$[h_i, e_j] = a_{ij}e_j, \quad [h_i, f_j] = -a_{ij}f_j,$$

$$\underbrace{[e_i, [e_i, \ldots [e_i, e_j] \ldots]]}_{-a_{ij}+1} = 0 \qquad (i \neq j),$$

$$\underbrace{[f_i, [f_i, \ldots [f_i, f_j] \ldots]]}_{-a_{ij}+1} = 0 \qquad (i \neq j).$$

In the algebra $\mathfrak{g}^A$, we have the grading

$$\deg h = (\underbrace{0, \ldots, 0}_{n}), \quad \deg e_i = (\underbrace{0, \ldots, 0, \overset{(i)}{1}, 0, \ldots, 0}_{n}),$$

$$\deg f_i = (\underbrace{0, \ldots, 0, -\overset{(i)}{1}, 0, \ldots, 0}_{n}).$$

In the following Theorems 2.5.9-2.5.12, we assume that the matrix A is irreducible (cannot be transformed by interchanging rows and columns into block-diagonal form with proper blocks). The proof of these theorems is contained in the articles by Kac [53] and Moody [69].

Theorem 2.5.9. (i) If the matrix A is nondegenerate, then the algebra $\mathfrak{g}^A$ is simple. If $\text{rang } A = n - k$ $(k > 0)$, then $\mathfrak{g}^A$ has a k-dimensional center $Z \subset \mathfrak{g}^A_{(0, \ldots, 0)}$, and the algebra $\mathfrak{g}^A/Z$ possesses no graded ideals, i.e., ideals I, such that $I = \bigoplus_{k_1, \ldots, k_n} I \cap \mathfrak{g}^A_{(k_1, \ldots, k_n)})$.

(ii) $\mathfrak{g}^A = N^-(\mathfrak{g}^A) \oplus H \oplus N^+(\mathfrak{g}^A)$, where $N^-(\mathfrak{g}^A)$ and $N^+(\mathfrak{g}^A)$ are subalgebras of algebra $\mathfrak{g}^A$ generated, respectively, by $f_1, \ldots, f_n$ and $e_1, \ldots, e_n$, while H is the (commutative) subalgebra generated by $h_1, \ldots, h_n$. In particular, $\mathfrak{g}^A_{(0, \ldots, 0)} = H$ and $\mathfrak{g}^A_{(k_1, \ldots, k_n)} = 0$, if among the numbers $k_1, \ldots, k_n$ there are positive ones and negative ones.

(iii) dim $H = n$; the system of relations for the gener-
ators $f_1, \ldots, f_n$ of the algebra $N^- (\mathfrak{g}^A)$ is constituted by
the relations

$$[\underbrace{f_i, [f_i, \ldots [f_i, f_j]}_{-a_{ij}+1} \ldots]] = 0,$$

and a similar fact is true for the algebra $N^+ (\mathfrak{g}^A)$.

The class of Kac—Moody algebras includes algebras with
very different properties. It is convenient to divide them
into three groups: algebras for which the symmetrized Cartan
matrix $bA = \| b_i a_{ij} \|$ is positive definite, algebras for which
this matrix is nonnegative semidefinite and of rank $n - 1$,
and the other algebras. The simplest are the algebras of
the first group.

Theorem 2.5.10. The class of algebras $\mathfrak{g}^A$ with positive-
definite matrices bA coincides with the class of simple finite-dimensional Lie algebras.

Here n is the rank of the algebra $\mathfrak{g}^A$, the decomposition
$\mathfrak{g}^A = N^- (\mathfrak{g}^A) \oplus H \oplus N^+ (\mathfrak{g}^A)$ is the classical Cartan decomposition, and the spaces $\mathfrak{g}^A_{(k_1, \ldots, k_n)}$ are the root subspaces of the
algebra $\mathfrak{g}^A$ (see [94]). In particular, dim $\mathfrak{g}^A_{(0, \ldots, 0)} = n$ and
dim $\mathfrak{g}^A_{(k_1, \ldots, k_n)} \leqslant 1$, if $(k_1, \ldots, k_n) \neq (0, \ldots, 0)$. The Cartan
matrices of finite-dimensional simple Lie algebras are con-
tained in the tables presented at the end of the appropriate
part of the Bourbaki treatise. These matrices are obtained
from Gram matrices of systems of simple roots (from Dynkin
diagrams) by multiplying the rows by numbers such that the
diagonal elements become equal to 2.

The information known at present about the algebras of
the third group is more or less limited to the following statement.

Theorem 2.5.11. If the matrix bA has negative eigenvalues and is of rank $\leqslant n - 2$, then the dimension of the space $\bigoplus_{k_1 + \ldots + k_n = k} \mathfrak{g}^A_{(k_1, \ldots, k_n)}$ grows exponentially with k.

The most interesting algebras $\mathfrak{g}^A$ are those of the second group.

Theorem 2.5.12. If the matrix bA is nonnegative definite and of rank $n - 1$, then the algebra $\mathfrak{g}^A$ has a one-dimensional center Z, and the algebra $\mathfrak{g}^A/Z$ is isomorphic to one of the following algebras:

(i) the algebra $(\mathfrak{g}^{S^1})^{\mathrm{pol}} = \mathfrak{g} \otimes \mathbb{C}[t, t^{-1}]$, where $\mathfrak{g}$ is a simple finite-dimensional Lie algebra;

(ii) the subalgebra $\bigoplus_{l=-\infty}^{\infty} \mathfrak{g}(l) \otimes t^l$ of the previous algebra, where $\mathfrak{g}(l)$ is the eigensubspace of a certain automorphism $\mathfrak{g} \rightarrow \mathfrak{g}$ of finite order m, corresponding to the eigenvalue $e^{2\pi i l/m}$ (in fact m can only equal 2 or 3).

In the case (i), the Cartan matrix of the algebra $\mathfrak{g}^A$ is obtained from the extended Dynkin diagram (see [94]) of the algebra $\mathfrak{g}$ in the same way as the Cartan matrix of the algebra $\mathfrak{g}$ itself is obtained from its ordinary Dynkin diagram. The grading in $\mathfrak{g}^A$ is described by the formula

$$\mathfrak{g}^A_{(k_1,\ldots,k_n)}/Z \cap \mathfrak{g}^A_{(k_1,\ldots,k_n)} = \mathfrak{g}_{(k_1-l_1k_n,\ldots,k_{n-1}-l_{n-1}k_n)} \otimes t^{k_n},$$

where $(l_1, \ldots, l_{n-1})$ are coordinates of the minimal root of the algebra $\mathfrak{g}$, and $\mathfrak{g}_{(\ldots)}$ are its root subspaces. [Note that $(l_1, \ldots, l_{n-1}, 1)$ is an eigenvector of the Cartan matrix corresponding to the zero eigenvalue.]

It follows from the above that

$$\dim g^A_{(k_1, \dots, k_n)} \begin{cases} = n, & \text{if } (k_1, \dots, k_n) = (0, \dots, 0), \\ = n - 1, & \text{if } (k_1, \dots, k_n) \text{ is an} \\ & \text{eigenvector of the Cartan matrix} \\ & \text{corresponding to the zero eigenvalue,} \\ \leqslant 1 & \text{otherwise.} \end{cases}$$

The last statement is valid for the algebras from part (ii) of Theorem 2.5.12 as well.

B. The cohomology of the algebras $N^+ (g^A)$. Let

$$Q_A (x_1, \dots, x_n) = -\frac{1}{2} \sum b_i a_{ij} x_i x_j + \sum b_i x_i.$$

Theorem 2.5.13 (see [62]). If $Q_A (k_1, \dots, k_n) \neq 0$, then

$$H_q^{(k_1, \dots, k_n)} (N^+ (g^A)) = 0$$

for any q.

In the case when the matrix A satisfies the conditions of Theorems 2.5.10 or 2.5.12, then Theorem 2.5.13 may be proved just as we proved Theorem 2.5.2 in Subsection 1: in the algebra $N^+ (g^A)$ it is easy to indicate a metric with respect to which the Laplace operator in $C_*^{(k_1, \dots, k_n)} (N^+ (g^A))$ is the multiplication by $Q_A (k_1, \dots, k_n)$. [It goes without saying that part (i) of Theorem 2.5.2 is a particular case of Theorem 2.5.13: in view of Sub-subsection A, the algebra

$T (n)$ has the Cartan matrix $\begin{Vmatrix} 2 & -2 \\ -2 & 2 \end{Vmatrix}$ for $n = 2$ and, for

$n > 2$, the following Cartan matrix:

$$\begin{Vmatrix} 2 & -1 & 0 & \dots & 0 & -1 \\ -1 & 2 & -1 & \dots & 0 & 0 \\ 0 & -1 & 2 & \dots & 0 & 0 \\ \dots & \dots & \dots & \dots & \dots & \dots \\ 0 & 0 & 0 & \dots & 2 & -1 \\ -1 & 0 & 0 & \dots & -1 & 2 \end{Vmatrix}.$$

(In both cases we can set $b_1 = \ldots = b_n = 1$.)

For algebras of the first and second group, we have the analog of part (ii) of Theorem 2.5.2: If $Q_A (k_1, \ldots, k_n) = 0$, then there exists a $q = q (k_1, \ldots, k_n)$, such that

$$\dim H_r^{(k_1,\ldots,k_n)} (N^+(\mathfrak{g}^A)) = \begin{cases} 1 & \text{for } r = q, \\ 0 & \text{for } r \neq q. \end{cases}$$

I do not know if a similar statement is true for algebras of the third group.

C. The passage to the algebras $\mathfrak{g}^A$ and current algebras. This passage may be carried through by means of the Goncharova spectral sequence (see Subsection 2) which, in the general case, gives about as much information as in the case of the algebras $T(n)$ presented in Subsection 2. Fuller results on the cohomologies of the algebras $\mathfrak{g}^{S^1}$ may be obtained by using methods similar to those of §§2 and 4: first we compute the cohomology of the algebras of "formal currents" $\mathfrak{g} \otimes \mathbb{R} [[t]]$, and then pass on to the algebra $\mathfrak{g}^{S^1}$ by using the additive techniques of Bott and Segal described in §4. This approach gives the following result (see [14]).

Theorem 2.5.14. Suppose $\mathfrak{g}$ is a real semisimple Lie algebra and G the corresponding Lie group. Suppose that the complexification of the algebra $\mathfrak{g}$ coincides with the complexification of the Lie algebra $\bar{\mathfrak{g}}$ of some complex Lie group $\bar{G}$. Then

$$H^* (\mathfrak{g}^{S^1}) \cong H^*(\bar{G}^{S^1}; \mathbb{R}).$$

In particular, $H^* (\mathfrak{sl} (n, \mathbb{R})^{S^1}) \cong H^* (SU (n)^{S^1}; \mathbb{R})$.

[The cohomology ring of the space $SU(n)^{S^1} = SU(n) \times$ $\Omega SU(n)$ is the tensor product of the exterior algebra in the generators of dimension $3, 5, \ldots, 2n-1$ and the polynomial ring in generators of dimension $2, 4, \ldots, 2n-2$.]

If the group G is compact itself, then Theorem 2.5.14 has the following specification.

Theorem 2.5.15. If the group G is compact, then the canonical homomorphism

$$H^*(\mathfrak{g}^{S^1}) \to H^*(G^{S^1}; \mathbb{R})$$

(see 1.3.3) is an isomorphism.

In conclusion, note that if we replace the circle by a manifold of higher dimension, then the cohomology of the current algebra apparently becomes rather meaningless. In particular, if $\dim M \geqslant 2$, then already the space $H^2(\mathfrak{g}^M)$ is infinite dimensional for any semisimple algebra $\mathfrak{g}$ (see [14]).

§6. COMPUTATIONS FOR LIE SUPERALGEBRAS

In this section we present results obtained by Leites and myself in the cohomology of finite-dimensional Lie superalgebras [61].

Unlike finite-dimensional Lie algebras, finite-dimensional Lie superalgebras may possess infinite-dimensional cohomology; the simplest example of the situation is given by an arbitrary finite-dimensional Lie superalgebra with trivial operation and nontrivial odd parts. However, in the examples considered below, the cohomology always turns out to be finite dimensional.

The definitions of the Lie superalgebras considered here are presented in §1.6.

1. The superalgebras $\mathfrak{gl}\,(m,\,n)$ and $\mathfrak{sl}\,(m,\,n)$. Theorem 2.6.1. If $m \geqslant n$, then the natural inclusions

$$\mathfrak{gl}\,(m,\,\mathbb{K}) \to \mathfrak{gl}\,(m,\,n)_0 \subset \mathfrak{gl}\,(m,\,n),$$
$$\mathfrak{sl}\,(m,\,\mathbb{K}) \to \mathfrak{sl}\,(m,\,n)_0 \subset \mathfrak{sl}\,(m,\,n)$$

induce an isomorphism in cohomology with trivial coefficients.

Proof. We limit ourselves to the first inclusion. Consider the Serre–Hochschild spectral sequence associated with the pair $(\mathfrak{gl}\,(m,\,n),\ \mathfrak{gl}\,(m,\,n)_0)$. For this spectral sequence, we have

$$E_1^{p,\,q} = H^q\,(\mathfrak{gl}\,(m,\,n)_0;\ S^p\,(\mathfrak{gl}\,(m,\,n)_1))$$
$$= H^q\,(\mathfrak{gl}\,(m,\,\mathbb{K})\,\oplus\,\mathfrak{gl}\,(n,\,\mathbb{K});\ S^p\,(\mathrm{Hom}\,(V_0,\,V_1)\,\oplus\,\mathrm{Hom}\,(V_1,\,V_0)),$$

where $V_0,\,V_1$ are the spaces $\mathbb{K}^m,\,\mathbb{K}^n$ with the standard action of the algebras $\mathfrak{gl}\,(m,\,\mathbb{K}),\ \mathfrak{gl}\,(n,\,\mathbb{K})$. It follows from Theorem 2.1.2 that

$$E_1^{p,\,q} = H^q\,(\mathfrak{gl}\,(m,\,\mathbb{K})\,\oplus\,\mathfrak{gl}\,(n,\,\mathbb{K}))$$
$$\otimes\,\mathrm{Inv}_{\mathfrak{gl}(m,\,\mathbb{K})}\mathrm{Inv}_{\mathfrak{gl}(n,\,\mathbb{K})}S^p\,(\mathrm{Hom}\,(V_0,\,V_1)\,\oplus\,\mathrm{Hom}\,(V_1,\,V_0)).$$

Let us compute the ring of invariants

$$\mathrm{Inv}_{\mathfrak{gl}(m,\,\mathbb{K})}\mathrm{Inv}_{\mathfrak{gl}(n,\,\mathbb{K})}S^*\,(\mathrm{Hom}\,(V_0,\,V_1)\,\oplus\,\mathrm{Hom}\,(V_1,\,V_0))$$
$$= \mathrm{Inv}_{\mathfrak{gl}(m,\,\mathbb{K})}\mathrm{Inv}_{\mathfrak{gl}(n,\,\mathbb{K})}\,(S^*\mathrm{Hom}\,(V_0,\,V_1)\,\otimes\,S^*\mathrm{Hom}\,(V_1,\,V_0)). \qquad (1)$$

To do this, define the functional

$$\eta_k\colon S^k\,\mathrm{Hom}\,(V_0,\,V_1)\,\otimes\,S^k\,\mathrm{Hom}\,(V_1,\,V_0) \to \mathbb{K}$$

by the formula

$$\eta_k\,(\varphi_1,\,\ldots,\,\varphi_k;\,\psi_1,\,\ldots,\,\psi_k) = \sum_{\sigma\in\mathrm{Symm}(k)}\mathrm{Tr}\,[(\varphi_1\circ\psi_{\sigma(1)})\circ\ldots\circ(\varphi_k\circ\psi_{\sigma(k)})]$$

$$= \sum_{\sigma \in \mathrm{Symm}(k)} \mathrm{Tr}\left[(\psi_1 \circ \varphi_{\sigma(1)}) \circ \ldots \circ (\psi_k \circ \varphi_{\sigma(k)})\right],$$

in which $\varphi_i \in \mathrm{Hom}(V_0, V_1)$, $\psi_i \in \mathrm{Hom}(V_1, V_0)$. Obviously η_k is an invariant.

Lemma. If $m \geqslant n$, then (1) is the polynomial ring in $\eta_1, \ldots, \eta_n$.

Proof. We begin with the remark that $\eta_{n+1}, \eta_{n+2}, \ldots$ are polynomials in $\eta_1, \ldots, \eta_n$, while $\eta_1, \ldots, \eta_n$ are independent. Indeed, in view of the symmetry of the functional η_k with respect to φ and ψ, it is defined by its values on the family of arguments of the form $(\varphi, \ldots, \varphi; \psi, \ldots, \psi)$. But

$$\eta_k(\varphi, \ldots, \varphi; \psi, \ldots, \psi) = \lambda_1(\varphi \circ \psi)^k + \ldots + \lambda_n(\varphi \circ \psi)^k$$
$$= \lambda_1(\psi \circ \varphi)^k + \ldots + \lambda_m(\psi \cdot \varphi)^k,$$

where $\lambda_i(\varphi \circ \psi)$, $\lambda_i(\psi \circ \varphi)$ are the eigenvalues of the operators $\varphi \circ \psi$, $\psi \circ \varphi$. If $m \geqslant n$, then the eigenvalues of the composition $\varphi \circ \psi$ are n arbitrary numbers (the composition $\psi \circ \varphi$ has $m - n$ zero eigenvalues). The required statement thus becomes a well-known proposition from the theory of symmetric polynomials (compare with the proof of Theorem 2.1.5).

Further,

$$S^p(\mathrm{Hom}(V_0, V_1) \oplus \mathrm{Hom}(V_1, V_0)) = S^p((V_0' \otimes V_1) \oplus (V_1' \otimes V_0))$$

$$= \bigoplus_{p_1 + p_2 = p} [S^{p_1}(V_0' \otimes V_1) \otimes S^{p_2}(V_1' \otimes V_0)].$$

It follows from Theorem 2.1.4 that for $p_1 \neq p_2$ there are no invariants in the space $S^{p_1}(V_0' \otimes V_1) \otimes S^{p_2}(V_1' \otimes V_0)$ and an arbitrary element ξ of the space

$$\mathrm{Inv}\,\mathrm{Inv}\,(S^k(V_0' \otimes V_1) \otimes S^k(V_1' \otimes V_0))$$
$$= \mathrm{Inv}\,[S^k(V_0 \otimes V_1') \otimes S^k(V_1 \otimes V_0')]'$$

is given by the following formula:

$$\xi(\alpha_1, \ldots, \alpha_k; \beta_1, \ldots, \beta_k; \gamma_1, \ldots, \gamma_k; \delta_1, \ldots, \delta_k)$$

$$= \sum_{\sigma_1, \sigma_2 \in \text{Symm}(k)} a(\sigma_1, \sigma_2) \prod_i \delta_i(\alpha_{\sigma_1(i)}) \beta_i(\gamma_{\sigma_2(i)})$$

$(\alpha_i \in V_0, \ \beta_i \in V_1', \ \gamma_i \in V_1, \ \delta_i \in V_0')$.   The symmetry conditions
which guarantee the inclusion of the functional ξ in the
space $[S^k(V_0 \otimes V_1') \otimes S^k(V_1 \otimes V_0')]'$ are of the form

$$a(\sigma_1, \sigma_2) = a(\sigma\sigma_1, \sigma_2\sigma^{-1}) = a(\sigma_1\sigma^{-1}, \sigma\sigma_2)$$

for $\sigma \in \text{Symm}(k)$,   i.e., $a(\sigma_1, \sigma_2) = b(\sigma_2\sigma_1) = b(\sigma_1\sigma_2)$, where
$b(\sigma) = a(\sigma, 1)$.   It follows from the previous equalities that
the function b is constant on classes of conjugate elements,
i.e., the given coefficients $a(\sigma_1, \sigma_2)$ depend only on the
lengths $k_1, \ldots, k_s$ of the cycles in the product $\sigma_1\sigma_2$.   We
see that

$$\xi = \sum a(k_1, \ldots, k_s)\eta_{k_1} \ldots \eta_{k_s}.$$

The lemma is proved.

Thus, E_1 is the tensor product of the exterior algebra
in the generators

$$\varphi_1' \in E_1^{0,1}, \ \varphi_2' \in E_1^{0,3}, \ldots, \varphi_m' \in E_1^{0,2m-1},$$
$$\varphi_1'' \in E_1^{0,1}, \ \varphi_2'' \in E_1^{0,3}, \ldots, \varphi_n'' \in E_1^{0,2n-1}$$

and the polynomial algebra in the generators

$$\eta_1 \in E_1^{2,0}, \ldots, \eta_n \in E_1^{2n,0}.$$

We conclude the proof of the theorem by showing that the generators φ_i', φ_i'' are transgressive and that transgression sends
φ_i' into η_i and φ_i'' into η_i. To do this, we consider the map
of the Weyl algebras $W(\mathfrak{gl}(m, \mathbb{K}))$, $W(\mathfrak{gl}(n, \mathbb{K}))$ into $C^*(\mathfrak{gl}(m, n))$, extending the natural inclusions

$$\Lambda^1\mathfrak{gl}\,(m,\mathbb{K})' = \mathfrak{gl}\,(m,\mathbb{K})' \to \mathfrak{gl}\,(m,n)_0' \subset C^1\,(\mathfrak{gl}\,(m,n)),$$

$$\Lambda^1\mathfrak{gl}\,(n,\mathbb{K})' = \mathfrak{gl}\,(n,\mathbb{K})' \to \mathfrak{gl}\,(m,n)_0' \subset C^1\,(\mathfrak{gl}\,(m,n))$$

(see 1.3.5.2°). What we presented in 1.3.5 enables us to compute automatically the values of these maps on $S^1\mathfrak{gl}\,(m,\mathbb{K})'$, $S^1\mathfrak{gl}\,(n,\mathbb{K})'$: the functionals $\alpha \in S^1\mathfrak{gl}\,(m,\mathbb{K})' = \mathfrak{gl}\,(m,\mathbb{K})'$, $\beta \in S^1\mathfrak{gl}\,(n,\mathbb{K})' = \mathfrak{gl}\,(n,\mathbb{K})'$ are transformed into two-dimensional cochains

$$S^2\mathfrak{gl}\,(m,n)_1 \to \mathbb{K},$$

acting according to the formulas

$$((\varphi_1,\psi_1),(\varphi_2,\psi_2)) \mapsto \alpha\,(\psi_1\circ\varphi_2 + \psi_2\circ\varphi_1),\ \beta\,(\varphi_1\circ\psi_2 + \varphi_2\circ\psi_1).$$

This implies, first of all, that each of these maps sends the invariants ζ_k from Theorem 2.1.5 into η_k and, second, that both of these maps are compatible with filtrations. The required statement now follows from the fact that the spectral sequence in 1.3.5 has invariants ζ_k which are the images under transgression of the exterior generators of the algebras $E_1^{0,*}$ (see 2.1.3).

2. The superalgebra $Q(n)$. Theorem 2.6.2. The ring $H^*\,(Q\,(n))$ is isomorphic to the quotient ring of the ring of symmetric polynomials in n variables by the ideal generated by symmetric polynomials in the squares of these variables.

Proof. In the Serre–Hochschild spectral sequence associated with the pair $(Q\,(n),\ Q\,(n)_0)$, we have

$$E_1 = H^*\,(Q\,(n)_0;\ S^*Q\,(n)_1') = H^*\,(\mathfrak{gl}\,(n,\mathbb{K})) \otimes S^*\,(\mathfrak{gl}\,(n,\mathbb{K})'),$$

and Theorem 2.1.5 shows that E_1 is the tensor product of the exterior algebra in n variables contained in $E_1^{0,1}$, $E_1^{0,3}$, $\ldots$, $E_1^{0,2n-1}$, and the polynomial algebras in n variables contained in $E_1^{1,0}$, $E_1^{2,0}$, $\ldots$, $E_1^{n,0}$. This last polynomial algebra can be conveniently identified with the algebra of symmetric polynomials in n variables, by assigning Newton polynomials (sums of powers of the variables) to the corresponding generators indicated in Theorem 2.1.5. Just as in the proof one establishes that the exterior generators are transgressive and then one can compute their images under transgression. The latter turn out to be Newton polynomials of even degrees. This implies the required statement.

Remark. One should not be surprised that the ring $H^*(Q(n))$ is commutative and not skew-commutative as usually happens with cohomology rings. Simply, as shown by this commutativity,

$$H^q(Q(n)) = \begin{cases} H_0^q(Q(n)) & \text{for even } q, \\ H_1^q(Q(n)) & \text{for odd } q \end{cases}$$

(even-dimensional cohomology consists of even elements, odd dimensional — of odd elements).

3. Other results. Theorem 2.6.3. The ring $H^*(P(n))$ is the exterior algebra in $[(n+1)/2]$ generators of dimensions $1, 5, \ldots, 4[(n+1)/2] - 3$.

Theorem 2.6.4.

$$H^*(\mathfrak{osp}(m, n)) = \begin{cases} H^*(\mathfrak{o}(m)) & \text{for } m \geqslant 2n, \\ H^*(\mathfrak{sp}(n, \mathbb{R})) & \text{for } m < 2n. \end{cases}$$

Theorem 2.6.5.

$$\dim H^q(W(0, n)) = \begin{cases} 0 & \text{for } q \neq 0, \ 2n - 1, \\ 1 & \text{for } q = 0, \ 2n - 1. \end{cases}$$

I will leave the proofs of these theorems to the reader, limiting myself to the following hints. Theorems 2.6.3 and

2.6.4 are proved exactly like Theorems 2.6.1 and 2.6.2, but the references to Theorem 2.1.4 and its corollaries must be replaced by references to the corresponding theorems about orthogonal and symplectic invariants (see [93]). The calculation of the cohomology of the algebra $W(0, n)$ may be carried out by means of the Serre—Hochschild spectral sequence corresponding to the pair $(W(0, n), \mathfrak{gl}(n, \mathbb{K}))$, where the inclusion $\mathfrak{gl}(n, \mathbb{K}) \to W(0, n)$ is defined by the formula $\| a_{ij} \| \mapsto \sum a_{ij} y_i \partial/\partial y_j$. The beginning of the proof follows the proof of Theorem 2.2.4, but the relations between the analogs of the invariants Ψ_k are completely different from the relations between the latter.

Chapter 3. Applications

§1. CHARACTERISTIC CLASSES OF FOLIATIONS

The theory of characteristic classes of foliations is
one of the most brilliant applications of the cohomology of
Lie algebras of vector fields. Here we develop only some se-
lected results of this theory. The reader who would like to
acquaint himself in more detail with foliations and their
characteristic classes is referred to the books by Tamura [88],
Lawson [60], and Pittie [73], to Lawson's survey article [59],
and to my own survey articles [19, 21].

   1. Foliations. A. Fundamental Definitions. Suppose
X is a smooth n-dimensional manifold without boundary and assume $0 \leqslant p \leqslant n$. We shall say that X is supplied with a
(smooth) foliation of dimension p if X is partitioned into
(pathwise) connected subsets F_α possessing the following
property: each point $x \in X$ can be covered by a chart φ:
$U \to \mathbb{R}^n$ from the structural atlas of the manifold X, such
that the components (with respect to pathwise connectedness)
of the intersections $U \cap F_\alpha$ are mapped by φ into p-dimensional
planes of the space $\mathbb{R}^n$, parallel to the plane of the first p
coordinate axes. The sets F_α are called leaves of the folia-

tion; X itself is called the <u>total manifold</u>. The number
$n - p$ is known as the <u>codimension</u> of the foliation; the codimension of the foliation $\mathcal{F}$ is denoted by codim $\mathcal{F}$.

These definitions (like most of the following ones) have
an obvious complex version. They may be generalized to manifolds with boundary, and here there are two competing possibilities: we can require that the leaves be transversal to
the boundary, or we can require that the components of the
boundary be leaves (the last of course is meaninful only in
the case of codimension-1 foliations). Foliations on manifolds
with boundary will sometimes appear in this book (and both of
the above possibilities will be needed), but their role is
purely technical. If no special mention is made we assume
that the total manifold of a foliation is a manifold without
boundary.

Trivial examples of foliations are given by smooth fibrations: if X is the total manifold of a smooth fibration with
$(n - p)$-dimensional base $(n = \dim X)$, then the partition of
the manifold X into the fibers of the bundle is a p-dimension-
al foliation. The next example in order of complexity is the
line foliation on the torus, i.e., the partition of the torus
$\mathbb{R}^2/\mathbb{Z}^2$ into the images of the lines $y = \lambda x + c$ for a fixed
λ. If λ is rational, then this foliation relates to the previous class, while in the case of an irrational λ each leaf
is dense in the torus. This last example illustrates the fundamental difference between foliations and fibrations: foliations have no base. (Of course it is possible to constitute
the quotient space of the total manifold by the partition into
leaves but, as shown by the previous example, the result may
turn out to be unsatisfactory both from the topological and
from the set theoretic point of view.)

The leaves of a foliation should not be supplied with
the induced topology, but rather the topology whose base consists of the components of the intersection of the leaf with
open subsets of the total manifold. With respect to this topology, a leaf is a p-dimensional manifold. It also possesses
a natural smooth structure with respect to which its inclusion
in the total manifold is an (injective) immersion. For example, the leaves of the torus foliation with irrational λ are
diffeomorphic to the line.

B. Tangent and normal bundles. The tangent bundle of
the total manifold of a foliation possesses a subbundle which
it is natural to call the tangent bundle of the foliation: it
consists of vectors which are tangent to the leaves. The corresponding quotient bundle is known as the normal bundle of
the foliation; its dimension equals the codimension of the
foliation. The notations for the tangent and normal bundles
of the foliation $\mathcal{F}$ are $\operatorname{tang} \mathcal{F}$ and $\operatorname{norm} \mathcal{F}$. It is appropriate
to mention at this point that, in the theory developed below,
the geometry of the leaves themselves plays a secondary role
as compared to the geometry of the transversals to the leaves.
In this connection, such objects as codimension and normal
bundle are more important for us than, say, the dimension and
the tangent bundle.

A foliation is called oriented if its normal bundle is
oriented. Note that neither the leaves nor the total manifold
of an oriented foliation need even be orientable. A foliation
is said to be framed if its normal bundle is trivial and supplied with a specific trivialization.

A smooth map f of a smooth manifold Y without boundary
into the total manifold X of a foliation $\mathcal{F}$ is said to be

transversal to $\mathcal{F}$, if f is transversal to the tangent bundle
of the foliation $\mathcal{F}$; of course this is possible only if
dim $Y \geqslant$ codim $\mathcal{F}$. If f is transversal to $\mathcal{F}$, then the components of the inverse images of the leaves of the foliation $\mathcal{F}$
constitute a foliation of manifold Y. The latter is said
to be induced from $\mathcal{F}$ by means of f and is denoted by $f^*\mathcal{F}$.
Clearly we have codim $f^*\mathcal{F} =$ codim $\mathcal{F}$ and norm $f^*\mathcal{F} = f^*$ norm $\mathcal{F}$.
If f is the inclusion of a submanifold into a manifold, then
$f^*\mathcal{F}$ is also called the restriction of $\mathcal{F}$ to Y and is also
denoted by $\mathcal{F} \mid Y$.

C. The determining system of forms. Suppose $\mathcal{F}$ is a
foliation of codimension q on the manifold X. A determining
system of forms of the foliation $\mathcal{F}$ in the open set $U \subset X$ is
a family $\omega_1, \ldots, \omega_q$ of smooth 1-forms defined in U and possessing two properties: (i) the forms $\omega_1, \ldots, \omega_q$ are linearly independent at each point of the set U; (ii) the restriction of each of these forms to any component of the intersection of any leaf with U is trivial. Obviously, to determine
a foliation, it suffices to specify a determining system of
forms in the open sets constituting a covering of the total
manifold. It is also clear that if $\{\omega_1, \ldots, \omega_q\}$, $\{\omega_1', \ldots, \omega_q'\}$
are determining systems of forms in U, U', then we have $\omega_i' = \sum_j \varphi_{ij} \omega_j$ (in $U \cap U'$), where the φ_{ij} are smooth functions for
which det $\| \varphi_{ij} \|$ does not vanish at any point of the intersec-
tion $U \cap U'$.

If $U = X$, then we are dealing with a global determining
system of forms. Obviously, for the existence of a global
determining system of forms, it is necessary and sufficient
that the normal bundle to the foliation be trivial, and the
choice of a global determining system of forms specifies the

choice of a definite homotopy class of the framing of this
foliation.

We stress that the system $\omega_1, \ldots, \omega_q$ of smooth 1-forms
given in the open set $U \subset X$ may not be a determining system
of forms for any foliation of codimension q, even if it satisfies condition (i): we must also require the "integrability
criterion." The latter can be stated as follows: for each
i the differential $d\omega_i$ must possess a representation in the
form $\sum_j \eta_j \wedge \omega_j$, where the η_j are certain smooth 1-forms; equivalent formulation [if we have (i)]: $\omega_1 \wedge \cdots \wedge \omega_q \wedge d\omega_i = 0$
for $i = 1, \ldots, q$. The fact that this criterion holds for any
determining system of forms of a foliation is quite obvious:
we may assume that the proposed system is given within the
charts described in the definition of foliations (see A), and
in local coordinates $x_1, \ldots, x_n$ of such a chart any determin-
ing system of forms can be written as $\sum_{j=1}^{q} f_{ij} dx_{p+j}$ $(i = 1, \ldots, q)$
with $\det \| f_{ij} \| \neq 0$. The converse statement, whose proof is
somewhat longer, is actually a well-known theorem of the calculus. (In the calculus this theorem is stated as follows:
suppose $\omega_1, \ldots, \omega_q$ is a system of smooth 1-forms defined in
a neighborhood of the point $\cdot 0 \in \mathbb{R}^n$ and linearly independent
at this point; if $\omega_1 \wedge \cdots \wedge \omega_q \wedge d\omega_i = 0$ for $i = 1, \ldots, q$,
then there exist smooth functions f_{ij}, q_j $(i, j = 1, \ldots, q)$ with
$\det \| f_{ij} \| \neq 0$ and $\omega_i = \sum_j f_{ij} dg_j$; for details, see [12].)

In the important case $q = 1$ the above reduces to the
following. A foliation of codimension 1 is given by a smooth
nonvanishing 1-form; this form ω must satisfy the integrabil-
ity criterion $\omega \wedge d\omega = 0$ (or $d\omega = \eta \wedge \omega$ for some smooth 1-
form η). A foliation of codimension 1 is defined by a 1-form

globally if and only if it is orientable and a choice of a
globally determining 1-form for the foliation determines a
choice of the orientation.

Note in conclusion that the integrability criterion holds
automatically if $q = n - 1$. For $q < n - 1$ this criterion
becomes a meaningful condition; for example, the 1-form
$x\, dy + dz$ in $\mathbb{R}^3$ does not satisfy it.

D. Concordance and homotopy equivalence. The foliations
$\mathcal{F}_0, \mathcal{F}_1$ of codimension q, given on the same manifold X, are
said to be concordant if there exists a foliation $\mathcal{H}$ in $X \times \mathbb{R}$
of the same codimension q, such that the inclusions i_0, $i_1 \colon X \to$
$X \times \mathbb{R}$, defined by the formulas $i_0(x) = (x, 0)$, $i_1(x) = (x, 1)$,
are transversal to $\mathcal{H}$ and $i_0^* \mathcal{H} = \mathcal{F}_0$, $i_1^* \mathcal{H} = \mathcal{F}_1$. Obviously,
the relation of concordance is reflexive and symmetric and
it is easy to see that it is also transitive; thus one can
speak of concordance classes of foliations. Clearly, concordant foliations have equivalent normal bundles [norm $\mathcal{F}_0 =$
i_0^* norm $\mathcal{H}$, norm $\mathcal{F}_1 = i_1^*$ norm $\mathcal{H}$ and i_0, i_1 are homotopic], while
their tangent bundles are stably equivalent. Moreover, if
the foliations $\mathcal{F}_0$, $\mathcal{F}_1$ are concordant, then the homotopy
class of the equivalence between norm $\mathcal{F}_0$ and norm $\mathcal{F}_1$ is
canonically defined.

Concordance should not be confused with homotopy equivalence, i.e., with the possibility of continuously deforming
one foliation into another (the topology in the set of foliations is induced from the $\mathcal{C}^\infty$-topology in the set of smooth
subbundles of the tangent bundle of the total manifold). For
example, the line foliations on the torus corresponding to
different values of λ (say λ_1 and λ_2) are obviously homotopic:
a homotopy is given by the foliations in which λ continuously

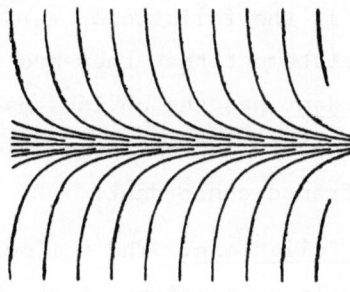

Fig. 7

changes from λ_1 to λ_2. However, it is easy to see that, using
the foliations with these different λ, it is not possible to
construct a foliation in the cylinder over the torus. Actually these foliations for different λ are all concordant.
[Construction of the foliation $\mathscr{H}$. Multiplying Fig. 7 by the
line, we obtain a foliation in $\mathbb{R}^3$, one of whose leaves will
be the plane. Turning the halfspace (into which $\mathbb{R}^3$ is divided by this plane) around a line perpendicular to this plane,
each by its own angle, we obtain a new foliation in $\mathbb{R}^3$ (in
order to make it smooth, it is necessary that the curves approach the horizontal line in Fig. 7 asymptotically very slow-
ly, say with the same speed as the graph of the function $y =$
$1/\log x$), and this foliation is still invariant with respect
to parallel translations preserving our plane. Taking the
quotient of $\mathbb{R}^3$ by the group $\mathbb{Z} \oplus \mathbb{Z}$ constituted by such translations, we obtain a foliation in the cylinder over the torus;
this is $\mathscr{H}$.] Examples of homotopic but nonconcordant foliations will appear in §2. The simplest examples of concordant
but nonhomotopic foliations are given by foliations of $\mathbb{R}^2 \setminus 0$,
the leaves of the first one being the circles with center 0,
and the leaves of the second — the two halves of the x-axis
and lines parallel to this axis.

Let us add that if the foliations $\bar{\mathcal{F}}_0, \mathcal{F}_1$ are oriented or framed, then the statement that they are <u>oriented-concordant</u> or <u>framed-concordant</u> has the obvious meaning. Clearly, if $\bar{\mathcal{F}}_0, \mathcal{F}_1$ are identical foliations possessing homotopic framings, then they are framed-concordant.

E. <u>Examples of foliations</u>. The following important example generalizes the line foliation of the torus described above. Suppose G is a Lie group, H its connected subgroup, and Γ its discrete subgroup. Right cosets of the subgroup H constitute a foliation (in G) which is invariant with respect to left translations by elements of the group G, and, in particular, by elements of the group Γ. Because of this last fact, G/Γ acquires a foliation whose dimension equals the codimension of the group II in G; we denote this by $\mathcal{F}\ (G, H, \Gamma)$. This foliation becomes the torus line foliation if $G = \mathbb{R} \oplus \mathbb{R}, H \cong \mathbb{R}, \Gamma = \mathbb{Z} \oplus \mathbb{Z}$. In the general case, this foliation is (real) analytical; its leaves are diffeomorphic to the quotient spaces $H/(H \cap g^{-1}\Gamma g)$ with $g \in G$; the fibers of the normal bundle of this foliation are canonically isomorphic to the space $\mathfrak{g}/\mathfrak{h}$, where $\mathfrak{g}, \mathfrak{h}$ are the Lie algebras of the groups $G, H,$ and the foliation becomes framed if a basis is fixed in $\mathfrak{g}/\mathfrak{h}$.

This example is perhaps not very typical of classical foliation theory. The list of examples at the disposal of this theory consists mainly of foliations of codimension 1, constructed at different times in order to prove the existence of codimension-1 foliations on some manifold or other. The actual problem of the existence of foliations of codimension 1 is in fact entirely solved: on an open manifold, such a foliation always exists and, in the case of a closed manifold,

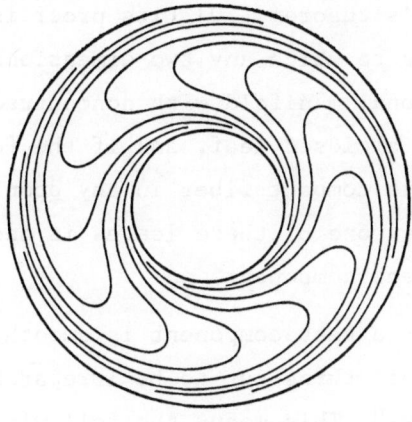

Fig. 8

the vanishing of the Euler characteristic of this manifold,
which is obviously necessary for the existence of a codimen-
sion-1 foliation, is also sufficient. This last fact was
proved by Thurston [90]; however, Thurston's proof can hardly
be called geometrically evident, so that the classical constructions of foliations have not lost their importance. The
best-known one among them is the codimension-1 foliation on
S^3, due to Reeb [75]; it is constructed in the following way.
First consider the foliation of the cylinder $\{(x, y, z) \in$
$\mathbb{R}^3 | x^2 + y^2 \leqslant 1\}$, whose leaves are the boundary of the cylinder
$x^2 + y^2 = 1$ and the surfaces $z = \exp(1 - x^2 - y^2)^{-1} + c$.   The
usual projection of the cylinder on the solid torus $S^1 \times D^2$
supplies the latter with a foliation of codimension 1. This
foliation in $S^1 \times D^2$ is known as a <u>Reeb component</u>; its axial
section is represented in Fig. 8. A foliation in the sphere
S^3 is obtained by the usual pasting together of the sphere
from two solid tori.

Note that all the leaves of the Reeb foliation of S^3,
except one, are diffeomorphic to the plane, while one is diffeomorphic to the torus. In this connection it is appropriate

to mention Novikov's theorem [70] (its proof is also presented
in [88]), according to which any two-dimensional foliation
on a three-dimensional manifold with noncontractible universal
covering possesses a closed leaf, and if the foliation is not
a smooth bundle with compact fiber in any open subset of the
total manifold, then one of these leaves is necessarily a
torus bounding a Reeb component.

Note also that a Reeb component is smooth, but not analytical. This defect turns out to be irreparable because of
"one-sided holonomy." This means the following. If we take
the meridian of the torus which is the leaf of our foliation
and consider the ring $S^1 \times \mathbb{R}$, obtained from the perpendiculars
to the leaf at the points of this meridian, then the foliation
will cut out a spiral on this ring on one side of the torus
and a family of concentric circles on the other side. In the
analytical situation this is of course impossible. In general,
there is a Haefliger theorem according to which there are no
analytical foliations of codimension 1 on closed simply connected manifolds (see [46] and also [70]).

Besides the Reeb construction, there are constructions
of codimension-1 foliations on a whole series of other manifolds, including all odd-dimensional spheres. We refer the
reader to Lawson's review article [59] for the details; here
I shall give an instructive construction of a codimension-1
foliation on S^5, also due to Lawson [58].

The construction is based on the following observation:
if the compact manifold X (with boundary in general) is fibered over the circle, then X possesses a foliation of codimension 1, one of whose leaves is the boundary. To prove
this, it suffices to construct a smooth function $X \to \mathbb{R}_+$,

which equals zero on ∂X, is positive in int X, and has no
critical values less than 1. This function, together with
the projection $X \to S^1$, constitutes the smooth map $X \to$
$S^1 \times \mathbb{R}$, transversal to the foliation pictured in Fig. 9 [in
$S^1 \times (1, \infty)$ the leaves of this foliation coincide with the
lines $t \times (1, \infty)$]. The foliation induced on X will be the
one required. This observation permits us to construct a foliation of codimension 1 on any closed manifold, cut by a submanifold of codimension 1 into two pieces, each of which is
fibered over a circle. But S^5 is so cut: the cut is made
by the surface $|z_1^3 + z_2^3 + z_3^3| = \varepsilon$, where ε is sufficiently
small (we are assuming that $S^5 \subset \mathbb{C}^3$). The piece $|z_1^3 + z_2^3 +$
$z_3^3| \geqslant \varepsilon$ is fibered over S^1 by the projection $(z_1, z_2, z_3) \mapsto$
arg $(z_1^3 + z_2^3 + z_3^3)$ (this is the so-called Milnor fibration), the
piece $|z_1^3 + z_2^3 + z_3^3| \leqslant \varepsilon$ is fibered over a nonsingular cubic
curve in $\mathbb{CP}^2$, which is diffeomorphic to the torus $S^1 \times S^1$.

   2.   The Godbillon–Vey class.   A.   Definition.   The construction below assigns to an arbitrary foliation $\mathcal{F}$ of codimension 1, a three-dimensional real cohomology class of the
total manifold, called, after its authors [36], the Godbillon–
Vey class of the foliation $\mathcal{F}$. For simplicity we begin with
the case when the foliation $\mathcal{F}$ is orientable; then it can be
defined globally by a 1-form ω, which does not vanish, and
this form is determined by the foliation up to multiplication
by a nonvanishing function (see 1C). The form ω satisfies an
integrability condition which is equivalent with the existence
of a 1-form η such that $\eta \wedge \omega = d\omega$. It turns out that: (i)
the form $\eta \wedge d\eta$ is closed; (ii) the cohomology class of this
form is independent of the arbitrariness in the construction
of the forms ω, η, and is determined by the foliation $\mathcal{F}$.

This class is called the <u>Godbillon—Vey class</u> of the foliation $\mathcal{F}$; it is denoted by gv $(\mathcal{F})$.

<u>Proof</u>. (i) Since $\eta \wedge \omega = d\omega$, one has $d(\eta \wedge \omega) = 0$. At the same time, $d(\eta \wedge \omega) = (d\eta \wedge \omega) - (\eta \wedge d\omega) = (d\eta \wedge \omega) - (\eta \wedge \eta \wedge \omega) = d\eta \wedge \omega$, and, therefore, $d\eta \wedge \omega = 0$. Hence $d\eta = \zeta \wedge \omega$, where ζ is a certain 1-form and $d(\eta \wedge d\eta) = d\eta \wedge d\eta = \zeta \wedge \omega \wedge \zeta \wedge \omega = 0$.

<u>Proof of (ii)</u>. If ω' is another form determining $\mathcal{F}$, then $\omega' = f\omega$, where f is a nonvanishing function; therefore,

$$d\omega' = df \wedge \omega + f \, d\omega = (d \log |f| + \eta) \wedge \omega'.$$

If, moreover, $d\omega' = \eta' \wedge \omega'$, then $(\eta' - \eta - d \log |f|) \wedge \omega = 0$ and, therefore, $\eta' - \eta - d \log |f| = g\omega$, where g is a function. Thus, $\eta' = \eta + g\omega + d \log |f|$ and

$$\eta' \wedge d\eta' = (\eta + g\omega + d\log |f|) \wedge (d\eta + dg \wedge \omega + g \, d\omega)$$
$$= \eta \wedge d\eta - dg \wedge \eta \wedge \omega + d \log |f| \wedge d\eta$$
$$+ d \log |f| \wedge dg \wedge \omega + \underline{g \, d} \log |f| \wedge d\omega$$

(we have taken into consideration the fact that $\eta \wedge d\omega = 0$ and $\omega \wedge d\eta = 0$). It remains to note that

$$- dg \wedge \eta \wedge \omega + d \log |f| \wedge d\eta$$
$$+ d \log |f| \wedge dg \wedge \omega + g \, d \log |f| \wedge d\omega$$
$$= -d \, (g \, d\omega + d \log |f| \wedge g\omega + d \log |f| \wedge \eta).$$

In the general case (when the foliation $\mathcal{F}$ is not necessarily orientable), we construct the double cover space over the total manifold — the lifting of the foliation on it will be an orientable foliation. Then we carry out the previous construction on the covering space, imposing a supplementary condition: the form ω must be odd (it must be multiplied by −1 under the action of the nontrivial covering automorphism),

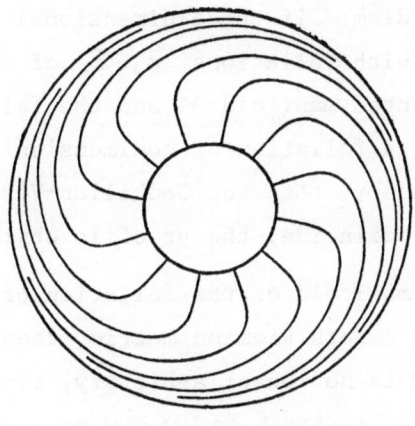

Fig. 9

while the form η must be even. The form $\eta \wedge d\eta$ will also
be even and will determine, therefore, a closed 3-form on the
given total manifold. The cohomology class of the latter is
by definition the Godbillon—Vey class; it is still denoted
by   gv $(\mathcal{F})$. The proof of the fact that it is well defined
repeats the previous definition with the only addition that
the function f may be even and the function g odd.

   B. Additions and commentary. First of all, note that
if $f: Y \to X$ is a smooth map transversal to the foliation $\mathcal{F}$
(of codimension 1) on X, then   gv $(f^*\mathcal{F}) = f^*$ (gv $(\mathcal{F})$). It follows from this obvious observation that concordant foliations
have identical Godbillon—Vey classes. The Godbillon—Vey
class, however, is not a homotopy invariant, as we shall see
below.

   Together with the Godbillon—Vey class, we can define the
Godbillon—Vey number. The latter appears if the codimension-
1 foliation is given in a three-dimensional closed oriented
manifold and equals the value of the Godbillon—Vey class on
the three-dimensional cycle. The Godbillon—Vey number is an

invariant of cobordism: if three-dimensional closed oriented
manifolds X_1, X_2 with foliations $\mathcal{F}_1$, $\mathcal{F}_2$ of codimension 1
both bound an oriented manifold Y and the foliations $\mathcal{F}_1$, $\mathcal{F}_2$
can be extended to a foliation of codimension 1 in Y, trans-
versal to the boundary, then the Godbillon—Vey numbers of the
foliations $\mathcal{F}_1$, $\mathcal{F}_2$ coincide; the proof is obvious.

If the total manifold of the foliation of codimension 1
is supplied with a smooth Riemann metric, then the construction of the form η is no longer arbitrary, since we can im-
pose the conditions $\| \omega \| = 1$, $\langle \eta, \omega \rangle = 0$ on ω and η. This en-
ables us to express the Godbillon—Vey form $\eta \wedge d\eta$ explicitly
in terms of elements of the Riemann metric. For example, if
we are dealing with a foliation on an oriented three-dimensional manifold, then

$$\eta \wedge d\eta = \chi^2 (\tau + l (N, Z)) \, dv,$$

where dv is the volume form, l the second quadratic form of
the leaf, while χ, τ, N, and Z are the curvature, torsion,
normal, and binormal of an integral curve of the vector field
perpendicular to the foliation (this is a result of Reinhart
and Wood [76]).

Nevertheless, despite the simplicity of the definition
of the Godbillon—Vey class and the apparent clarity of the
last formula, the geometric meaning of the Godbillon—Vey class
remains unclarified.

C. Computation for the Reeb foliation. The following
calculation shows that the Reeb foliation on S^3 (see Subsection 1E) has a zero Godbillon—Vey class. Certain details of
this calculation will be used later (in Sub-subsection E).

For $(x_1, x_2, x_3, x_4) \in S^3$, set

$$\varphi = \arg (x_1 + ix_2), \quad \psi = \arg (x_3 + ix_4),$$
$$t = x_1^2 + x_2^2 \ [= 1 - x_3^2 - x_4^2].$$

(Here φ and ψ are defined modulo 2π; φ is not defined at all if $x_1 = x_2 = 0$, and ψ , if $x_3 = x_4 = 0$.) The Reeb component may be defined by the formula

$$\omega = \begin{cases} \alpha (t) \, d\psi + \beta (t) \, dt, & \text{if} \quad t \leqslant 1/2, \\ \alpha (1-t) \, d\varphi + \beta (1-t) \, dt, & \text{if} \quad t \geqslant 1/2, \end{cases}$$

where α, β are certain smooth functions on the closed interval $[0, 1/2]$, possessing the following properties:

$$\alpha^2 + \beta^2 = 1, \quad \alpha (0) = 1, \tag{1}$$
$$\alpha (1/2) = \alpha' (1/2) = \alpha'' (1/2) = \ldots = 0.$$

(The functions α and β may be explicitly calculated by using the definition of the Reeb foliation, but the answer is cumbersome and I will not provide it.) A direct verification shows that $d\omega = \eta \wedge \omega$, where

$$\eta = \begin{cases} \alpha' (t) \, (-\beta (t) \, d\psi + \alpha (t) \, dt), & \text{if} \quad t \leqslant 1/2, \\ \alpha' (1-t) \, (-\beta (1-t) \, d\varphi + \alpha (1-t) \, dt), & \text{if} \quad t \geqslant 1/2, \end{cases}$$

and that $\eta \wedge d\eta \equiv 0$.

Thus, the Godbillon–Vey number is equal to zero for the classical Reeb foliation, as well as for any foliation defined by the form ω with functions α, β, satisfying conditions (1). Actually, all these foliations are cobordant to zero, but the proof of this statement meets with considerable difficulties (see [82]).

D. Nontriviality. Let $G = \mathrm{SL} (2, \mathbb{R})$, and denote by H the subgroup of the group G, consisting of matrices of the form $\left\| \begin{matrix} a & b \\ 0 & a^{-1} \end{matrix} \right\|$ with $a > 0$, and fix a discrete subgroup Γ of the group G with compact quotient space G/Γ (the fact that such a subgroup Γ exists is well known; actually, an explicit construction will be provided further). We shall show that the foliation $\mathcal{F}(G, H, \Gamma)$ in G/Γ (see Subsection 1E) has a nontrivial Godbillon–Vey class.

Denote by ω_{-1}, ω_0, ω_1 the right-invariant forms on SL (2, $\mathbb{R}$), corresponding to the cochains

$$\left\| \begin{matrix} x & y \\ z & -x \end{matrix} \right\| \mapsto z, \quad \left\| \begin{matrix} x & y \\ z & -x \end{matrix} \right\| \mapsto x, \quad \left\| \begin{matrix} x & y \\ z & -x \end{matrix} \right\| \mapsto y$$

of the Lie algebra $\mathfrak{sl}(2, \mathbb{R})$ of matrices with zero trace (see 1 1.3). Obviously, $d\omega_{-1} = 2\omega_{-1} \wedge \omega_0$, $d\omega_0 = \omega_{-1} \wedge \omega_1$, $d\omega_1 = 2\omega_0 \wedge \omega_1$ [since similar relations hold in the cochain complex of the Lie algebra $\mathfrak{sl}(2, \mathbb{R})$ and $\omega_{-1} \wedge \omega_0 \wedge \omega_1$ is a biinvariant volume form]. The forms ω_{-1}, ω_0, ω_1 are obtained by lifting certain forms ω'_{-1}, ω'_0, ω'_1 defined on G/Γ, and, clearly, the form ω'_{-1} is integrable (the relations between the forms ω and their differentials listed above remain valid when we pass to the forms ω') and determines precisely the foliation $\mathcal{F}(G, \ H, \ \Gamma)$ in G/Γ. Since $d\omega'_{-1} = -2\omega'_0 \wedge \omega'_{-1}$, the Godbillon–Vey class is represented by the form $-2\omega'_0 \wedge d(-2\omega'_0) = -4\omega'_{-1} \wedge \omega'_0 \wedge \omega'_1$, which is not exact since $\omega'_{-1} \wedge \omega'_0 \wedge \omega'_1$ is the volume form.

In conclusion, note that for an appropriate choice of the group Γ the previous foliation has a clear geometric description. Suppose P is a surface of genus $\geqslant 2$. It is well known that P possesses a metric of constant negative curvature and that the universal cover over P with this metric is the Lobachevski plane L. The group Γ_0 of automorphisms

of the covering space $L \to P$ is the subgroup of the isometry
group of the Lobachevski plane, i.e., of the group G/Z, where
G has the previous meaning, while $Z(\cong \mathbb{Z}_2)$ is the center of
the group G. For the group Γ we take the inverse image of
the group Γ_0 under the projection $G \to G/Z$. The group G/Z
acts in the manifold $\mathrm{Tang}_1 L$ of unit tangent vectors to L,
and this action is free and transitive. This enables us to
identify G/Z with $\mathrm{Tang}_1 L$, fixing some point in $\mathrm{Tang}_1 L$.
Under this identification, G/Γ becomes $\mathrm{Tang}_1 P$ (the manifold
of unit tangent vectors to P); right cosets of the group H
are transformed by the projection $G \to G/Z = \mathrm{Tang}_1 L$ into the
set of parallel vectors of the Lobachevski plane L, and, under
the projection $G \to G/\Gamma = \mathrm{Tang}_1 P$, into the set of parallel vectors of the surface P (two vectors are called parallel if
the geodesics which they determine asymptotically converge in
the positive direction). In other words, the leaves of the
foliation $\mathcal{F}(G, H, \Gamma)$ and the leaves of the inverse image of
this foliation in G/Z are horocycles of the surface P and
the Lobachevski plane L. Note also that the Riemann metric
determined in $G/\Gamma = \mathrm{Tang}_1 P$ by the invariant metric of the
group G and the Lobachevski metric are proportional, so that
the Godbillon–Vey form of the foliation $\mathcal{F}(G, H, \Gamma)$ computed
above is proportional to the volume form determined in G/Γ
by the Lobachevski metric. The last proportionality coefficient, as shown by a simple calculation, is actually equal
to 1.

E. Variability. Here we present a construction due to
Thurston [89], whose result is a smooth one-parameter family
of codimension-1 foliations on S^3, whose Godbillon–Vey numbers
increase with the increase of the parameter. This shows that

all the foliations of this family are pairwise nondiffeomor-
phic, pairwise nonconcordant, and even pairwise noncobordant.

Suppose $m \geqslant 3$. Choose an arbitrary polygon Q of m
sides on the Lobachevski plane and the symmetric polygon Q^*
with respect to some line. Throw out the vertices from Q
and Q^*, and denote by R, R^* the inverse images of the re-
maining parts of the polygons Q, Q^* under the projection
$\mathrm{Tang}_1 \, L \to L$. In Sub-subsection D, we have constructed a
certain foliation $\mathcal{F}_0$ in $\mathrm{Tang}_1 L$ of codimension 1, its determining form ω_0, and the form η_0 satisfying $\eta_0 \wedge \omega_0 = d\omega_0$
(the notations $\mathcal{F}_0$, ω_0, η_0 were not used there); all these objects are invariant with respect to isometries of the Loba-
chevski plane, and the form $\eta_0 \wedge d\eta_0$ equals the standard volume form on $\mathrm{Tang}_1 L$. Restrict $\mathcal{F}_0$ and ω_0, η_0 to R and R^*
and glue together R with R^*, identifying the component of
the boundary of the manifold R corresponding to the side AB
of the polygon Q with the component of the boundary of the
manifold R^* corresponding to the symmetric side A^*B^* of
the polygon Q^* (A^* is symmetric to A and B^* is symmetric
to B), according to the diffeomorphism determined by the isometry of the Lobachevski plane which sends A into A^* and B
into B^*. Since $\mathcal{F}_0$ and ω_0, η_0 are invariant with respect
to isometry, after a natural smoothing of the constructed man-
ifold $R \cup R^*$, the foliations which exist in R and R^* and
the corresponding forms constitute a certain foliation $\mathcal{F}_1$
and yield certain forms ω_1, η_1. Further, the constructed manifold can be naturally compactified and becomes a manifold
with boundary consisting of m tori (corresponding to the m
vertices of the polygon Q). This compact manifold will be
denoted by X. The foliation $\mathcal{F}_1$ and the forms ω_1, η_1 can
be extended to a foliation $\mathcal{F}_2$ and to the forms ω_2, η_2 in X

such that ω_2 determines $\mathcal{F}_2$, $\eta_2 \wedge \omega_2 = d\omega_2$ and, moreover, the following statements hold: (i) $\mathcal{F}_2$ is transversal to ∂X; (ii) the restriction of the foliation $\mathcal{F}_2$ to each component of the boundary is a line foliation of the torus (the slope of the lines has a simple relationship with the corresponding angle of the polygon Q); (iii) the restriction $\omega_2 \,|\, \partial X$ is closed; (iv) the restriction $\eta_2 \,|\, \partial X$ is trivial; (v) $\int_X \eta_2 \wedge d\eta_2 = 4\pi$ area Q (the latter follows from the calculation of the Godbillon–Vey form carried out in the previous subsection).

Consider the smooth map $f\colon X \to X$, coinciding in some tubular neighborhood U of the boundary ∂X with the natural projection $U \to \partial X$ and mapping $X \setminus U$ diffeomorphically into int X. Obviously, such a map f may be assumed transversal to $\mathcal{F}_2$; set $\mathcal{F}_3 = f^*\mathcal{F}_2$, $\omega_3 = f^*\omega_2$, $\eta_3 = f^*\eta_2$. The foliation $\mathcal{F}_3$ and the forms ω_3, η_3 still have the properties of the foliation $\mathcal{F}_2$ and of the forms ω_2, η_2 and, moreover, the form ω_3 is closed in U, while the form η_3 is trivial in U. Now, fix a smooth function $t\colon X \to I$, equaling 0 on ∂X, greater than $1/2$ on $X \setminus U$, and possessing no singularities on the intersections of the leaves of the foliation $\mathcal{F}_3$ with U; also fix smooth functions $\alpha, \beta\colon I \to I$ satisfying $\alpha^2 + \beta^2 = 1$, $\alpha \,|\, [0,\ 1/4] = 0$, $\alpha \,|\, [1/2,\ 1] = 1$. Then we can define the forms ω_4, η_4 (on X) by setting

$$\omega_4 \,|\, X \setminus U = \omega_3 \,|\, X \setminus U, \quad \eta_4 \,|\, X \setminus U = \eta_3 \,|\, X \setminus U,$$
$$\omega_4 \,|\, U = (\alpha \circ t)\, \omega_3 \,|\, U + (\beta \circ t)\, dt,$$
$$\eta_4 \,|\, U = (\alpha' \circ t)\, [-(\beta \circ t)\, \omega_3 \,|\, U + (\alpha \circ t)\, dt]$$

(the prime denotes derivation). Obviously, the form ω_4 does not vanish anywhere on X and we have $\eta_4 \wedge \omega_4 = d\omega_4$ (the fact that the form $\omega_3 \,|\, U$ is closed is required to check this re-

lation). Clearly, $(\eta_4 \mid U) \wedge d\,(\eta_4 \mid U) = 0,$ so that $\eta_4 \wedge d\eta_4 =$
$\eta_3 \wedge d\eta_3$ on all of X . The form ω_4 determines a foliation
$\mathcal{F}_4$ on X such that every boundary component is a leaf. Let
us paste on m solid tori to X , so as to obtain the sphere S^3 ,
and supplement the foliation $\mathcal{F}_4$ by adding Reeb components
in the solid tori, obtaining a certain foliation $\mathcal{F} = \mathcal{F}\,(Q)$
of S^3 . The forms ω_4, η_4 will be supplemented to obtain forms
ω, η on S^3 by adding the forms described in Sub-subsection
D on the solid tori. It is important that we have $\eta \wedge d\eta = 0$
in the solid tori, so that

$$\int_{S^3} \eta \wedge d\eta = \int_X \eta_4 \wedge d\eta_4 = \int_X \eta_3 \wedge d\eta_3 = \int_X \eta_2 \wedge d\eta_2 = 4\pi \text{ area } Q.$$

Thus, the Godbillon–Vey number of the foliation $\mathcal{F}\,(Q)$ equals
4π area Q . If we now change the polygon Q so that its area
changes from 0 (not included) to its maximal value $\pi\,(m-2)$,
then the Godbillon–Vey number of the foliation $\mathcal{F}\,(Q)$ will
change from 0 to $4\pi^2\,(m-2)$.

   This construction leads to the following question. Does
there exist a family of foliations of codimension 1 on S^3 for
which the Godbillon–Vey number increases unboundedly when the
parameter increases?

   3. Digression: $\mathfrak{g}$-Structures. The notion of $\mathfrak{g}$-struc-
ture, which will be defined below, establishes a relationship
between the Godbillon–Vey class and the cohomology of the Lie
algebra of formal vector fields and defines a large number
of other characteristic classes of foliations.

   A. Definition of $\mathfrak{g}$-structure. Suppose $\mathfrak{g}$ is a Lie algebra. By a $\mathfrak{g}$-structure on a smooth manifold X we mean a
smooth 1-form ω on X with values in $\mathfrak{g}$, satisfying the
"Maurer–Cartan equation"

$$d\omega = -\frac{1}{2}[\omega, \omega].$$

The latter means that for any vector fields ξ_1, ξ_2 on X

$$d\omega\,(\xi_1, \xi_2) = -[\omega\,(\xi_1), \omega\,(\xi_2)].$$

(this definition and terminology can be found, for example, in the Bernstein–Rozenfeld article [5]).

If f is a smooth map of the manifold Y into the manifold X supplied with a $\mathfrak{g}$-structure ω, then $f^*\omega$ is a $\mathfrak{g}$-structure on Y ("the induced $\mathfrak{g}$-structure"). If $\varphi\colon \mathfrak{g} \to \mathfrak{g}_1$ is a homomorphism of one Lie algebra into another, then the 1-form $\varphi \circ \omega$, defined from the $\mathfrak{g}$-structure ω on X, is a $\mathfrak{g}_1$-structure on X. Concordance between $\mathfrak{g}$-structures ω_0, ω_1 on X is defined as a $\mathfrak{g}$-structure Ω on $X \times \mathbb{R}$, such that $i_0^*\Omega = \omega_0$, $i_1^*\Omega = \omega_1$, where i_0, i_1 are defined by the formulas $x \mapsto (x, 0)$, $x \mapsto (x, 1)$. The operations f^* and $\varphi \circ$ described above are obviously invariant with respect to concordance.

B. Examples. Trivial example: If $\mathfrak{g}$ is a one-dimensional commutative Lie algebra, then a $\mathfrak{g}$-structure on X is simply a closed 1-form on X. More serious examples:

1°. If $\mathfrak{g}$ is the Lie algebra of some Lie group G, then a $\mathfrak{g}$-structure on X is none other than a flat connection in the standard trivial G-bundle $X \times G \to X$, i.e., is a right-invariant foliation in $X \times G$ of codimension $\dim G$ transversal to the fibers of the projection $X \times G \to X$. This foliation is easily described directly: since the tangent space $\mathrm{Tang}_{(x,\,g)}\,(X \times G)$ is $\mathrm{Tang}_x\,X \times \mathrm{Tang}_g\,G = \mathrm{Tang}_x\,X \times \mathfrak{g}$ (the identification of $\mathrm{Tang}_g\,G$ with $\mathfrak{g}$ is carried out by means of the differential of right translation), the graph of the (linear) map ω_x: $\mathrm{Tang}_x\, X \to \mathfrak{g}$ is a $\dim X$-dimensional plane in $\mathrm{Tang}_{(x',g)}\, (X \times G)$, transversal to $\mathrm{Tang}_{(x,g)}\, (x \times G)$; the condition $d\omega = -\frac{1}{2}[\omega, \omega]$ means exactly that these planes constitute an integrable system, i.e., a system of tangent planes to a certain foliation; the latter is G-invariant, since the system of planes is G-invariant.

The most important carrier of such a $\mathfrak{g}$-structure is the group G itself; its tangent space is identified with $\mathfrak{g}$ by means of right translations, and the isomorphisms $\mathrm{Tang}_g\, G \to \mathfrak{g}$ which thus arise constitute a 1-form on G with values on $\mathfrak{g}$ satisfying the Maurer–Cartan condition. Then (right) quotient space G/Γ of the group G by any of its discrete subgroups Γ also possesses a canonical $\mathfrak{g}$-structure (the description of the structure repeats the previous one).

2°. Now suppose $\mathfrak{g}$ is the Lie algebra $\mathrm{Vect}\, M$ of smooth vector fields on the manifold M. If M is compact, then $\mathrm{Vect}\, M$ is in a certain sense the Lie algebra of the group $\mathrm{Diff}\, M$ of diffeomorphisms of the manifold M and the previous description of $\mathfrak{g}$-structure retains its meaning. We shall give a simpler description of $\mathrm{Vect}\, M$-structures here, including the noncompact case as well. Namely, a $\mathrm{Vect}\, M$-structure on X is a foliation of codimension $\dim M$ in $X \times M$ transversal to the fibers of the projection $X \times M \to X$.   Indeed, a $\mathrm{Vect}\, M$-structure ω consists of the linear maps ω_x: $\mathrm{Tang}_x\, X \to \mathrm{Vect}\, M$ $(x \in X)$,   i.e., linear maps $\omega_{x,m}$: $\mathrm{Tang}_x\, X \to \mathrm{Tang}_m\, M$ $(x \in X, m \in M)$. The graph of the last map is a $\dim X$-dimensional plane in $\mathrm{Tang}_x\, X \times \mathrm{Tang}_m\, M = \mathrm{Tang}_{(x,m)}\, (X \times M)$ transversal to $\mathrm{Tang}_{(x,m)}\, (x \times M)$, while the condition $d\omega = -\frac{1}{2}[\omega, \omega]$ means exactly that these planes constitute an integrable system.

3°. If we are given a framed foliation of codimension n on X , then a W_n-structure is defined on X canonically up to concordance (here and further in this section we understand W_n to be $\mathbb{R}W_n$). The construction is the following. The trivialization of the normal bundle determines a determining system of forms for the foliation, say $\omega_1, \ldots, \omega_n$. The integrability of this system implies the existence of 1-forms $\omega_{ij}\ (1 \leqslant i, j \leqslant n)$ satisfying

$$d\omega_i = \sum_j \omega_j \wedge \omega_{ji}.$$

Taking differentials in this equality, we see that

$$0 = \sum_j d\omega_j \wedge \omega_{ji} - \sum_j \omega_j \wedge d\omega_{ji} = \sum_j \omega_j \wedge \left(\sum_s \omega_{js} \wedge \omega_{si} - d\omega_{ji}\right),$$

and, since the forms $\omega_1, \ldots, \omega_n$ are linearly independent, there exist forms $\omega_{ji_1i_2}$ satisfying

$$\sum_j \omega_j \wedge \omega_{ji_1i_2} = \sum_j \omega_{i_1j} \wedge \omega_{ji_2} - d\omega_{i_1i_2}.$$

Continuing in the same vein, we obtain a family of 1-forms $\omega_{ji_1\ldots i_s}$ satisfying

$$\sum_j \omega_j \wedge \omega_{ji_1\ldots i_s} = \sum_{p_1,\ldots,p_l} \sum_j \omega_{i_1 p_1 \ldots i_{p_l} j} \wedge \omega_{ji_{q_1}\ldots i_{q_m}} - d\omega_{i_1\ldots i_s}, \qquad (2)$$

where the first sum on the right-hand side is taken over all partitions $\{2,\ldots,s\} = \{p_1,\ldots,p_l\} \cup \{q_1,\ldots,q_m\}$ such that $p_1 < \ldots < p_l$, $q_1 < \ldots < q_m$. Now we can define a 1-form ω on X with values in $W_n = \bigoplus_{j=1}^n \mathbb{R}[[t_1, \ldots, t_n]]\, \partial/\partial t_j$ by setting, for any tangent vector ξ in X,

$$\omega(\xi) = \sum_s \sum_{j,\, i_1,\ldots,i_s} \omega_{ji_1\ldots i_s}(\xi)\, t_{i_1} \ldots t_{i_{s-1}} \frac{\partial}{\partial t_j}.$$

Relation (2) means precisely that $d\omega = -\frac{1}{2}[\omega, \omega]$.   Thus, ω is a W_n-structure.

Let us indicate an important property of this construction. If Y is a submanifold of the manifold X transversal
to the foliation and a W_n-structure has been chosen on Y already,

$$\omega^Y(\xi) = \sum_s \sum_{j,\, i_1,\ldots,\, i_s} \omega^Y_{ji_1\ldots i_s}(\xi)\, t_{i_1} \ldots t_{i_{s-1}} \frac{\partial}{\partial t_j}\,,$$

so that $\omega^Y_j = \omega_j \,|\, Y$, then a W_n-structure on X may be
chosen so as to satisfy the condition $\omega \,|\, Y = \omega^Y$.   Indeed,
assume that we have carried out part of our construction, obtaining $\omega_{ji_1\ldots i_t}$ for $t < s$, and that for $t < s$

$$\omega_{ji_1\ldots i_t} \,|\, Y = \omega^Y_{ji_1\ldots i_t}.$$

Let us find arbitrary solutions $\omega_{ji_1\ldots i_s}$ of Eqs. (2) and then
subtract Eqs. (2) for ω^Y from the restriction to Y of Eqs.
(2) for ω. We obtain

$$\sum_j (\omega_j \,|\, Y) \wedge (\omega_{ji_1\ldots i_s} \,|\, Y - \omega^Y_{ji_1\ldots i_s}) = 0,$$

i.e.,

$$\omega_{ji_1\ldots i_s} \,|\, Y - \omega^Y_{ji_1\ldots i_s} = \sum_k \varphi^Y_k (\omega_k \,|\, Y),$$

where the φ^Y_k are smooth functions on Y (depending on $j, i_1,$
$\ldots, i_s$).   We extend these functions to functions φ_k on X
and replace our form $\omega_{ji_1\ldots i_s}$ by the form

$$\omega_{ji_1\ldots i_s} + \sum_k \varphi_k \omega_k.$$

Adjusting the solution of Eq. (2) in this way at each step,
we obtain a W_n-structure on X with the required properties.

It follows from the above that the previous construction
determines a W_n-structure uniquely up to concordance: if

ω, ω' are two W_n-structures corresponding to the framed foliation $\mathcal{F}$ on X, then the W_n-structure Ω corresponding to
the foliation $\mathcal{F} \times \mathbb{R}$ on $X \times \mathbb{R}$ and coinciding with ω on
$X \times 0$ and with ω' on $X \times 1$ is a concordance between ω and
ω'.

Thus, we have constructed a function which assigns to
every class of framed concordant framed foliations of codimension n on the manifold X a class of concordance W_n-structures
on X, and this function is compatible with the pullback operation of foliations and W_n-structures.

In conclusion we note that the passage from a framed foliation to a W_n-structure may be described in more geometric
terms. To do this, it suffices to fix a smooth map $X \times$
$\mathbb{R}^n \to X$, which sends the space $x \times \mathbb{R}^n$ diffeomorphically,
for every $x \in X$, onto a small transversal surface element
to the foliation at the point x, sending $(x, 0)$ into x and compatible (in an obvious sense) with the framing of the foliation. Then for any point $y \in X$ sufficiently close to x the
leaves of the foliation determine a germ φ_y of a diffeomor-
phism $y \times \mathbb{R}^n \to x \times \mathbb{R}^n$, i.e., $\mathbb{R}^n \to \mathbb{R}^n$, at the point 0. For
$y = x$ this diffeomorphism is the identity. The differential
of the map $y \mapsto \varphi_y$ is a linear map $\mathrm{Tang}_x X \to W_n$, while the
family of all such maps constitutes a 1-form ω on X with
values in W_n; the fact that $d\omega = -\frac{1}{2}[\omega, \omega]$ can be checked
without much difficulty.

C. Characteristic classes of $\mathfrak{g}$-structures. The following construction assigns to each $\mathfrak{g}$-structure ω on X the homomorphism

$$\mathrm{Char}_\omega\colon H^* (\mathfrak{g}) \to H^* (X;\ \mathbb{R}),$$

so that if $f\colon Y \to X$ is a smooth map, then

$$\mathrm{Char}_{f^*\omega} = f^* \circ \mathrm{Char}_\omega,$$

while if $\varphi\colon \mathfrak{g} \to \mathfrak{g}_1$ is a homomorphism of Lie algebras, then

$$\mathrm{Char}_{\varphi \circ \omega} = \mathrm{Char}_\omega \circ \varphi^*.$$

Thus, to an arbitrary $\alpha \in H^*(\mathfrak{g})$ will correspond a function assigning to each $\mathfrak{g}$-structure ω on X the element $\mathrm{Char}_\omega^{\,\cdot}(\alpha)$ of the space $H^q(X;\mathbb{R})$ — "the characteristic class of the $\mathfrak{g}$-structure ω." These characteristic classes are natural with respect to X and to $\mathfrak{g}$ and are identical for concordant $\mathfrak{g}$-structures. Moreover, if $q = \dim X$, then the "characteristic number" — the value of the characteristic class on the fundamental cycle of the manifold X — is invariant with respect to cobordism.

Suppose $\mathfrak{g}$ is a (real) Lie algebra, and suppose the manifold X is supplied with a $\mathfrak{g}$-structure ω. For the cochain $\alpha \in C^q(\mathfrak{g})$ the formula

$$(v_1, \ldots, v_q) \mapsto \alpha(\omega(v_1), \ldots, \omega(v_q)),$$

in which $v_1, \ldots, v_q$ are tangent vectors to the manifold X at the same point, determines a certain q-form $\omega(\alpha)$ on X.

Lemma. The map $\alpha \mapsto \omega(\alpha)$ commutes with the differential

$$\omega(d\alpha) = d\omega(\alpha).$$

Proof. The condition $d\omega = -\frac{1}{2}[\omega, \omega]$ means that for any vector fields ξ, η on X we have

$$\omega([\xi, \eta]) - \xi\omega(\eta) - \eta\omega(\xi) = [\omega(\xi), \omega(\eta)]$$

(on the left- and right-hand sides of this relation we have functions on X with values in $\mathfrak{g}$). Therefore, for any vector fields $\xi_1, \ldots, \xi_{q+1}$ on X,

$$[d\omega\,(\alpha)]\,(\xi_1,\ldots,\xi_{q+1})$$
$$= \sum_{1\leqslant s<t\leqslant q+1}(-1)^{s+t-1}[\omega\,(\alpha)]\,([\xi_s,\xi_t],\xi_1,\ldots\hat{\xi}_s\ldots\hat{\xi}_t\ldots,\xi_{q+1})$$
$$-\sum_{1\leqslant s\leqslant q+1}(-1)^s\xi_s\,[\omega\,(\alpha)]\,(\xi_1,\ldots\hat{\xi}_s\ldots,\xi_{q+1})$$
$$=\sum_{1\leqslant s<t\leqslant q+1}(-1)^{s+t-1}\alpha\,(\omega\,([\xi_s,\xi_t]),\,\omega\,(\xi_1),\ldots$$
$$\ldots\omega\,(\hat{\xi}_s)\ldots\omega\,(\hat{\xi}_t)\ldots,\,\omega\,(\xi_{q+1}))-\sum_{s<t}(-1)^s\alpha\,(\omega\,(\xi_1),\ldots\omega\,(\hat{\xi}_s)$$
$$\ldots,\,\omega\,(\xi_{t-1}),\,\xi_s\omega\,(\xi_t),\,\omega\,(\xi_{t+1}),\ldots,\,\omega\,(\hat{\xi}_{q+1}))$$
$$-\sum_{s>t}(-1)^s\alpha\,(\omega\,(\xi_1),\ldots,\,\omega\,(\xi_{t-1}),\,\xi_s\omega\,(\xi_t),\,\omega\,(\xi_{t+1}),$$
$$\ldots\omega\,(\hat{\xi}_s)\ldots,\,\omega\,(\xi_{q+1}))$$
$$=\sum_{1\leqslant s<t\leqslant q+1}(-1)^{s+t-1}\alpha\,([\omega\,(\xi_s),\,\omega\,(\xi_t)],\,\omega\,(\xi_1),\ldots\omega\,(\hat{\xi}_s)$$
$$\ldots\omega\,(\hat{\xi}_t)\ldots,\,\omega\,(\xi_{q+1}))=[\omega\,(d\alpha)]\,(\xi_1,\ldots,\xi_{q+1}).$$

Thus, we have defined a homomorphism of the complex $C^{\cdot}\,(\mathfrak{g})$ into the de Rham complex $\Omega^{\cdot}(X)$ of the manifold X. The corresponding induced homomorphism $H^*\,(\mathfrak{g})\to H^*\,(X,\mathbb{R})$ is by definition Char_ω.

The homomorphism Char_ω is natural (in the sense described above) with respect to continuous maps $X\to Y$ and to homomorphisms $\mathfrak{g}\to\mathfrak{g}_1$. The first naturality property implies that characteristic classes of $\mathfrak{g}$-structures are invariant with respect to concordance, while characteristic numbers are invariant with respect to cobordism.

Example. If $\mathfrak{g}$ is the Lie algebra of a Lie group G, then characteristic classes of $\mathfrak{g}$-structures have the following simple description. As we know, a $\mathfrak{g}$-structure on X is in this case an invariant foliation of codimension dim G in $X\times G$ transversal to the fibers of the projection $X\times G\to X$. The leaves of this foliation determine (locally) maps $X\to G$, which differ only by right translations of the group G on the intersections of their domains of definition. A cochain of the Lie algebra $\mathfrak{g}$ is a right-invariant form on G.

Carrying over this form into X by means of these local maps we obtain a single form over all of X. The corresponding map $C^*(\mathfrak{g}) \to \Omega^*(X)$ obviously commutes with differentials, so that a homomorphism in cohomology is defined: $H^*(\mathfrak{g}) \to H^*(X; \mathbb{R})$. This is the homomorphism defining characteristic classes of $\mathfrak{g}$-structures.

For example, for a canonical $\mathfrak{g}$-structure on G/Γ, where Γ is a discrete subgroup of the group G, characteristic classes are determined by the canonical homomorphism $H^*(\mathfrak{g}) \to H^*(G/\Gamma; \mathbb{R})$, which sends the class of the cochain $\alpha \in C^q(\mathfrak{g})$ into the class of the q-form on G/Γ, which, when lifted to G, becomes the right-invariant form corresponding to α.

D. The relative case. Assume that $\mathfrak{h}$ is a finite-dimensional subalgebra of the algebra $\mathfrak{g}$, corresponding to the Lie group H, and that each canonical action of the algebra $\mathfrak{h}$ in $\mathfrak{g}$ is the differential of some action of the group H and $\mathfrak{g}$ extending the canonical action of H in $\mathfrak{h}$ (compare with Subsection 1.3.1). We define a $(\mathfrak{g}, H)$-structure on the manifold X as a pair consisting of a principal H-bundle $p: \tilde{X} \to X$ and a $\mathfrak{g}$-structure $\tilde{\omega}$ on $\tilde{X}$ and possessing the following properties: (i) $\tilde{\omega}(h\tilde{v}) = h(\tilde{\omega}\tilde{v})$ for $h \in H$, $\tilde{v} \in \mathrm{Tang}.\tilde{\Gamma}$; (ii) on the kernel of the differential $dp: \mathrm{Tang}_{\tilde{x}} \tilde{X} \to \mathrm{Tang}_{p(\tilde{x})} X$ $(\tilde{x} \in \tilde{X})$ the form $\tilde{\omega}$ coincides with the canonical isomorphism between this kernel and $\mathfrak{h}$. In the case $\mathfrak{h} = \mathfrak{g}$ this definition coincides with the definition of connection in the bundle p.

Suppose $\omega = (\tilde{\omega}, p: \tilde{X} \to X)$ is a $(\mathfrak{g}, H)$-structure on X. It is easy to check that for any cochain $\alpha \in C^q(\mathfrak{g}, H)$ the form $\tilde{\omega}(\alpha)$ is the lifting of a certain form $\omega(\alpha) \in \Omega^q(X)$; the assignment $\alpha \mapsto \omega(\alpha)$ commutes with differentials and determines a homomorphism

$$H^q(\mathfrak{g}, H) \to H^q(X; \mathbb{R})$$

whose image, by definition, consists of characteristic classes of the $(\mathfrak{g}, H)$-structure ω. The formal properties of these classes resemble the corresponding properties of characteristic classes of $\mathfrak{g}$-structures, and we shall not list them.

4. **General theory of characteristic classes of foliations.** The generalization of the Godbillon–Vey class to foliations of arbitrary codimension were independently carried through by Bernstein and Rozenfeld [4, 5] and Bott and Haefliger [8]. The construction described below resembles that of Bernstein–Rozenfeld.

A. **The framed case.** Comparing 3B and 3C, we see that for a manifold X supplied with a framed foliation $\mathcal{F}$ of codimension n, the homomorphism

$$\mathrm{Char}_{\mathcal{F}}: H^*(W_n) \to H^*(X;\; \mathbb{R})$$

is defined and satisfies $\mathrm{Char}_{f*\mathcal{F}} = f^* \circ \mathrm{Char}_{\mathcal{F}}$ for any smooth map $f: X \to Y$, transversal to $\mathcal{F}$. Thus, to every element of the space $H^q(W_n)$ corresponds a q-dimensional characteristic class of framed foliations of codimension n. The considerations developed in Subsection 3 make the computation of these classes possible in principle for framed foliations given by the determining system of forms $\omega_1, \ldots, \omega_n$. To do this, it suffices to find forms ω_{ji} satisfying the condition $d\omega_i = \sum_j \omega_j \wedge \omega_{ji}$, and the subsequent forms $\omega_{ji_1 \ldots i_s}$, satisfying condition (2) from Subsection 3B; then the characteristic class corresponding to the cohomology class of the cocycle $\alpha \in C^q(W_n)$ is represented by the differential form $v_1, \ldots, v_q \mapsto \alpha(\xi_1, \ldots, \xi_q)$, where

$$\xi_i = \sum_{j,\, i_1, \ldots, i_s} \omega_{ji_1 \ldots i_s}(v_i)\, t_{i_1} \ldots t_{i_s} \frac{\partial}{\partial t_j}\, .$$

This procedure may be simplified by using Theorem 2.2.4'.
Namely, in view of this theorem, the complex $C^{\cdot}(W_n)$ is homotopy equivalent to its subcomplex which is multiplicatively
generated by the cochains

$$\alpha_{ji} \in C^1(W_n), \quad \alpha_{ji}\left(\sum a_{ji_1\ldots i_s} t_{i_1}\ldots t_{i_s}\frac{\partial}{\partial t_j}\right) = a_{ji}$$

and their differentials. If our element of the space $H^q(W_n)$
is represented by the cocycle

$$\sum_{i,j,k,l} a_{j_1\ldots j_r i_1\ldots i_r l_1\ldots l_s k_1\ldots k_s}\alpha_{j_1 i_1}\ldots\alpha_{j_r i_r}d\alpha_{l_1 k_1}\ldots d\alpha_{l_s k_s},$$

then the corresponding characteristic class of our framed foliation is represented by the form

$$\sum_{i,j,k,l} a_{j_1\ldots j_r i_1\ldots i_r l_1\ldots l_s k_1\ldots k_s}\omega_{j_1 i_1}\wedge\cdots\wedge\omega_{j_r i_r}\wedge d\omega_{l_1 k_1}\wedge\cdots\wedge d\omega_{l_s k_s}.$$

We see that in order to compute the characteristic class of
the framed foliation it is not necessary to determine all the
forms constituting the corresponding W_n-structure: the forms
ω_{ji} alone suffice.

For example, $H^q(W_1) = 0$ for $q \neq 0, 3$ and $H^3(W_1) = \mathbb{R}$
(see Subsection 2.2.3A). Thus, for a framed, i.e., oriented,
foliation of codimension 1, our construction yields the only
characteristic class of positive dimension and this dimension
equals 3. An explicit computation of the cocycle representing
the nontrivial element from $H^3(W_1)$ (see 2.2.3D) shows that
this characteristic class is represented by the form $\omega_{11}\wedge$
$d\omega_{11}$. We see that this is the Godbillon–Vey class.

To framed codimension-2 foliations, our construction assigns five characteristic classes of dimensions 5, 5, 7, 8, 8 (again, compare 2.2.3A). A direct computation of the cocycles representing the cohomology of the algebra W_2 allows us to write out the forms representing these classes:

$$d\,(\omega_{11} + \omega_{22})^2 \wedge (\omega_{11} + \omega_{22});$$
$$(d\omega_{11} \wedge d\omega_{22} - d\omega_{12} \wedge d\omega_{21}) \wedge (\omega_{11} + \omega_{22});$$
$$(d\omega_{11} \wedge d\omega_{22} - d\omega_{12} \wedge d\omega_{21}) \wedge (\omega_{11} - \omega_{22}) \wedge \omega_{12} \wedge \omega_{21};$$
$$(d\omega_{11} \wedge d\omega_{22} - d\omega_{12} \wedge d\omega_{21}) \wedge \omega_{11} \wedge \omega_{12} \wedge \omega_{21} \wedge \omega_{22};$$
$$d\,(\omega_{11} + \omega_{22})^2 \wedge \omega_{11} \wedge \omega_{12} \wedge \omega_{21} \wedge \omega_{22}$$

(recall that the forms ω_{ji} are defined by the relation $d\omega_i = \sum_j \omega_j \wedge \omega_{ji}$).

Framed foliations of arbitrary codimension n possess characteristic classes of dimensions from $2n + 1$ to $n\,(n + 2)$; among them for any n the classes of the forms

$$\left(\sum_i d\omega_{ii}\right)^n \wedge \sum_i \omega_{ii},$$
$$\left(\sum_i d\omega_{ii}\right)^n \wedge \left(\bigwedge_{i,j} \omega_{ij}\right)$$

[of degrees $2n + 1$ and $n\,(n + 2)$] are present; each of these classes generalizes the Godbillon–Vey class.

B. The general case. Suppose $\mathscr{F}$ is an arbitrary foliation of codimension n on the manifold X, and let $p: \tilde{X} \to X$ be the principal $O(n)$-bundle associated with the bundle norm $\mathscr{F}$ supplied in an arbitrary metric. The foliation $p^*\mathscr{F}$ on $\tilde{X}$ possesses a canonical framing (a point of the space $\tilde{X}$ is a pair consisting of the point $x \in X$ and the linear isomorphism $\lambda: \mathrm{norm}_x\,\mathscr{F} \to \mathbb{R}^n$, where $\mathrm{norm}_x\,\mathscr{F}$ is the fiber of the bundle norm $\mathscr{F}$ over x; the compositions of maps

$$\mathrm{tang}_{(x,\,\lambda)}\tilde{X} \xrightarrow{\,dp\,} \mathrm{tang}_x X \xrightarrow{\,\mathrm{pr}\,} \mathrm{norm}_x\,\mathscr{F} \xrightarrow{\,\lambda\,} \mathbb{R}^n$$

constitute a $\mathbb{R}^n$-valued 1-form on $\tilde{X}$, i.e., n ordinary 1-forms on $\tilde{X}$; these 1-forms determine $p^*\mathcal{F}$ and constitute the framing of $p^*\mathcal{F}$). Clearly, the W_n-structure corresponding to the framed foliation $p^*\mathcal{F}$, together with the bundle p: $\tilde{X} \to X$, constitute a $(W_n, O(n))$-structure. We obtain the homomorphism

$$\text{Char}_{\mathcal{F}}: H^*(W_n, O(n)) \to H^*(X; \mathbb{R}),$$

which transforms elements of the space $H^*(W_n, O(n))$ into characteristic classes of nonframed foliations of codimension n. We know the cohomology $H^*(W_n, O(n))$ (see Theorem 2.2.6). In particular, $H^*(W_1, O(1)) = H^*(W_1)$, i.e., nonframed foliations of codimension 1 also have a unique characteristic class; this is again the Godbillon–Vey class. Nonframed foliations of codimension 2 have three characteristic classes of dimension 4, 5, 5; the first of these is the real Pontryagin class of the bundle norm $\mathcal{F}$.

In general, real Pontryagin classes of the bundle norm $\mathcal{F}$ are part of the characteristic classes of a nonframed foliation $\mathcal{F}$. The corresponding elements of the space $H^*(W_n, O(n)) = H^*(X_n/O(n); \mathbb{R})$ (see Theorem 2.2.6) are simply real Pontryagin classes of the $O(n)$-bundle $\text{pr}: X_n \to X_n/O(n)$, i.e., images of even real universal Chern classes under the projection $X_n/O(n) \to \text{sk}_{2n}BU(n) \subset BU(n)$. Incidentally, comparing the last remark with the fact that $H^q(\text{sk}_{2n}BU(n); \mathbb{R}) = 0$ for $q > 2n$, we obtain the following well-known Bott theorem: polynomials in real Pontryagin classes of the bundle norm $\mathcal{F}$ of degree $> 2 \operatorname{codim} \mathcal{F}$ equal 0.

Let us add that, similarly to the characteristic classes considered above, we can define characteristic classes of foliations supplied with transversal structures of the most varied kind. For example, elements of the space $H^*(W_n,$ SO $(n)) = H^*(W_n, \mathfrak{o}(n))$ are the characteristic classes of oriented foliations of codimension n, while elements of the space $H^*(\mathbb{C}W_n)$ are the characteristic classes of complex folia- tions of codimension n.

C. Characteristic classes of foliated bundles. Since a Vect M-structure on the manifold X is a foliation of co- dimension dim M on $X \times M$, transversal to the fibers of the projection $X \times M \to X$, elements of the space H^q (Vect M) may be viewed as characteristic classes of such foliations with values in the q-dimensional cohomology of the space X. These classes, however, do not give any new invariants by com- parison with the ordinary characteristic classes of foliations. For example, elements of the space H^*(Vect S^1) correspond to characteristic classes of foliations of codimension 1 in the space $X \times S^1$ transversal to the circles — fibers of the pro- jection $X \times S^1 \to X$. The ring H^* (Vect S^1) is a free skew- commutative algebra with two generators in dimensions 2 and 3 (see Theorem 2.4.2). The corresponding characteristic class- es assign to foliations $\mathcal{F}$ in $X \times S^1$ elements of the spaces $H^2(X; \mathbb{R})$, $H^3(X; \mathbb{R})$, and are obtained from gv $(\mathcal{F}) \in H^3(X \times S^1; \mathbb{R})$ by integration over fibers and restriction to $X = X \times 1 \subset X \times S^1$.

5. The variability of characteristic classes. As we saw in Subsection 2E, if foliations of codimension 1 are grouped together in a one-parameter family, the Godbillon—Vey class may depend nontrivially on this parameter. The object of study of the present subsection is the range of other characteristic classes of $\mathfrak{g}$-structures.

The main ideas of this subsection are contained in the article by Gelfand, Feigin, and myself [34], mentioned previously (see Subsection 2.2.5). Another exposition of the theory of variation of characteristic classes of foliations and some other $\mathfrak{g}$-structures is contained in the works of Kamber and Tondeur (see, e.g., [56]).

A. General theory. Assume that the manifold X is supplied with a one-parameter family of $\mathfrak{g}$-structures ω_t $(t \in \mathbb{R})$, which depends smoothly on the parameter t. Then to each tangent vector $v \in \operatorname{Tang} X$ corresponds a one-parameter family of elements of the algebra $\mathfrak{g}$, i.e., a smooth map $\mathbb{R} \to \mathfrak{g}$. Obviously such maps constitute a Lie algebra themselves. We denote the latter algebra by $\tilde{\mathfrak{g}}$ (this is the "current algebra" that we know) and remark that a smooth one-parameter family of $\mathfrak{g}$-structures is a $\tilde{\mathfrak{g}}$-structure. In principle we can apply the general theorem from Subsection 3 to $\tilde{\mathfrak{g}}$-structures, i.e., elements of the space $H^*(\tilde{\mathfrak{g}})$ may be interpreted as characteristic classes of $\tilde{\mathfrak{g}}$-structures (= smooth one-parameter families of $\mathfrak{g}$-structures). However, the space $H^*(\tilde{\mathfrak{g}})$ is very large; therefore, we shall single out certain especially important classes among the characteristic classes of $\tilde{\mathfrak{g}}$-structures.

First of all, note that to each $\tilde{\mathfrak{g}}$-structure $\{\omega_t\}$ corresponds the $\mathfrak{g}$-structure ω_0 (we can take ω_{t_0} for any $t_0 \in \mathbb{R}$); therefore, each characteristic class of the $\mathfrak{g}$-structure is a characteristic class of the $\tilde{\mathfrak{g}}$-structure. This correspondence comes from the map $H^*(\mathfrak{g}) \to H^*(\tilde{\mathfrak{g}})$, induced by the homomorphism $\tilde{\mathfrak{g}} \to \mathfrak{g}$, sending each map $\varphi: \mathbb{R} \to \mathfrak{g}$ into the element $\varphi(0)$ of the algebra $\mathfrak{g}$. Actually, $H^*(\mathfrak{g})$ is a direct summand in the space $H^*(\tilde{\mathfrak{g}})$: the projection $H^*(\tilde{\mathfrak{g}}) \to H^*(\mathfrak{g})$ is induced by the map $\mathfrak{g} \to \tilde{\mathfrak{g}}$, sending the element g of the algebra $\mathfrak{g}$ into the constant map $\mathbb{R} \to \mathfrak{g}$ with value g.

More serious examples of characteristic classes of $\tilde{\mathfrak{g}}$-structures (particularly important for the problem of variability) are the derivatives of the characteristic classes of $\mathfrak{g}$-structures with respect to the parameter. In more detail, suppose $\alpha \in H^q(\mathfrak{g})$. If $\{\omega_t\}$ is a $\tilde{\mathfrak{g}}$-structure on the manifold X, then $\mathrm{Char}_{\omega_t}(\alpha)$ is a smooth one-parameter family of elements of the space $H^q(X; \mathbb{R})$, and we can compute the derivative $\frac{d}{dt}\mathrm{Char}_{\omega_t}(\alpha)|_{t=0}$; this is also an element of the space $H^q(X; \mathbb{R})$, and the function which sends this element into the $\tilde{\mathfrak{g}}$-structure $\{\omega_t\}$ is a characteristic class of $\tilde{\mathfrak{g}}$-structures. It is easy to see that this characteristic class corresponds to a definite element of the space $H^q(\tilde{\mathfrak{g}})$, and this element is determined as follows. Suppose $\gamma \in C^q(\mathfrak{g})$ is an arbitrary cochain. Define the cochain $\tilde{\gamma} \in C^q(\tilde{\mathfrak{g}})$ by the formula

$$\tilde{\gamma}(\varphi_1, \ldots, \varphi_q) = \sum_{i=1}^{q} \gamma\left(\varphi_1(0), \ldots, \varphi_{i-1}(0), \frac{d}{dt}\varphi_i(t)\Big|_{t=0}, \varphi_{i+1}(0), \ldots, \varphi_q(0)\right)$$

$[\varphi_1, \ldots, \varphi_q \in \tilde{\mathfrak{g}}]$. The assignment $\gamma \mapsto \tilde{\gamma}$ defines a homomorphism $C^q(\mathfrak{g}) \to C^q(\tilde{\mathfrak{g}})$; such homomorphisms are defined for all q and, as shown by a simple verification, commute with differentials; therefore, we get the homomorphism $H^q(\mathfrak{g}) \to H^q(\tilde{\mathfrak{g}})$, which we shall denote by Var. Obviously, the characteristic class $\{\omega_t\} \mapsto \frac{d}{dt}\mathrm{Char}_{\omega_t}(\alpha)|_{t=0}$ corresponds precisely to the class $\mathrm{Var}\,\alpha \in H^q(\tilde{\mathfrak{g}})$. Thus, to vary the characteristic class of the $\mathfrak{g}$-structure corresponding to α (i.e., for the existence of a smooth one-parameter family of $\mathfrak{g}$-structures with variable values of the characteristic class corresponding to α) it is necessary that the class $\mathrm{Var}\,\alpha$ be nonzero. In other words, the kernel of the homomorphism Var consists of nonvarying ("rigid") classes.

The homomorphism Var is not very convenient to us because we do not know the cohomology of the algebra $\tilde{\mathfrak{g}}$. In order to overcome this difficulty, note that Var can be represented as the composition

$$H^q(\mathfrak{g}) \xrightarrow{\text{var}} H^{q-1}(\mathfrak{g};\, \mathfrak{g}') \to H^q(\tilde{\mathfrak{g}}),$$

where var is the homomorphism defined in 2.2.6A and the second arrow denotes the homomorphism defined on the cochain level by the formula

$$\gamma \mapsto \{(\varphi_1, \ldots, \varphi_q) \mapsto \sum_{i=1}^{q} [\gamma(\varphi_1(0), \ldots \hat{\varphi}_i(0) \ldots, \varphi_q(0))]\left(\frac{d}{dt}\varphi_i(t)\Big|_{t=0}\right)\}$$

$[\gamma \in C^{q-1}(\mathfrak{g};\, \mathfrak{g}');\ \varphi_1, \ldots, \varphi_q \in \tilde{\mathfrak{g}}]$. The last homomorphism is a monomorphism and its image is a direct summand: the right inverse homomorphism $H^q(\tilde{\mathfrak{g}}) \to H^{q-1}(\mathfrak{g};\, \mathfrak{g}')$ is defined on the cochain level by the formula

$$\beta \mapsto \{(g_1, \ldots, g_{q-1}) \mapsto \{g \mapsto \beta(\bar{g}_1, \ldots, \bar{g}_{q-1}, \overline{tg})\}\},$$

where $\beta \in C^q(\mathfrak{g})$; $g_1, \ldots, g_{q-1}, g \in \mathfrak{g}$; we denote by $\bar{g}_1, \ldots, \bar{g}_{q-1}$ the constant mapping $\mathbb{R} \to \mathfrak{g}$ with values $g_1, \ldots, g_{q-1}$; we denote by $\overline{tg}$ the map $\mathbb{R} \to \mathfrak{g}$, which carries t to tg. [In fact, the direct summand $H^{q-1}(\mathfrak{g};\, \mathfrak{g}')$ of the space $H^q(\tilde{\mathfrak{g}})$ does not intersect the preceding direct summand $H^q(\mathfrak{g})$, so that $H^q(\tilde{\mathfrak{g}}) = H^q(\mathfrak{g}) \oplus H^{q-1}(\mathfrak{g};\, \mathfrak{g}') \oplus \ldots$; moreover, one can complete this direct decomposition to a direct decomposition

$$H^q(\tilde{\mathfrak{g}}) = H^q(\mathfrak{g}) \oplus H^{q-1}(\mathfrak{g};\, \mathfrak{g}') \oplus H^{q-2}(\mathfrak{g};\, \Lambda^2\mathfrak{g}') \oplus \ldots \oplus H^0(\mathfrak{g};\, \Lambda^q\mathfrak{g}') \oplus \ldots;$$

of course these remarks are of no value in what follows.] Thus, the characteristic classes of a $\tilde{\mathfrak{g}}$-structure, defined as derivatives with respect to the parameter of the characteristic classes of the g-structure, lie in $H^{q-1}(\mathfrak{g}; \mathfrak{g}')$, and the passage from a characteristic class of a g-structure to its derivative with respect to the parameter is effected by the homomorphism var: $H^q(\mathfrak{g}) \to H^{q-1}(\mathfrak{g}; \mathfrak{g}')$. In particular, for the variability of the class corresponding to $\alpha \in H^q(\mathfrak{g})$, it is necessary that the class var α be nonzero, and the kernel Ker var consist of nonvarying (rigid) classes.s.

B. The case of finite-dimensional algebras. This case has not been considered much from the point of view of the general theory described: for example, for any semisimple Lie algebra. $\mathfrak{g}$ the cohomology $H^*(\mathfrak{g}; \mathfrak{g}')$ is trivial (see [94] and §2.1) so that for a semisimple algebra $\mathfrak{g}$ all the characteristic classes of g-structures are rigid. The simplest example of a nontrivial homomorphism var is given by the (commutative) algebra $\mathbb{R}$: obviously, $\mathbb{R}'$ is a trivial $\mathbb{R}$-module so that

$$H^q(\mathbb{R}; \mathbb{R}') = H^q(\mathbb{R}) = \begin{cases} \mathbb{R}, & \text{if } q = 0, 1, \\ 0 & \text{otherwise}; \end{cases}$$

as shown by a lightning-like computation, var: $H^1(\mathbb{R}) \to H^0(\mathbb{R}; \mathbb{R})$ is an isomorphism [exercise: var: $H^1(\mathfrak{g}) \to H^0(\mathfrak{g}; \mathfrak{g}')$ is always an isomorphism]. Recall that an $\mathbb{R}$-structure on a manifold is a closed 1-form. Obviously, the characteristic class of $\mathbb{R}$-structure corresponding to the natural generator of the space $H^1(\mathbb{R})$ is simply the cohomology class of this form; there is no doubt that this class varies. A more complicated example:

$$H^*(\mathfrak{gl}(n, \mathbb{R}); \mathfrak{gl}(n, \mathbb{R})') = H^*(\mathfrak{gl}(n, \mathbb{R})) \otimes \text{Inv}(\mathfrak{gl}(n, \mathbb{R}))' = H^*(\mathfrak{gl}(n, \mathbb{R}))$$

[the trace is the only invariant functional on $\mathfrak{gl}(n, \mathbb{R})$ up to a constant factor]. The homomorphism var: $H^q(\mathfrak{gl}(n, \mathbb{R})) \to$

$H^{q-1}(\mathfrak{gl}(n, \mathbb{R}); \mathfrak{gl}(n, \mathbb{R})') = H^{q-1}(\mathfrak{gl}(n, \mathbb{R}))$ can be written in terms of the canonical multiplicative basis $\varphi_1, \ldots, \varphi_n$ of the ring $H^*(\mathfrak{gl}(n, \mathbb{R}))$ by using the formula (in which $i_1 < \ldots < i_s$)

$$\mathrm{var}(\varphi_{i_1} \ldots \varphi_{i_s}) = \begin{cases} \varphi_{i_2} \ldots \varphi_{i_s}, & \text{if } i_1 = 1, \\ 0, & \text{if } i_1 > 1. \end{cases}$$

The fact that characteristic classes of $\mathfrak{gl}(n, \mathbb{R})$-structures corresponding to elements of the ring $H^*(\mathfrak{gl}(n, \mathbb{R}))$, not belonging to the kernel of the homomorphism var, actually do vary has a proof, which it will be useful for the reader to find as an exercise.

C. The case of framed foliations. Since the homomorphism var: $H^q(W_n) \to H^{q-1}(W_n; W_n')$ has already been calculated (Theorem 2.2.11) it remains to state the result.

Theorem 3.1.1. If the class $\alpha \in H^q(W_n)$ belongs to the image of the homomorphism $H^q(W_{n+1}) \to H^q(W_n)$ induced by the inclusion $W_n \to W_{n+1}$, then the characteristic class of framed foliations of codimension n corresponding to α is rigid.

It should be mentioned that Theorem 3.1.1 actually does not require the general theory from Sub-subsection A for its proof. It was first proved by Heitsch [50] and then proved again many times by different authors. Here is a very simple proof, due to Bernstein and Rozenfeld [5]. Suppose $\{\mathcal{F}_t\}$ is a smooth one-parameter family of framed foliations of codimension n on X. Together they constitute a framed foliation of codimension $n+1$ on $X \times \mathbb{R}$ (the leaves of the new foliation are sets of the form $F_t \times t$, where F_t are leaves of the foliation $\mathcal{F}_t$). Obviously, the foliation $\mathcal{F}_t$ may be represented by a family $\{\omega_t\}$ of W_n-structures smoothly depending on t. It is easy to check that the 1-form

$$\omega = \omega_{x_{n+1}} + \left(\frac{\partial}{\partial t_{n+1}}\right) dx_{n+1}$$

on $X \times \mathbb{R}$ with values in W_{n+1} [by t_{n+1} we denote the $(n + 1)$-st coordinate in the space whose formal vector fields constitute the algebra W_{n+1}; by x_{n+1} we denote the coordinate in $X \times \mathbb{R}$ along the factor $\mathbb{R}$] determines precisely the framed foliation $\mathcal{F}$ on $X \times \mathbb{R}$. Denote by η the inclusion $W_n \to W_{n+1}$ and by i_t the inclusion $X \to X \times \mathbb{R}$, defined by the formula $x \mapsto (x, t)$. Obviously, $i_t^* \omega = \eta \circ \omega_t$. It follows that if $\alpha \in H^*(W_{n+1})$, then the value of the characteristic class corresponding to $\eta^* \alpha$ on the foliation $\mathcal{F}_t$ equals $i_t^* \mathrm{Char}_\omega(\alpha)$; this last expression does not depend on t, since all the maps i_t are homotopic.

The following proposition will be harder to prove without using the general theory of Sub-subsection A (and the computations from 2.2.6).

Theorem 3.1.2. Suppose α, $\beta \in \bigoplus_{q>0} H^q(W_n)$, and let $\mathcal{F}_t$ be a smooth family of framed foliations of codimension n on some manifold X. Suppose, further, α_t, β_t are the values of the characteristic classes corresponding to α, β on $\mathcal{F}_t$. Then

$$\alpha_0\left(\frac{d}{dt}\beta_t\Big|_{t=0}\right) = 0.$$

This immediately follows from the triviality of the $H^*(W_n)$-module structure in $H^*(W_n; W_n')$ (see Theorem 2.2.10).

D. The case of codimension-1 foliations. I do not have any meaningful additions to the general theory corresponding to the case of codimension 1; simply, in this case, the theory

may be developed along elementary lines, like the Godbillon–
Vey theory.

Suppose ω is a nonvanishing integrable 1-form on the manifold X. Suppose that we are given a deformation of this form
in the class of 1-forms; this deformation will be considered
in the infinitesimal first-order neighborhood of the trivial
value of the parameter; in other words, we write

$$\omega_t = \omega + t\alpha + \ldots,$$

where the dots stand for the summand $o(t)$ which we ignore.
The integrability of this form means that

$$0 = \omega_t \wedge d\omega_t = (\omega + t\alpha + \ldots) \wedge d(\omega + t\alpha + \ldots)$$
$$= \omega \wedge d\omega + t(\omega \wedge d\alpha + \alpha \wedge d\omega) + \ldots,$$

i.e., that $\omega \wedge d\omega = 0$ (which we already know) and $\omega \wedge d\alpha +$
$\alpha \wedge d\omega = 0$. The relation $\omega \wedge d\omega = 0$, as we mentioned a number of times, means that there exists a 1-form η satisfying
$d\omega = \eta \wedge \omega$; the relation $\omega \wedge d\alpha + \alpha \wedge d\omega = 0$ shows, in view
of this, that

$$\omega \wedge (d\alpha + \alpha \wedge \eta) = 0,$$

i.e., that there exists a 1-form β satisfying

$$d\alpha + \alpha \wedge \eta = \beta \wedge \omega.$$

[The meaning of the form β is clarified by the relation
$d(\omega + t\alpha + \ldots) = (\eta + t\beta + \ldots) \wedge (\omega + t\alpha + \ldots).$] It follows from the relations between the forms $\omega, \eta, \alpha, \beta$ that the
forms

$$\eta \wedge d\eta, \quad \beta \wedge d\eta, \quad \beta \wedge \eta \wedge d\eta$$

(of degrees 3, 3, 4) are closed and that their cohomology
classes are determined by the given family of foliations (even

by the 1-jets of this family with respect to the parameter in zero). These three forms represent three characteristic classes of the family of foliations of codimension 1: the Godbillon–Vey class, its derivative with respect to the parameter, and one more class of dimension 4. These classes exhaust the part $H^*(W_1) \oplus H^*(W_1; W_1')$ of the space $H^*(\mathscr{W}_1)$: according to Theorems 2.2.4 and 2.2.10,

$$H^q(W_1) = \begin{cases} \mathbb{R} & \text{for } q = 3, \\ 0 & \text{for other } q > 0, \end{cases}$$

$$H^q(W_1; W_1') = \begin{cases} \mathbb{R} & \text{for } q = 2, 3, \\ 0 & \text{for other } q > 0. \end{cases}$$

The fact that the first two classes are nontrivial was proved in Subsections 2D, E. It would be interesting to construct an example of a family of foliations for which the third (four-dimensional) characteristic class is nontrivial.

The codimension-1 case is interesting from the point of view of Theorem 3.1.2 as well. This theorem shows that any one-parameter family of foliations of codimension 1 is such that the product of the Godbillon–Vey class by its derivative with respect to the parameter equals zero [this directly follows from the above: $(\eta \wedge d\eta) \wedge (\beta \wedge d\eta) = 0$, since $d\eta \wedge d\eta = 0$ — see Subsection 2A]. This shows, for example, that if $\mathscr{F}$ is a foliation of codimension 1 on $S^3 \times S^3$ with a nonzero God-billon–Vey class (t_1, t_2) [we identify $H^3(S^3 \times S^3; \mathbb{R})$ with $\mathbb{R} \oplus \mathbb{R}$], then the quotient $t_1 : t_2$ does not change under deformations of $\mathscr{F}$ [to be more precise, it can change only when we pass through $(0, 0)$]. In connection with this observation, it is reasonable to ask: What values can the Godbillon–Vey class on $S^3 \times S^3$ assume? The only known construction of such foliations is to take the codimension-1 foliation on

S^3 (which can have any Godbillon–Vey class) and carry it over to $S^3 \times S^3$ by means of some map $S^3 \times S^3 \to S^3$. However, in this way we can only obtain foliations whose Godbillon–Vey class is (t_1, t_2) with rationally comeasurable t_1, t_2. Possibly the Godbillon–Vey class of foliations on $S^3 \times S^3$ cannot assume any other values: this would be in good agreement with the previous remark on deformations.

E. The relative case. The theory of the variations of characteristic classes for $(\mathfrak{g}, H)$-structures resembles the absolute analog so much that it would be tiresome even to list its main statements. I limit myself to mentioning the fact that characteristic classes of oriented foliations of codimension n, corresponding to the elements of the images of the homomorphism

$$H^* (W_{n+1}, \; O \, (n + 1)) \to H^* (W_n, \; O \, (n)),$$
$$H^* (W_{n+1}, \; \mathfrak{o} \, (n + 1)) \to H^* (W_n, \; \mathfrak{o} \, (n)),$$

induced by inclusions are rigid.

6. Characteristic classes of foliations of the form $\mathcal{F} \, (G, H, \Gamma)$. Recall that the foliation $\mathcal{F} \, (G, H, \Gamma)$ is defined on the (right) quotient space G/Γ and possesses a canonical framing if a basis is chosen in $\mathfrak{g}/\mathfrak{h}$.

Already the calculation of the Godbillon–Vey class for foliations of this form, carried through in Subsection 2D, may lead the reader to think that characteristic classes of such foliations can always be computed without much difficulty. This is indeed the case, and the calculation of these classes is the subject of a rather large number of papers (see the bibliography in [22]). There is not much sense in listing all the specific results of such computations here and in this

subsection we only point out some of the more important facts;
the contents of this subsection are covered by my article
[20] and by the article by Kamber and Tondeur [55].

It follows from 1.1.3 that to the pair (G, H) with a discrete quotient group $(\mathrm{Norm}\, H)/H$ corresponds the inclusion
$\mathfrak{g} \to W_n$, where $n = \dim G/H$, while according to Subsection
3D (see example 2°) the quotient space G/Γ possesses a canonical $\mathfrak{g}$-structure (the condition that the group $(\mathrm{Norm}\, H)/H$
is discrete holds trivially in the examples considered below
and we ignore it). Combining this $\mathfrak{g}$-structure and the above
inclusion, we obtain a W_n-structure on G/Γ, and clearly this
is precisely the W_n-structure which corresponds to the foliation $\mathcal{F}(G, H, \Gamma)$ in the sense of 3B (see example 3°). Thus,
the homomorphism

$$\mathrm{Char:}\ H^*(W_n) \to H^*(G/\Gamma; \mathbb{R}),$$

corresponding to the foliation $\mathcal{F}(G, H, \Gamma)$, splits up into the
composition of homomorphisms

$$H^*(W_n) \to H^*(\mathfrak{g}), \quad H^*(\mathfrak{g}) \to H^*(G/\Gamma;\ \mathbb{R}),$$

of which the first has nothing to do with Γ and is induced
by our inclusion $\mathfrak{g} \to W_n$, while the second has nothing to do
with H and corresponds, in the sense of Subsection 3C, to
the canonical $\mathfrak{g}$-structure on G/Γ (the second homomorphism
is explicitly described at the end of Subsection 3C).

We begin by considering the second homomorphism. If the
algebra $\mathfrak{g}$ is unitary (see 1.3.6B) and the quotient space G/Γ
is compact, then this homomorphism is a monomorphism. In
fact, in this case there is a Poincaré duality in the cohomology of the algebra $\mathfrak{g}$. In the top dimension our homomorphism
is an isomorphism (the invariant volume form goes into the

invariant volume form), and the monomorphic property needed
follows from this: if the nonzero class $\alpha \in H^q(\mathfrak{g})$ goes to
0, then we choose $\beta \in H^{\dim G - q}(\mathfrak{g})$ with $\alpha\beta \neq 0$ (one can do this
by virtue of the Poincaré duality) and we have arrived at a
contradiction with the fact that under our homomorphism $\alpha\beta$
does not go into 0 but our homomorphism is multiplicative.
In addition, if the group G is semisimple, then the algebra
$\mathfrak{g}$ is unitary (see [94] and 1.3.6B) and G has a discrete subgroup Γ with compact G/Γ (see [6] and 1D). Thus, if the
group G is semisimple, then for an appropriately chosen group
Γ the kernel of the homomorphism

$$\text{Char: } H^*(W_n) \to H^*(G/\Gamma; \mathbb{R})$$

coincides with the kernel of the homomorphism $H^*(W_n) \to H^*(\mathfrak{g})$;
now in general the second kernel is contained in the first.
Consequently, to construct a normalized foliation of the form
$\mathcal{F}(G, H, \Gamma)$ with nontrivial characteristic class corresponding
to $\alpha \in H^*(W_n)$, it is first of all necessary to find a finite-
dimensional subalgebra $\mathfrak{g}$ of the algebra W_n, such that the inclusion homomorphism $H^*(W_n) \to H^*(\mathfrak{g})$ does not carry α to 0.

B. Finite-dimensional subalgebras of the algebra W_n.
In order to determine for which cohomology classes α such an
algebra $\mathfrak{g}$ exists, it would be useful to list all the finite-
dimensional subalgebras of the algebra W_n. However, we can
only do this for $n = 1$.

Proposition 3.1.3. Every finite-dimensional subalgebra
of the algebra W_1 is of dimension $\leqslant 3$. The only three-dimensional subalgebra (up to a formal change of variable on the
line) of the algebra W_1 is (additively) generated by the vector fields $d/dt, t(d/dt), t^2(d/dt)$. Every two-dimensional subalgebra of the algebra W_1 is transformed by such a change of

variable into the subalgebra generated by the fields $t \, (d/dt)$
and $t^k \, (d/dt)$ for some k.

Proof. Suppose $\mathfrak{g}$ is a finite-dimensional subalgebra of
the algebra W_1. Choose a ("triangular") basis in $\mathfrak{g}$ of the
form $e_i = t^{k_i} \, (d/dt) + \ldots$ (here the dots stand for higher degree terms) such that $0 \leqslant k_1 < k_2 < \ldots$ Since $[t^k \, (d/dt),$
$t^l \, (d/dt)] = (l - k) \, t^{k+l-1} \, (d/dt)$, the set $\{k_1, k_2, \ldots\}$ contains all
the numbers of the form $k_i + k_j - 1$ with $i \neq j$. It is easy
to check that finite sets of nonnegative integers possessing
this property are exhausted by the sets $\{k\}, \{1, k\}, \{0, 1, 2\}$.
It remains to note that the vector field $t \, (d/dt) + \ldots$ can
be given the form $t \, (d/dt)$ by a formal change of variable and
that the subalgebra of the algebra W_1, containing $t \, (d/dt)$,
possesses a basis consisting of monomial fields.

The subalgebra of the algebra W_1 spanning the elements
$d/dt, \, t \, (d/dt), \, t^2 \, (d/dt)$ is isomorphic to $\mathfrak{sl} \, (2, \mathbb{R})$. Its cohomology is isomorphic to the cohomology of the algebra W_1, and,
as can be easily checked, the inclusion homomorphism $H^* \, (W_1) \to$
$H^* \, (\mathfrak{sl} \, (2, \mathbb{R}))$ is an isomorphism. This fact explains the
nontriviality of the Godbillon–Vey class of the foliation
$\mathcal{F} \, (G, H, \Gamma)$ for $G = \mathrm{SL} \, (2, \mathbb{R})$ – see 2D.

However, for arbitrary n the classification of finite-
dimensional subalgebras of the algebra W_n is hardly possible:
indeed, any locally effective transitive action of a Lie group
on an n-dimensional manifold determines the inclusion of its
Lie algebra in W_n. Actually, already in W_2 there are very
many finite-dimensional subalgebras and among them there are
transitive subalgebras of arbitrarily large dimensions: for
example, the vector fields

$$\frac{\partial}{\partial t_1}, \quad \frac{\partial}{\partial t_2}; \quad t_1 \frac{\partial}{\partial t_1}, \quad t_2 \frac{\partial}{\partial t_1}, \quad t_2 \frac{\partial}{\partial t_2};$$

$$t_2^2 \frac{\partial}{\partial t_1}, \quad t_2^2 \frac{\partial}{\partial t_2} + mt_1t_2 \frac{\partial}{\partial t_1}; \quad t_2^3 \frac{\partial}{\partial t_1}; \quad \ldots; \quad t_2^m \frac{\partial}{\partial t_1}$$

for any integer m additively generate a subalgebra of the algebra W_2. Hence we shall attack our problem without possessing a classification of finite-dimensional subalgebras of the algebra W_n.

C. Calculations for the pair $(W_n, \mathfrak{sl}\,(n+1),\,\mathbb{R}))$. From the point of view of characteristic classes, the most useful finite-dimensional subalgebra of the algebra W_n is the algebra $\mathfrak{sl}\,(n+1,\mathbb{R})$, included in W_n by means of the homomorphism described in 1.1.3. Probably, the dimension $n\,(n+2)$ of the algebra $\mathfrak{sl}\,(n+1,\mathbb{R})$ is the largest possible dimension of a semisimple and perhaps even a unitary subalgebra of the algebra W_n. It is interesting that the number $n\,(n+2)$ coincides with the maximum dimension of nontrivial cohomology of the algebra W_n (see 2.2.3A).

Theorem 3.1.4. The sequence

$$H^q(W_n) \to H^q\,(\mathfrak{sl}\,(n+1,\,\mathbb{R})) \to H^q\,(\mathfrak{gl}\,(n,\,\mathbb{R})),$$

in which the first homomorphism is induced by the previous inclusion while the second is induced by the inclusion $\mathfrak{gl}\,(n,\,\mathbb{R}) \to \mathfrak{sl}\,(n+1,\,\mathbb{R})$, defined by the formula $A \mapsto A - (\operatorname{tr} A)e_{n+1}^{n+1}$, is exact for $q > 0$.

Proof. The composition of the two inclusions indicated above coincides, up to an automorphism of the algebra $\mathfrak{gl}\,(n,\,\mathbb{R})$, with the canonical inclusion $\mathfrak{gl}\,(n,\,\mathbb{R}) \to W_n$ (see 1.1.3), and the triviality of the corresponding homomorphism in cohomology (in positive dimensions) is already known to us. [It follows from the triviality of the spaces $E_\infty^{0;\,q}\,(q > 0)$ of the Serre–Hochschild spectral sequence corresponding to the pair $(W_n, \mathfrak{gl}\,(n,\,\mathbb{R}))$, and this triviality was established when we proved Theorem 2.2.4 — see, e.g., Lemma 3 in 2.2.2.] Thus, we must prove that the dimension of the image of the homomorphism

$$H^{>0}(W_n) \to H^{>0}(\mathfrak{sl}(n+1, \mathbb{R}))$$

is greater than or equal to the dimension of the kernel of the homormophism $H^*(\mathfrak{sl}(n+1, \mathbb{R})) \to H^*(\mathfrak{gl}(n, \mathbb{R}))$; this last kernel coincides with the ideal generated by φ_{n+1}, and its dimension equals $\frac{1}{2}$ dim $H^*(\mathfrak{sl}(n+1, \mathbb{R})) = 2^{n-1}$.

Consider the Serre–Hochschild spectral sequence corresponding to the pairs $(\mathfrak{sl}(n+1, \mathbb{R}), \mathfrak{gl}(n, \mathbb{R})), (W_n, \mathfrak{gl}(n, \mathbb{R}))$, and the homomorphism of the second sequence into the first induced by our inclusion $\mathfrak{sl}(n+1, \mathbb{R}) \to W_n$. For convenience, we shall denote the terms of these spectral sequences by $'E_r^{p,q}$, $''E_r^{p,q}$, and denote the components of the homomorphism by $\eta_r^{p,q}$. Obviously, $'E_2^{*,0}$ is the truncated polynomial algebra in one variable of degree 2 $['E_2^{*,0} = H^*(\mathbb{C}P^n, \mathbb{R})]$, and $'E_2^{0,*} = H^*(\mathfrak{gl}(n, \mathbb{R}))$. It is also clear that $\eta_2^{0,*}$ is an isomorphism, and since η commutes with differentials, it follows that $\eta_2^{2,0}: \,''E_2^{2,0} \to 'E_2^{2,0}$ is also an isomorphism. Therefore, the homomorphism $\eta_2^{2k,0}: \,''E_2^{2k,0} \to 'E_0^{2k,0}$ sends the element ψ_1^k, for $k = 1$, ..., n into a nonzero element, so that $\eta_2: \,''E_2 \to 'E_2$ is an epimorphism. Thus, $\eta_r^{2n,*}: \,''E_r^{2n,*} \to 'E_r^{2n,*}$ is an epimorphism for any r, including ∞ ($''E_r^{2n,q}$, $'E_r^{2n,q}$ are quotient spaces of the spaces $''E_2^{2n,q}$, $'E_2^{2n,q}$), and, therefore, the dimension of the image of the homomorphism $H^{>0}(W_n) \to H^{>0}(\mathfrak{sl}(n+1, \mathbb{R}))$ is no less than dim $'E_\infty^{2n,*}$. But the ring $'E_\infty^{2n,*}$ is clearly the exterior algebra in generators of dimension $3, 5, \ldots, 2n-1, 2n+1$, contained in $'E_\infty^{0,3}, 'E_\infty^{0,5} \ldots, 'E_\infty^{0,2n-1}, 'E_\infty^{2n,1}$ and giving rise to $(2n+1)$-dimensional generators, and, therefore, dim $'E_\infty^{2n,*} = 2^{n-1}$.

As we pointed out in 1.1.3, the inclusion $\mathfrak{sl}(n+1, \mathbb{R}) \to W_n$ considered above corresponds to the subgroup $H = \{\|a_{ij}\| \in$

$SL\ (n + 1,\ \mathbb{R})\ |\ a_{1,n+1} = \ \ldots = a_{n,n+1} = 0\}$ of the group $SL\ (n + 1,\ \mathbb{R})$. Thus, if Γ is a discrete subgroup of the group $SL\ (n + 1,\ \mathbb{R})$ with compact quotient space $SL\ (n + 1,\ \mathbb{R})/\Gamma$, then the image of the homomorphism

$$\text{Char:}\ H^{>0}\ (W_n) \to H^{>0}\ (SL\ (n + 1,\ \mathbb{R})/\Gamma;\ \mathbb{R}),$$

corresponding to the foliation $\mathcal{F}\ (SL\ (n + 1,\ \mathbb{R}),\ H,\ \Gamma)$, is of dimension 2^{n-1}. This result may be viewed as the generalization of the theorem on the nontriviality of the Godbillon—Vey class of the foliation $\mathcal{F}\ (SL\ (2,\ \mathbb{R}),\ H,\ \Gamma)$, proved in 2D.

Note also that $H^*\ (\mathfrak{sl}\ (n + 1,\ \mathbb{R});\ \mathfrak{sl}\ (n + 1,\ \mathbb{R})') = 0$, and, therefore, characteristic classes of $\mathfrak{sl}\ (n + 1,\ \mathbb{R})$-structures do not change under deformations. Therefore, the characteristic classes of the foliation $\mathcal{F}\ (SL\ (n + 1,\ \mathbb{R}),\ H,\ \Gamma)$ remain unchanged when it is deformed in the class of foliations of the same type. [In the last statement the group $SL\ (n + 1,\ \mathbb{R})$ may be replaced by an arbitrary semisimple Lie group.] This theorem clarifies a classical result in the theory of discrete isometry groups in the Lobachevski plane. Namely, it is easy to construct a nontrivial family $\{\Gamma_t\}$ of groups continuously depending on the parameter t and possessing a compact fundamental domain. The Godbillon—Vey number of the foliation $\mathcal{F}\ (SL\ (2,\ \mathbb{R}),\ H,\ \Gamma_t)$ is proportional to the area of the fundamental domain of the group Γ_t and, by the above, this number does not depend on t. Therefore, the area of the fundamental domain of the group Γ_t does not depend on t. We have obtained a new proof of this classical fact.

D. In conclusion, let us mention some other results. In [20] it is proved that on the kernel of the inclusion homomorphism

$$Ji^{n(n+2)}(W_n) \to H^{n(n+2)}(\mathfrak{sl}(n+1), \mathbb{R})$$

[which has dimension $p(n) - 1$] a similar homomorphism $H^{n(n+2)}(W_n) \to H^{n(n+2)}(\mathfrak{g})$ is trivial for any finite-dimensional subalgebra $\mathfrak{g}$ of the algebra W_n. It is also conjectured there that any number may be taken in this statement instead of $n(n+2)$. This naive conjecture was shown to be false by Baker [3] [in his example, $n = 2k \geqslant 6$, $\mathfrak{g} = \mathfrak{sl}(k+2)$ and the inclusion $\mathfrak{g} \to W_n$ corresponds to the subgroup H of the group $SL(k+2, \mathbb{R})$, for which $SL(k+2, \mathbb{R})/H$ is the Grassmann manifold $G(k+2, 2)$]. However, Pittie [73] later proved the following statement, which may be viewed as a weakened version of the above-mentioned conjecture.

Theorem 3.1.5. There exists a space $A \subset H^{>0}(W_n)$ of dimension 2^{n-1}, such that for any semisimple group G and its parabolic subgroup P of codimension n, the image of the corresponding homomorphism $H^{>0}(W_n) \to H^{>0}(\mathfrak{g})$ coincides with the image of the space A under this homomorphism.

§2. COMBINATORIAL IDENTITIES

1. Introduction. The source of the identities with which we shall be concerned in this section is the famous Euler identity

$$\prod_{n=1}^{\infty}(1-t^n) = 1 + \sum_{n=1}^{\infty}(-1)^n(t^{\frac{3n^2-n}{2}} + t^{\frac{3n^2+n}{2}})$$
$$= 1 - t - t^2 + t^5 + t^7 - t^{12} - t^{15} + t^{22} + t^{26} \ldots,$$

discovered at the end of the 1740s. It is interesting to note that, although this identity has a very simple elementary proof (see [23] or [49]), Euler did not succeed in proving it (in this connection see [74], Chapter 6). Nevertheless,

Euler used his identity, which he viewed as a conjecture, to prove two remarkable corollaries, which I hazard to mention here, despite the fact that they are very well known.

Proposition 3.2.1. Suppose that, as before, $p(n)$ denotes the number of decompositions of the number n into a sum of natural summands. Put $p(0) = 1$, $p(n) = 0$ for $n < 0$; then for $n > 0$

$$p(n) = p(n-1) + p(n-2) - p(n-5) - p(n-7)$$
$$+ p(n-12) + p(n-15)\ldots$$
$$= \sum_{k=1}^{\infty} (-1)^{k-1} \left[p\left(n - \frac{3k^2 - k}{2}\right) + p\left(n - \frac{3k^2 + k}{2}\right) \right].$$

Proof.   Let

$$P(t) = 1 + \sum_{n=1}^{\infty} p(n) t^n.$$

Obviously, $P(t) = \prod_{m=1}^{\infty} (1 + t^m + t^{2m} + t^{3m} + \ldots) = \prod_{m=1}^{\infty} (1 - t^m)^{-1}$.   Thus we have $P(t) \prod_{m=1}^{\infty} (1 - t^m) = 1$, i.e.,

$$(1 + p(1) t + p(2) t^2 + \ldots)(1 - t - t^2 + t^5 + t^7 - t^{12} - t^{15} \ldots) = 1.$$

The required identity is obtained by setting the coefficient of t^n on the left-hand side equal to zero.

Proposition 3.2.2.   Suppose $d(n)$ denotes the sum of all positive integer divisors of the number n. Let $d(n) = 0$ for $n < 0$. Then for $n > 0$

$$d(n) = d(n-1) + d(n-2) - d(n-5) - d(n-7)$$
$$+ d(n-12) + d(n-15)\ldots$$

$$= \sum_{k=1}^{\infty} (-1)^{k-1} \left[d \left(n - \frac{3k^2 - k}{2} \right) + d \left(n - \frac{3k^2 + k}{2} \right) \right],$$

where, instead of d (0), if such a summand appears in the sum, one should take n.

Proof.   Let

$$D(t) = \sum_{n=1}^{\infty} d(n) t^n.$$

Obviously,

$$D(t) = \sum_{m=1}^{\infty} m \left(t^m + t^{2m} + t^{3m} + \dots \right) = \sum_{m=1}^{\infty} m t^m (1 - t^m)^{-1}$$

$$= -t \sum_{m=1}^{\infty} \frac{d}{dt} \log (1 - t^m) = -t \frac{d}{dt} \log \prod_{m=1}^{\infty} (1 - t^m)$$

$$= -t \frac{d}{dt} \prod_{m=1}^{\infty} (1 - t^m) \Big/ \prod_{m=1}^{\infty} (1 - t^m),$$

i.e.,

$$(d(1) t + d(2) t^2 + \dots) (1 - t - t^2 + t^5 + t^7 - \dots)$$
$$= t + 2t^2 - 5t^5 - 7t^7 + 12t^{12} + 15t^{15} - \dots$$

The required identity is obtained by setting the coefficients of t^n on the right- and left-hand sides equal to each other.

These two elegant recurrent formulas already indicate that the Euler identity is, per se, a deep mathematical fact. And, indeed, it is the starting point of a whole branch of combinatorics, which has attracted the attention of leading mathematicians of all generations.

Seventy years after Euler's discovery, the identity

$$\prod_{m=1}^{\infty} (1 - t_1^m t_2^m)(1 - t_1^m t_2^{m-1})(1 - t_1^{m-1} t_2^m)$$

$$= 1 + \sum_{k=1}^{\infty} (-1)^k (t_1^{\frac{k(k+1)}{2}} t_2^{\frac{k(k-1)}{2}} + t_1^{\frac{k(k-1)}{2}} t_2^{\frac{k(k+1)}{2}}),$$

now known as the Gauss—Jacobi identity, was discovered. At
first it was proved by methods from the theory of elliptic
functions, but at the present time an elementary proof is also
known (see [49]). Note that if we put $t_1 = t^2$, $t_2 = t$ in the
Gauss—Jacobi identity, we recover the Euler identity; if we
take its derivative with respect to t_2 and then set $t_1 = t$,
$t_2 = 1$, we obtain the formula

$$\prod_{n=1}^{\infty} (1 - t^n)^3 = \sum_{k=1}^{\infty} (-1)^{k-1} (2k-1) t^{\frac{k(k-1)}{2}}.$$

The fact that the cube of the "Euler function" $\prod (1 - t^n)$ can
be represented by a series hardly less remarkable, perhaps,
than the series for the Euler function itself, is quite unexpected, because the decomposition into series for the square
of this function is not attractive in any way.

Years went by and new identities appeared, relating infinite products with infinite sums. The simplest of them,
like the previous identity, gave expressions for different
powers of the Euler function. Here, for example, are two such
identities, which are due, respectively, to Klein and, sur-
prisingly enough, to the physicist Dyson:

$$\prod_{n=1}^{\infty} (1 - t^n)^8 = \sum \left[\frac{1}{3} + \frac{3}{2} (3klm - kl - km - lm) \right] t^{-(kl+km+lm)},$$

where the sum on the right-hand side is taken over all triples k, l, m of integers satisfying $k + l + m = 1$;

$$\prod_{n=1}^{\infty} (1 - t^n)^{24} = \frac{1}{1!\,2!\,3!\,4!} \sum \begin{vmatrix} 1 & 1 & 1 & 1 & 1 \\ a & b & c & d & e \\ a^2 & b^2 & c^2 & d^2 & e^2 \\ a^3 & b^3 & c^3 & d^3 & e^3 \\ a^4 & b^4 & c^4 & d^4 & e^4 \end{vmatrix} t^{\frac{a^2+b^2+c^2+d^2+e^2}{10} - 1},$$

where the sum on the right-hand side is taken over all quintuples a, b, c, d, e of integers whose residues modulo 5 are 1, 2, 3, 4, 0, respectively, and which satisfy $a + b + c + d + e = 0$. Here, as we have already seen in the example of the first power, the square, and the cube of the Euler function, only for some chosen values of the power k can the function $\prod (1 - t^n)^k$ be represented by a "nice" series.

In his emotional article [13], Dyson tells us that "as a relief from the serious business of physics" he wrote out a list of these distinguished powers k, and only since "his mind was so well compartmentalized" he did not recognize the beginning of the sequence of dimensions of simple Lie algebras in his list. (Additional arguments about "distinguished powers" are given in [23] and in Subsections 3 and 4.)

The relationship between Lie algebras and combinatorial identities was first discovered by Macdonald [67], who proved a universal identity containing the long sequence of all the previous ones. We now know several proofs of the Macdonald identities, and all of them are related to Lie algebras in one way or another. In Subsection 3, we develop the homological proof due to Garland and Lepowsky [24] of one of the sequences of Macdonald identities (with certain improvements due to Feigin). As explained in Subsection 4, this proof may be modified to get the other Macdonald identities.

2. General outline for proving combinatorial identities.

A universal method for obtaining identities is the Euler—
Poincaré formula: If

$$0 \leftarrow C_0 \leftarrow C_1 \leftarrow \ldots \leftarrow C_N \leftarrow 0$$

is a complex consisting of finite-dimensional linear spaces,
say, over $\mathbb{C}$, and $H_0, \ldots, H_N$ is its homology, then

$$\sum_{q=0}^{N} (-1)^q c_q = \sum_{q=0}^{N} (-1)^q h_q,$$

where $c_q = \dim C_q$, $h_q = \dim H_q$.   We shall apply this formula
to chain complexes of Lie algebras.   Actually, our algebras,
as a rule, will be infinite dimensional; their chain spaces
will be infinite dimensional as well.   Matters can be fixed
up by considering graded Lie algebras, say $\mathfrak{g} = \bigoplus_{m=1}^{\infty} \mathfrak{g}_{(m)}$ with
dim $\mathfrak{g}_{(m)} < \infty$.   Then the complex $C_{\cdot}(\mathfrak{g})$ can be decomposed into
the sum of complexes

$$0 \leftarrow C_0^{(m)}(\mathfrak{g}) \leftarrow C_1^{(m)}(\mathfrak{g}) \ldots \leftarrow 0$$

(see 1.3.7) and, for every m, the complex $C_{\cdot}^{(m)}(\mathfrak{g})$ consists
of a finite number of finite-dimensional spaces.   Applying
the Euler—Poincaré formula, we obtain

$$\sum_{q} (-1)^q c_q^{(m)} = \sum_{q} (-1)^q h_q^{(m)} \qquad (m = 0, 1, 2, \ldots),$$

where $c_q^{(m)} = \dim C_q^{(m)}(\mathfrak{g}), h_q^{(m)} = \dim H_q^{(m)}(\mathfrak{g})$.   It is convenient to
group this sequence of identities into one identity by intro-
ducing the formal variable t:

$$\sum_{q,m} (-1)^q c_q^{(m)} t^m = \sum_{q,m} (-1)^q h_q^{(m)} t^m$$

(the left- and right-hand sides of the identity are viewed
as formal power series).   The left-hand side of the latter
identity may be rewritten:

Lemma.

$$\sum_{q,m} (-1)^q c_q^{(m)} t^m = \prod_j (1 - t^j)^{d_j},$$

where $d_j = \dim \mathfrak{g}_{(j)}$.

Indeed,

$$C_q^{(m)}(\mathfrak{g}) = \bigoplus_{\substack{m_1+m_2+m_3+\ldots=q,\\ m_1+2m_2+3m_3+\ldots=m}} \bigotimes_j (\Lambda^{m_j} \mathfrak{g}_{(j)}),$$

and, therefore,

$$\sum_{q,m} (-1)^q c_q^{(m)} t^m = \sum_{q,m} (-1)^q \sum_{\substack{m_1+m_2+m_3+\ldots=q,\\ m_1+2m_2+3m_3+\ldots=m}} \prod_j \binom{d_j}{m_j} t^m$$

$$= \sum_{m_1,m_2,m_3,\ldots} \prod_j (-1)^{m_j} \binom{d_j}{m_j} t^{jm_j} = \prod_j \sum_{m_j} (-1)^{m_j} \binom{d_j}{m_j} t^{jm_j} = \prod_j (1 - t^j)^{d_j}.$$

Thus,

$$\prod_j (1 - t^j)^{d_j} = \sum_{q,m} (-1)^q h_q^{(m)} t^m,$$

and, in order to write the identity in its final form, it suffices to compute the homology $H_q^{(m)}(\mathfrak{g})$.

Further we shall apply this outline in a somewhat more general situation — when the algebra $\mathfrak{g}$ possesses a "poly-grading" $\mathfrak{g} = \oplus \mathfrak{g}_{(m_1,\ldots,m_k)}$. Repeating the above with obvious modifications, we obtain the following result.

Theorem 3.2.3. Suppose

$$\mathfrak{g} = \bigoplus_{\substack{m_1 \geqslant 0,\ldots,m_k \geqslant 0,\\ m_1+\ldots+m_k > 0}} \mathfrak{g}_{(m_1,\ldots,m_k)}$$

is the (poly)graded Lie algebra satisfying $d_{m_1 \ldots m_k} =$ dim $\mathfrak{g}_{(m_1, \ldots, m_k)} < \infty$. If

$$\dim H_q^{(m_1, \ldots, m_k)} = h_q^{(m_1, \ldots, m_k)},$$

we have the identity

$$\prod_{j_1, \ldots, j_k} (1 - t_1^{j_1} \ldots t_k^{j_k})^{d_{j_1 \ldots j_k}} = \sum_{q, m_1, \ldots, m_k} (-1)^q h_q^{(m_1, \ldots, m_k)} t_1^{m_1} \ldots t_k^{m_k}$$

(the left- and right-hand sides are viewed as formal power series in $t_1, \ldots, t_k$).

3. The generalized Gauss—Jacobi identity (the simplest of series of Macdonald identities). The application of Theorem 3.2.3 to the algebras $T^+(n)$ from Subsection 2.5.1, together with Theorem 2.5.2 and the formulas for dim $T^+_{(k_1, \ldots, k_n)}(n)$ given in Subsection 2.5.1A, immediately yield one of the series of Macdonald identities containing the Gauss—Jacobi identity.

Theorem 3.2.4. For any $n \geqslant 2$, we have the identity

$$\prod_{m=1}^{\infty} \left\{ (1 - t_1^m \ldots t_n^m)^{n-1} \prod_{1 \leqslant j < i \leqslant n} \left[\left(1 - t_1^{m-1} \ldots t_n^{m-1} \prod_{s=i}^{j-1} t_s \right) \right.\right.$$

$$\left.\left. \times \left(1 - t_1^m \ldots t_n^m \Big/ \prod_{s=i}^{j-1} t_s \right) \right] \right\} = \sum_{Q(j_1, \ldots, j_n) = 0} (-1)^{q \, (j_1, \ldots, j_n)} t_1^{j_1} \ldots t_n^{j_n}, \qquad (1)$$

where $Q(j_1, \ldots, j_n) = -j_1^2 - \ldots - j_n^2 + j_1 j_2 + \ldots + j_{n-1} j_n + j_n j_1 + j_1 + \ldots + j_n$, while $q(j_1, \ldots, j_n)$ is defined by formulas (7) and (8) from 2.5.1.

Remark. Define the transformations $s_i \colon \mathbb{Z}^n \to \mathbb{Z}^n$ by the formulas

$$s_1(k_1, \ldots, k_n) = (k_n - k_1 + k_2 + 1, k_2, \ldots, k_n),$$

$$s_i(k_1, \ldots, k_n) = (k_1, \ldots, k_{i-1}, k_{i-1} - k_i + k_{i+1} + 1, k_{i+1}, \ldots, k_n) \text{ for } 1 < i < n,$$

$$s_n(k_1, \ldots, k_n) = (k_1, \ldots, k_{n-1}, k_{n-1} - k_n + k_1 + 1).$$

A direct verification shows that for any i

(i) $s_i^2 = \mathrm{id}$;

(ii) $Q \circ s_i = Q$;

(iii) $q \circ s_i (k_1, \ldots, k_n) = q(k_1, \ldots, k_n) \pm 1$;

(iv) if $Q(j_1, \ldots, j_n) = 0$, then there exist r and $i_1, \ldots,$ i_r, such that $(j_1, \ldots, j_n) = s_{i_r} \circ \ldots \circ s_{i_1} (0, \ldots, 0)$. [For the proof of statement (iv), it is necessary to use information on the solutions of the equation $Q(j_1, \ldots, j_n) = 0$ contained in the proof of Theorem 2.5.2.] Obviously, in the situation of statement (iv), we have $(-1)^{q(j_1, \ldots, j_n)} = (-1)^r$, and this enables us to find the coefficients of the right-hand side of the identity (1) without using formulas (7) and (8) from 2.5.1.

For $n = 2$, the identity (1) becomes the Gauss–Jacobi identity.

If we apply the operator

$$\frac{\partial^{n(n-1)/2}}{\partial t_2 \partial t_3^2 \ldots \partial t_n^{n-1}}$$

to the left- and right-hand sides of identity (1) and then set $t_1 = t, t_2 = \ldots = t_n = 1$, we obtain the following identity:

$$\prod_{m=1}^{\infty} (1 - t^m)^{n-1} = \sum_{Q(j_1, \ldots, j_n) = 0} (-1)^{q(j_1, \ldots, j_n)} \binom{j_2}{1} \binom{j_3}{2} \ldots \binom{j_n}{n-1} t^{j_1}. \quad (2)$$

For $n = 3$ and $n = 5$ the reader might attempt to transform this identity into the Klein and Dyson ones, written out in Subsection 1.

If in identity (1) we set $t_1 = t^2$, $t_2 = \ldots = t_n = t$, we obtain the identity

$$\prod_{m=1}^{\infty} (1 - t^m)^{n-1} = \sum_{Q(j_1,\ldots,j_n)=0} (-1)^{q(j_1,\ldots,j_n)} t^{2j_1+j_2+\ldots+j_n}. \qquad (3)$$

We see that, despite what we said in Subsection 1, our theory gives certain formulas for <u>all</u> powers of the Euler function $\prod (1 - t^m)$.   For example, the identity (3) for $n = 3$ can be put in the form

$$\prod_{m=1}^{\infty} (1 - t^m)^2 = \sum_{r \geqslant 0,\, s \geqslant 0} t^{1(r^2+s^2+rs)} (t^r + t^s + t^{-r-s})$$

$$- \sum_{\substack{r \geqslant 0,\, s \geqslant 0,\\ r+s \neq 1}} t^{1(r^2+s^2+rs-r-s)} (t^{r+1} + t^{s+1} + t^{-r-s+2}).$$

Thus, the statement made in Subsection 1, claiming that no reasonable formula for the k-th power of the Euler function exists for all k, is not exactly correct. Such formulas exist for all k, and, for certain k, many such formulas exist (we shall see this again in Subsection 4). Simply the complexity of the simplest formula for $\prod (1 - t^m)^k$ does not increase monotonically with k. For example, (2) and (3) for the $(n^2 - 1)$-st and the $(n - 1)$-st power of the Euler function are, roughly speaking, equally complicated; in particular, the formula for the square of the Euler function written out above is about as complicated as Klein's formula for the 8th power of the Euler function; but this result, which we feel is sufficiently good for the 8th power, does not satisfy us in the case of squares.

    4. <u>Other identities</u>. This subsection is based on 2.5.3. In it, just as in 2.5.3, we shall not give complete proofs

of the theorems stated and we shall use the terminology of classical Lie algebra theory more widely than usual. Details can be found by the reader in the articles by Macdonald and Garland–Lepowsky, cited above, as well as in Macdonald's book [68].

Interesting combinatorial identities may be obtained by applying Theorem 3.2.3 to other graded Lie algebras. The most convenient subject matter for this is given by the sub-algebras $N^+(\mathfrak{g}^A)$ of Kac–Moody algebras, whose homology was computed in 2.5.3B. Recall that the algebra $N^+(\mathfrak{g}^A)$ is constructed from the Cartan matrix A, i.e., by the square integer matrix $A = \|a_{ij}\|_{i,\,j=1}^n$, in which $a_{11} = \ldots = a_{nn} = 2$, $a_{ij} \leqslant 0$ for $i \neq j$, and for which there exist positive numbers $b_1, \ldots, b_n$, such that the matrix $bA = \|b_i a_{ij}\|$ is symmetric. The algebra itself is generated by n elements $e_1, \ldots, e_n$, which satisfy the relations

$$\underbrace{[e_i, [e_i, \ldots [e_i, e_j] \ldots]]}_{-a_{ij}+1} = 0.$$

It possesses a $\mathbb{Z}^n$-grading, defined by the condition

$$\deg e_i = (0, \ldots, 0, \overset{(i)}{1}, 0, \ldots, 0).$$

It follows from 2.5.3 that

$$H_*^{(k_1,\ldots,k_n)}(N^+(\mathfrak{g}^A)) = 0, \quad \text{if } Q_A(k_1, \ldots, k_n) \neq 0,$$

where

$$Q_A(k_1, \ldots, k_n) = \sum_{i,j} b_i a_{ij} k_i k_j - 2 \sum_i b_i k_i.$$

Thus, Theorem 3.2.3 gives us the identity

$$\prod_{\substack{k_1 \geqslant 0, \ldots, k_n \geqslant 0, \\ k_1 + \ldots + k_n > 0}} (1 - t_1^{k_1} \ldots t_n^{k_n})^{\dim g^A(k_1, \ldots, k_n)} = \sum_{Q(j_1, \ldots, j_n) = 0} L(j_1, \ldots, j_n) t_1^{j_1} \ldots t_n^{j_n},$$

(4)

where $L(j_1, \ldots, j_n)$ are certain coefficients. In the computation of these coefficients, the following additional considerations turn out to be useful. Define the transformation s_i: $\mathbf{Z}^n \to \mathbf{Z}^n$ by the formula

$$s_i(k_1, \ldots, k_n) = (k_1, \ldots, k_{i-1}, -\sum_{j \neq i} a_{ij}k_j - k_i + 1, k_{i+1}, \ldots, k_n).$$

The transformation group of the lattice $\mathbf{Z}^n$ generated by these transformations is denoted by W^A. (It is isomorphic to the so-called <u>Weyl group</u> of the algebra g^A.) As shown by brief calculation,

$$Q_A \circ s_i(k_1, \ldots, k_n) = Q_A(k_1, \ldots, k_n),$$

and it turns out that $Q_A(k_1, \ldots, k_n) = 0$ implies that for any i

$$L \circ s_i(k_1, \ldots, k_n) = -L(k_1, \ldots, k_n).$$

In many cases, in particular in all cases covered by Theorems 2.5.10 and 2.5.12, the group W^A acts transitively on the set of integer points of the surface $Q_A(k_1, \ldots, k_n) = 0$. Since we always have $L(0, \ldots, 0) = 1$, in these cases $L(k_1, \ldots, k_n) = \pm 1$ for all $k_1, \ldots, k_n$. I do not know whether the latter is true in the general case or not.

The best-known identities are the ones which correspond to the algebras $N^+(g^A)$ with Cartan matrices of rank $n - 1$ with nonnegative eigenvalues (these are the identities which we usually call Macdonald identities). According to Theorem 2.5.12, there exist two different kinds of algebras g^A with such A (as in 2.5.3, we always limit ourselves to irreducible matrices A). The first kind includes algebras obtained by central extension of the algebras $(\mathfrak{g}^{S^1})^{\text{pol}} = \mathfrak{g} \otimes C [t,\ t^{-1}]$, where $\mathfrak{g}$ is a finite-dimensional simple (complex) Lie algebra (of rank $n-1$). For such an algebra, our identity acquires the form

$$\prod_{m \geqslant 1} \left\{ (1 - T^m)^{n-1} \prod_{(k_1,\ldots,k_{n-1}) \in \Delta^+} [(1 - T^{m-1} t_1^{k_1} \ldots t_{n-1}^{k_{n-1}}) \right.$$

$$\left. \times (1 - T^m / t_1^{k_1} \ldots t_{n-1}^{k_{n-1}})] \right\} = \sum_{w \in W^A} \text{sgn} (w)\, t^{w(0,\ldots,0)}, \qquad (5)$$

where Δ^+ is the set of coordinates of positive roots of the algebra $\mathfrak{g}$, $t^{w(0,\ldots,0)} = t_1^{w_1} \ldots t_n^{w_n}$ and $T = t_1^{l_1} \ldots t_{n-1}^{l_{n-1}} t_n$, where, in its turn, $(w_1, \ldots, w_n) = w (0, \ldots, 0)$ and $(l_1, \ldots, l_{n-1})$ is the greatest positive root of the algebra $\mathfrak{g}$. If $\mathfrak{g} = \mathfrak{sl}\,(n-1,\ \mathbb{C})$, i.e., if

$$A = \begin{Vmatrix} 2 & -1 & 0 & \ldots & 0 & -1 \\ -1 & 2 & -1 & \ldots & 0 & 0 \\ 0 & -1 & 2 & \ldots & 0 & 0 \\ \ldots & \ldots & \ldots & \ldots & \ldots & \ldots \\ 0 & 0 & 0 & \ldots & 2 & -1 \\ -1 & 0 & 0 & \ldots & -1 & 2 \end{Vmatrix},$$

we obtain identity (1) from Subsection 1.

Just as we deduced identities (2) and (3) from (1) in Subsection 3, it is possible to use identity (5) to obtain certain formulas for $\prod (1 - t^m)^{\dim \mathfrak{g}}$ and $\prod (1 - t^m)^{\text{rang} \mathfrak{g}}$. (This conclusively explains the phenomenon of "distinguished powers" — see Subsection 1.)

The second kind of algebra is constituted by the algebras constructed from exterior automorphisms of simple algebras possessing a finite order. The right-hand side of the identity corresponding to these algebras looks just like the right-hand side of identity (5) and, in order to find their left-hand sides, one must find the dimensions of the spaces $H^q_{(k_1,\ldots,k_n)} \times (N^r(\mathfrak{g}^A))$. This is not too difficult to do, but the answer turns out to be rather cumbersome and we limit ourselves here to the simplest particular case — the algebra corresponding to the automorphism "minus transposition" of the algebra $\mathfrak{sl}(3,\mathbb{C})$. This algebra has the Cartan matrix $\begin{Vmatrix} 2 & -1 \\ -4 & 2 \end{Vmatrix}$, and the corresponding identity has the following form:

$$\prod_{m=1}^{\infty} (1 - t_1^m t_2^{2m})(1 - t_1^m t_2^{2m-1})(1 - t_1^{m-1}t_2^{2m-1}) \times (1 - t_1^{m-1}t_2^{4m-4})(1 - t_1^{2m-1}t_2^{4m})$$

$$= \sum_{k=-\infty}^{\infty} t_1^{\frac{3k^2+k}{2}}(t_2^{3k^2-2k} - t_2^{3k^2-4k+1}).$$

(This identity was known long before the theory developed here arose. It was discovered in 1928 by Watson.)

Identities corresponding to Kac–Moody algebras with other Cartan matrices are also of interest. If the matrix A is positive definite, then the algebra $\mathfrak{g}^A$ is finite dimensional (see Theorem 2.5.10), and the series standing on the left- and right-hand sides of the identity (4) become polynomials. The actual identity acquires the form

$$\prod_{(k_1,\ldots,k_n)\in\Delta^+} (1 - t_1^{k_1}\ldots t_n^{k_n}) = \sum_{Q_A(j_1,\ldots,j_n)=0} (\pm 1) t_1^{j_1}\ldots t_n^{j_n},$$

where Δ^+ is the set of coordinates of positive roots of the algebra $\mathfrak{g}^A$. Note that the equation $Q_A = 0$ determines, in this case, an ellipsoid and, therefore, has a finite number of integer solutions (the number of these solutions equals the order of the Weyl group of the algebra $\mathfrak{g}^A$). For example, if $\mathfrak{g} = \mathfrak{sl}(n+1,\mathbb{C})$, then the identity may be written in the form

$$\prod_{1 \leqslant i < j \leqslant n} (1 - x_i x_{i+1} \ldots x_j) = \sum_{\sigma \in \mathrm{Symm}(n)} \mathrm{sgn}\,(\sigma) \prod_{k=1}^{n} x_k^{\sigma(1)+\ldots+\sigma(k)-k(k-1)/2}.$$

Finally, in the case when the matrix A possesses negative eigenvalues or is of rank $< n - 1$, we can say something definite only about the right-hand side of identity (4). To be sure, this right-hand side sometimes looks rather attractive. For example, for

$$A = A_1 = \begin{pmatrix} 2 & -1 \\ -5 & 2 \end{pmatrix}, \quad A = A_2 = \begin{pmatrix} 2 & -3 \\ -3 & 2 \end{pmatrix}$$

the identity (4) acquires the form

$$\prod (1 - t_1^{k_1} t_2^{k_2})^{\dim \mathfrak{g}_{(k_1, k_2)}^{A_1}} = \sum_{j=-\infty}^{\infty} t_1^{a_j} (t_2^{b_j} - t_2^{b_j+1}),$$

$$\prod (1 - t_1^{k_1} t_2^{k_2})^{\dim \mathfrak{g}_{(k_1, k_2)}^{A_2}} = 1 + \sum_{j=1}^{\infty} (-1)^j (t_1^{c_2 j-2-1} t_2^{c_2 j-1} + t_1^{c_2 j-1} t_2^{c_2 j-2-1}),$$

$$(6)$$

where the sequences $\{a_j\}$, $\{b_j\}$ are defined by the equations

$$a_j - 4a_{j-1} + 4a_{j-2} - a_{j-3} = 0, \quad b_j - 4b_{j-1} + 4b_{j-2} - b_{j-3} = 0$$

with "initial data" $a_{-1} = 1$, $a_0 = 0$, $a_1 = 0$, $b_{-1} = 2$, $b_0 = 0$, $b_1 = 1$, while c_i denotes the i-th Fibonacci number ($c_0 = 1$, $c_1 = 1$, $c_2 = 2$, ...). Concerning the exponents $\dim \mathfrak{g}_{(k_1, k_2)}^{A_i}$ we know only that they increase rapidly. Possibly, someday explicit formulas will be found for them, and this will make the identities (6) no less meaningful than the previous ones.

§3. INVARIANT DIFFERENTIAL OPERATORS

1. Introduction. In this section, we consider invariant differential operators in spaces of tensor expressions. In order to determine such an operator, it is necessary to fix the number of its arguments r (this number, known as the "arity" of the operator, comes from the words "unary," "binary," "r-nary," etc.), the number of its independent variables

n, r types of tensor or generalized tensor fields (see 1.2.2) in n-dimensional space, to which the arguments of the operator belong, and one more type of (generalized) tensor fields, to which the values of the operator will belong. The operator must transform r tensor fields of the specified types into a tensor field of the specified type; in local coordinates, it must look like a differential operator and, most important, it must be invariant with respect to changes of variables. In the subsequent subsections of this section, we shall be mainly interested in the case $n = 1$, and will only touch upon the general theory of invariant differential operators. The reader interested in details may find them in the very well-written survey article by Kirillov [57].

The simplest example of an invariant differential operator is the differential of exterior differential forms. This is a unary operator of the first order and, as can be shown, there are no other invariant unary operators of positive order (see [79]). However, the number of binary invariant operators is already very large. The most popular examples are the commutation of vector fields and Lie derivatives; the latter is a binary operator which assigns to a vector field and to a (generalized) tensor field of arbitrary type a tensor field of the same type — the Lie derivative of the given tensor field along the given vector field. Here are three more examples, each of which generalizes the commutation of vector fields. The commutation of polyvector fields (the antisymmetric Schouten concomitant) is defined by the formula

$$[\xi_1 \wedge \cdots \wedge \xi_l, \eta_1 \wedge \cdots \wedge \eta_m] = \sum_{i,j} (-1)^{i+j} [\xi_i, \eta_j] \wedge \xi_1$$

$$\wedge \cdots \hat{\xi}_i \cdots \wedge \xi_l \wedge \eta_1 \wedge \cdots \hat{\eta}_j \cdots \wedge \eta_m,$$

in which $\xi_1, \ldots, \xi_l, \eta_1, \ldots, \eta_m$ are vector fields. This is an
invariant binary operation supplying the space of all smooth
polyvector fields of the manifold by a Lie superalgebra structure. The <u>symmetric Schouten concomitant</u> is defined in a
similar way; it sends sections of the l-th and m-th symmetric
power of the tangent bundle of the manifold into sections of
power $(l + m - 1)$ of this bundle. (Another description of
this operator is the following: Sections of the symmetric
power of the tangent bundle are fiberwise polynomial functions
on the space of the cotangent bundle and to these functions
the Poisson bracket is applied.) Finally, the <u>Nijenhuis</u>
<u>bracket</u> can be applied to two vector-valued differential forms
on the manifold and its result is also a vector-valued form.
The simplest description of this operation is based on the
remark that vector-valued forms may be viewed as (super)der-
ivation in the ring of ordinary exterior differential forms.
Such superderivations constitute, in a natural way, a Lie
superalgebra, and the operation in this superalgebra is the
Nijenhuis bracket. (Note that all the operators listed above
are of first order.)

A complete classification of binary invariant differential operators was recently obtained by Grozman [40]. In particular, it turned out that all invariant binary operators
are of order $\leqslant 3$. Grozman's list is rather long, but with one
exception we can say, without stretching the point too much,
that each of his operators was known to the classics. The
exception is a third-order operator acting in generalized tensor fields on the line. Both of its arguments are of the
form $\varphi(x)\, dx^{-2/3}$, and it ranges over fields of the form $\psi(x) \times$
$dx^{5/3}$. Here is the formula which defines it:

$$\varphi_1(x)\,dx^{-2/3} \otimes \varphi_2(x)\,dx^{-2/3} \mapsto \psi(x)\,dx^{4/3},$$

where

$$\psi(x) = 2\begin{vmatrix} \varphi_1(x) & \varphi_2(x) \\ \varphi_1'''(x) & \varphi_2'''(x) \end{vmatrix} + 3\begin{vmatrix} \varphi_1'(x) & \varphi_2'(x) \\ \varphi_1''(x) & \varphi_2''(x) \end{vmatrix}.$$

This simple operator, whose invariance seems to be a miracle, has heightened the interest to invariant differential operators precisely on the line. Some interesting results concerning this case are presented in [57].

From our point of view, the problem of listing invariant differential operators is that of computing the zero-dimensional cohomology of the algebra W_n with coefficients in certain special modules. As we already pointed out in Chapter 2, in the case $n = 1$ we can move ahead much further in the computation of this cohomology than in the general case, so we should not be surprised by the fact that this case will also be considered separately in the theory of invariant differential operators.

2. Skew-symmetric invariant differential operators on the line. A. Statement of the problem. In Subsection 1, we did not specify to which smoothness class our tensor fields belong. In this connection, there is a considerable freedom of choice, and this choice does not influence the end result. It is most convenient to assume that our tensor fields have polynomial components.

As was pointed out in 1.2.2, the type of generalized tensor field on the line is determined by one complex number λ. The corresponding fields are of the form $\varphi(x)\,dx^{-\lambda}$, where $\varphi(x)$ is a polynomial; they constitute a W_1-module $F\lambda$ (as in

2.3.2, to which the present subsection is closely connected, we denote by W_1 the algebra of <u>polynomial</u> vector fields and understand in a similar sense the algebras and modules L_0, L_1, F_λ, etc.). Thus, an n-ary differential operator on the line is simply a W_1-homomorphism

$$F_{\lambda_1} \otimes \ldots \otimes F_{\lambda_n} \to F_{\lambda_0} \quad (\lambda_0, \lambda_1, \ldots, \lambda_n \in \mathbb{C}). \tag{1}$$

To make this definition agree with Subsection 1, we should have imposed the additional requirement on the homomorphism (1) which would make it a differential operator. However, as we shall see further (see Lemma 3 of this subsection), a W_1-homomorphism (1) is automatically a differential operator and has the differential order $\lambda_1 + \ldots + \lambda_n - \lambda_0$ [in particular, it follows from the existence of the nontrivial invariant operator (1) that $\lambda_1 + \ldots + \lambda_n - \lambda_0$ is a nonzero integer; actually this also follows from the next lemma].

<u>Lemma 1</u>. Every W_1-homomorphism (1) sends $f_{j_1} \otimes \ldots \otimes f_{j_n}$ into a vector which is a multiple of $f_{j_1 + \ldots + j_n - \lambda_1 - \ldots - \lambda_n + \lambda_0}$ (we use the notation introduced in 2.3.2A).

This is an obvious corollary to the fact that the homomorphism (1) commutes with the transformation effected by the element e_0 of the algebra W_1.

Now we note that the W_1-homomorphism $\varphi : F_{\lambda_1} \otimes \ldots \otimes F_{\lambda_n} \to F_{\lambda_0}$ induces a W_1-homomorphism $\varphi' : F'_{\lambda_0} \to F'_{\lambda_1} \otimes \ldots \otimes F'_{\lambda_n}$. (Note that, as in 2.3.2A, we understand F'_λ as the space of <u>finite</u> linear functionals $F_\lambda \to \mathbb{C}$, so that it is necessary to check that the homomorphism φ' is well defined; this verification, however, is obvious in view of Lemma 1.)

Let $s_\varphi = \varphi'(f'_0) \in F'_{\lambda_1} \otimes \ldots \otimes F'_{\lambda_n}$.

Lemma 2. (i) s_φ is a homogeneous element of degree $\lambda_1 + \ldots + \lambda_n - \lambda_0$; (ii) $e_i s_\varphi = 0$ for $i > 0$; (iii) $\varphi \leftrightarrow s_\varphi$ is a bijection between W_1-homomorphisms $F_{\lambda_1} \otimes \ldots \otimes F_{\lambda_n} \to F_{\lambda_0}$ and elements of the space

$$(F'_{\lambda_1} \otimes \ldots \otimes F'_{\lambda_n})_{(\lambda_1 + \ldots + \lambda_n - \lambda_0)},$$

which are annihilated by all the e_i with $i > 0$.

Indeed, (i) follows from Lemma 1, (ii) follows from the fact that $e_i f'_0 = 0$ for $i > 0$, and, in order to prove (iii), it suffices to note that for any i we have

$$f'_i = \frac{(-1)^i}{i!} \underbrace{e_{-1} \ldots e_{-1}}_{i} f'_0,$$

and, therefore, $\varphi'(f'_i) = (-1)^i e^i_{-1} s_\varphi / i!$, so that the homomorphism φ', as well as φ, is well determined by the vector s_φ.

[Note that statement (iii) of Lemma 2 also follows from Theorem 2.2.8.]

A nonzero element of a W_1-, L_0-, or L_1-module annihilated by all the e_i with $i > 0$ is said to be a <u>singular vector</u> of this module.

Lemma 3. Every W_1-homomorphism (1) is a differential operator with constant coefficients of order $\lambda_1 + \ldots + \lambda_n - \lambda_0$.

Proof. Suppose

$$s = \sum_{j_1 + \ldots + j_n = \lambda_1 + \ldots + \lambda_n - \lambda_0} a_{j_1 \ldots j_n} f'_{j_1} \otimes \ldots \otimes f'_{j_n}$$

is a singular vector corresponding to the homomorphism (1) in accordance with Lemma 2 (iii). Consider the map $F_{\lambda_1} \otimes \ldots \otimes F_{\lambda_n} \to F_{\lambda_0}$, given by the formula

$$(\varphi_1 \, dx^{-\lambda_1}) \otimes \ldots \otimes (\varphi_n \, dx^{-\lambda_n})$$

$$\mapsto \sum_{j_1 + \ldots + j_n = \lambda_1 + \ldots + \lambda_n - \lambda_0} \frac{a_{j_1 \ldots j_n}}{j_1! \ldots j_n!} \frac{d^{j_1} \varphi_1}{dx^{j_1}} \ldots \frac{d^{j_n} \varphi_n}{dx^{j_n}} \, dx^{-\lambda_0}. \tag{2}$$

A direct verification shows that the homomorphism (2) commutes with the transformation effected by the element $e_{-1} = d/dx$ of the algebra W_1. Therefore, the same is true for the homomorphism $F_{\lambda_0}' \to F_{\lambda_1}' \otimes \ldots \otimes F_{\lambda_n}'$, adjoint to (2). Obviously, the latter sends f_0' into s, i.e., homomorphisms adjoint to (1) and (2) act in the same way on f_0'. But this means they coincide, since both commute with e_{-1}, while $f_i' = (-1)^i e_{-1}^i f_0'/i!$. Therefore, the homomorphisms (1) and (2) also coincide. The lemma is proved.

Further we limit ourselves to skew-symmetric operators, i.e., W_1-homomorphisms $\Lambda^n F_\lambda \to F_{\lambda_0}$. By Lemma 2, such homomorphisms correspond to homogeneous singular vectors of degree $n\lambda - \lambda_0$ of the module $\Lambda^n F_\lambda'$, while by Lemma 3 they are represented by the formula

$$(\varphi_1 \, dx^{-\lambda}) \wedge \ldots \wedge (\varphi_n \, dx^{-\lambda})$$

$$\mapsto \sum_{\substack{0 \leqslant j_1 < \ldots < j_n, \\ j_1 + \ldots + j_n = N}} b_{j_1 \ldots j_n} \begin{vmatrix} \varphi_1^{(j_1)} & \ldots & \varphi_1^{(j_n)} \\ \cdots & \cdots & \cdots \\ \varphi_n^{(j_1)} & \ldots & \varphi_n^{(j_n)} \end{vmatrix} dx^{-(n\lambda - N)} \tag{3}$$

with $b_{j_1 \ldots j_n} \in \mathbb{C}$. It is already clear from this formula that if an invariant operator $\Lambda^n F_\lambda \to F_{\lambda_0}$ is nontrivial, then its order $N = n\lambda - \lambda_0$ is no less than $\frac{n(n-1)}{2}$. The difference $N - \frac{n(n-1)}{2}$ shall be called the <u>true order</u> of our operator.

B. <u>Statement of the main theorem</u>. The following statement is proved in [18].

<u>Theorem 3.3.1</u>. For all $\lambda \in \mathbb{C}$, $n > 0$, $k \in \mathbb{Z}$ there exists no more than 1 (up to proportionality) invariant differential operator

$$\Lambda^n F_\lambda \rightarrow F_{n\lambda - \frac{n(n-1)}{2} - k}$$

(of true order k). For its existence it is necessary and sufficient to have one of the following conditions:

(i) $k = 0$;

(ii) $0 < k \leqslant n$ and λ satisfies the quadratic equation

$$\left[\left(\lambda + \frac{1}{2}\right)(k' + 1) - n\right]\left[(\lambda + \frac{1}{2})(k'' + 1) - n\right] = \frac{(k'' - k')^2}{2}, \qquad (4)$$

in which k', k'' are arbitrary complementary divisors of the number k (i.e., $k' \in \mathbb{Z}_+$, $k'' \in \mathbb{Z}_+$, $k'k'' = k$).

The part of this theorem corresponding to the case $k \leqslant 0$ is covered by what we mentioned in Sub-subsection A. Indeed, the remark made at the end of that subsection shows that the true order of a skew-symmetric invariant differential operator cannot be negative. As to operators of true order 0, they correspond to singular vectors of the module $\Lambda^n F_\lambda'$, of degree $n(n-1)/2$; but

$$\dim (\Lambda^n F_\lambda')_{(r)} = \begin{cases} 0 & \text{for } r < n(n-1)/2, \\ 1 & \text{for } r = n(n-1)/2, \end{cases}$$

and, therefore, for any λ there is a singular vector in $(\Lambda^n F_\lambda')_{(n(n-1)/2)}$, and it is unique up to proportionality. The corresponding operator

$$\Lambda^n F_\lambda \rightarrow F_{n\lambda - \frac{n(n-1)}{2}}$$

acts in accordance with the formula

$$(\varphi_1\, dx^{-\lambda}) \wedge \cdots \wedge (\varphi_n\, dx^{-\lambda}) \;\mapsto\; \begin{vmatrix} \varphi_1 & \varphi_1' & \cdots & \varphi_1^{(n-1)} \\ \varphi_2 & \varphi_2' & \cdots & \varphi_2^{(n-1)} \\ \cdots & \cdots & \cdots & \cdots \\ \varphi_n & \varphi_n' & \cdots & \varphi_n^{(n-1)} \end{vmatrix} dx^{-n\lambda + \frac{n(n-1)}{2}}. \tag{5}$$

It is denoted by $\Delta_{\lambda,\,n}$ and called <u>general position operator</u>.

The proof of the other parts of Theorem 3.3.1 are considerably less trivial. They require a deep understanding of projective representations of Lie algebras of vector fields on the circle which I shall not give here. A small extract from this proof, which explains the role played by the divisors of the true order of an operator, is given below in Sub-subsection D. The next subsection is devoted for the most part to specific examples.

C. <u>Commentary to Theorem 3.3.1</u>. First of all, Theorem 3.3.1 shows that all nontrivial skew-symmetric invariant differential operators on the line are of true order $\leqslant n$, i.e., of order $\leqslant \frac{n(n+1)}{2}$. In this connection, it should be noted that at present no n-ary invariant differential operators of order $> \frac{n(n+1)}{2}$ are known, neither skew-symmetric nor others, neither on the line nor on spaces of higher dimension.

If $0 < k \leqslant n$, then n-ary skew-symmetric invariant differential operators on the line exist for specially chosen values of λ (this explains the term "general position operators"). These λ can be found from Eq. (4). As shown by a computation, under condition (ii) of Theorem 3.3.1, Eq. (4) has real roots which coincide if and only if $k'' = k'$. These roots are positive, except for a single exception: if $k = n$ and $\min(k', k'') = 1$, then one of the roots is 0. Further, suppose d is the number of positive integer divisors of the number k and $k_1 = 1, k_2, \ldots, k_{d-1}, k_d = k$ is the increasing sequence of these divisors. For indices i, j, such that $i \leqslant j$ and $k_i k_j = k$, denote by λ_i, λ_j the roots of Eq. (4) (in which $k' = k_i$, $k'' = k_j$) disposed in nondecreasing order. The λ_1, $\ldots, \lambda_d$ is also an increasing sequence. This sequence, of course, depends on n, since Eq. (4) depends on n. The numbers $\lambda_1, \ldots, \lambda_d$ for some k and n are presented in Table 2.

Some of our operators have a fairly simple description. For $n = 1$ we have the general position operator

$$\Delta_{\lambda, 1}: F_\lambda \rightarrow F_\lambda ,$$

which is the identity, and one first-order operator $d: F_0 \rightarrow F_{-1}$, corresponding to the divisor 1 of the number 1; it is the derivation of functions.

For $n = 2$ the general position operator

$$\Delta_{\lambda, 2}: \Lambda^2 F_\lambda \rightarrow F_{2\lambda-1}$$

is of order 1. Moreover, there is one operator of true order 1, i.e., order 2,

$$\Lambda^2 F_{1/2} \rightarrow F_{-1},$$

corresponding to the divisor 1 of the number 1: it is the composition $d \circ \Delta_{1/2, 2}$ and two operators of true order 2, i.e., of order 3 (corresponding to the divisors 1 and 2 of the number 2)

$$\Lambda^2 F_0 \rightarrow F_{-3}, \quad \Lambda^2 F_{2/3} \rightarrow F_{-5/3}.$$

The first of these operators is the composition $\Delta_{-1,2} \circ \Lambda^2 d$, the second is none other than the Grozman operator mentioned in Subsection 1. For any n the only n-ary operator of true order 1 is the composition of the general position operator $\Delta_{(n-1)/2, n}$, assuming values in F_0, and the operator $d: F_0 \rightarrow F_{-1}$.

TABLE 2

n	k	$\lambda_1, ..., \lambda_d$	n	k	$\lambda_1, ..., \lambda_d$
1	0	0	4	3	$(2 \pm \sqrt{2})/2$
2	1	1/2	4	4	0, 5/6, 9/5
2	2	0, 2/3	...	...	....
3	1	1			$\dfrac{n-1}{2}$
3	2	$(9 \pm \sqrt{21})/12$	n	1	
3	3	0, 5/4	...	...	....
4	1	3/2			$0, ..., \dfrac{(n-1)(n+2)}{2(n+1)}$
4	2	$(7 \pm \sqrt{7})/6$	n	n	
			...	...	....

It is just as simple to construct for any n the operator of true order n corresponding to the divisor 1 of the number n. This is the operator $\Lambda^n F_0 \to F_{-n(n+1)/2}$, which is the composition $\Delta_{-1,n} \circ \Lambda^n d$.

A simple direct description also exists for the n-ary operator of true order n corresponding to the divisor n of the number n and directly generalizing the Grozman operator. In this case, $\lambda = \dfrac{(n-1)(n+2)}{2(n+1)}$; denote this fraction by $\lambda(n)$. As can be seen directly, the operator $\Delta_{\lambda(n),n+1}$ assumes values in F_{-1}. Since $\Delta_{\lambda(n),n+1}$ is a differential operator, it can be extended to a W_1-homomorphism $\Lambda^{n+1} \mathcal{F}_{\lambda(n)} \to \mathcal{F}_{-1}$ (see 2.3.2). The composition of this homomorphism with the W_1-homomorphism $\mathcal{F}_{-1} \to \mathbb{C}$, sending a differential form into its residue, is the W_1-homomorphism $\Lambda^{n+1} \mathcal{F}_{\lambda(n)} \to \mathbb{C}$. The latter determines a W_1-homomorphism $\Lambda^n \mathcal{F}_{\lambda(n)} \to \mathcal{F}'_{\lambda(n)}$, and since $\mathcal{F}'_{\lambda(n)} = \mathcal{F}_{-1-\lambda(n)}$ (see 2.3.2A), we obtain the W_1-homomorphism $\Lambda^n \mathcal{F}_{\lambda(n)} \to \mathcal{F}_{-1-\lambda(n)}$. Obviously the latter also is a differential operator with constant coefficients and therefore maps $\Lambda^n F_{\lambda(n)}$ into $F_{-1-\lambda(n)}$. We have obtained a W_1-homomorphism $\Lambda^n F_{\lambda(n)} \to F_{-1-\lambda(n)}$, that is,

$$\Lambda^n F_{\frac{(n-1)(n+2)}{2(n+1)}} \rightarrow F_{-\frac{n(n+3)}{2(n+1)}} \cdot$$

This is precisely our operator.

A few more operators from Theorem 3.3.1, unfortunately far from all of them, can be given by rather simple explicit formulas (see [16, 18]). For example, for any $k = 1, \ldots, n$ the operators

$$\Lambda^n F_\lambda \rightarrow F_{n\lambda - \frac{n(n-1)}{2} - k},$$

corresponding to the divisors $1, k$ of the number k, are defined by formula (3), in which $b_{j_1 \ldots j_n} = b^{j_{n-k+1} - (n-k), \ldots, j_n - (n-1)}$,

$$b^{l_1, \ldots, l_k} = \frac{1}{l_1! \ldots l_k!} \begin{vmatrix} 1 & 1 & \cdots & 1 \\ -1 & l_1 & \cdots & l_k \\ \cdots & \cdots & \cdots & \cdots \\ (-1)^k & l_1^k & \cdots & l_k^k \end{vmatrix} \prod_{s=1}^{k} \prod_{t=s}^{l_s} \frac{(2\lambda - n + 1)(t - k) + k - 1}{2(t+1)} \cdot$$

In conclusion, note that the statements of Theorems 2.3.5a-e and 3.3.1 are very similar. In all of these theorems we are concerned with the cohomology of the algebra W_1 with coefficients in modules depending on parameters and, every time, it turns out that singular values of the parameters to which "jumps" of the cohomology correspond are located on families of second-order curves described by elegant formulas. One would like to hope that behind all this there is some general fact — perhaps relating to the structure of the Laplace operator spectrum in the chain complex of the algebra W_1 (see 2.3.1b) and to the theory of representations of Lie algebras of vector fields on the circle (see [18, 54]). However, at present, not even a hypothetical formulation of this fact has been found.

D. Hint of the proof of Theorem 3.3.1. Our problem consists in finding homogeneous singular vectors of degree
$\frac{n(n-1)}{2} + k$ in the module $\Lambda^n F'_\lambda$. Such a vector must be of
the form

$$\sum_{\substack{j_1 < \ldots < j_n, \\ j_1 + \ldots + j_n = \frac{n(n-1)}{2} + k}} a_{j_1 \ldots j_n} f'_{j_1} \wedge \cdots \wedge f'_{j_n}.$$

Its singularity condition, consisting of the fact that it is
annihilated by vector fields e_i for $i = 1, 2, \ldots, k,$ can be
written in the form of the system of equations

$$\sum_{s=1}^{n} (j_s - (i+1)\lambda)\, a_{j_1 \ldots j_s + i \ldots j_n} = 0 \qquad (6_i)$$

$$\left(j_1 < \cdots < j_n, j_1 + \cdots + j_n = \frac{n(n-1)}{2} + k - i, i = 1, 2, \ldots, k \right),$$

where it was assumed that $a_{m_1 \ldots m_n}$ is defined not only for
$m_1 < \ldots < m_n$, but depends on $m_1, \ldots, m_n$ in a skew-symmetric
way. In system (6), as a rule, there are more equations than
unknowns. We must find out for what λ the system has solutions and find them for these λ. Note that since $[e_1, e_i] =$
$(i - 1) e_{i+1},$ Eqs. (6_i) for $i \geqslant 3$ are consequences of Eqs.
(6_i) for $i = 1, 2,$ and it suffices to solve the system constituted by Eqs. $(6_1), (6_2)$.

If $k > n,$ it can be shown that our system has no solutions for any λ; this case will not be considered here. When
$k \leqslant n,$ it follows from $j_1 < \ldots < j_n, j_1 + \cdots + j_n = \frac{n(n-1)}{2} + k$
that $j_1 = 1, \ldots, j_{n-k} = n - k,$ and we put

$$a^{l_1 \ldots l_k} = a_{1, \ldots, n-k, \, n-k+1+l_1, \ldots, n+l_k}.$$

Our system acquires the form

$$\sum_{s=1}^{k} (n - k + s + l_s - (i + 1)\lambda) a^{l_1 \ldots l_k} = 0$$

$$(0 \leqslant l_1 \leqslant \ldots \leqslant l_k, \; l_1 + \ldots + l_k = k, \; i = 1, 2, \ldots, k). \qquad (7\mathrm{i})$$

As we see, the family of unknowns now depends only on k, while
n and λ appear in the coefficients and may be viewed as parameters, and the parameter n is now allowed to assume arbitrary
complex values. Denote by S_k the subset of the complex
plane $\mathbb{C}^2$, constituted by those (n, λ) for which the system
(7) has a nontrivial solution. In [18] it is proved that S_k
is the union of the curves (4) and also of a finite number
of isolated points for which n is either purely imaginary or
real but less than k, and this constitutes the main part of
the proof of Theorem 3.3.1. Here we limit ourselves to the
following simple proposition, whose proof comes through in
a rather unexpected way.

Proposition 3.3.2. Except for its isolated points, S_k
is an algebraic curve, determined by an equation whose principal term is a divisor of the product

$$(n - (k_1 + 1)\lambda) \ldots (n - (k_d + 1)\lambda),$$

where $k_1, \ldots, k_d$ are all the natural divisors of the number k.

Proof. Suppose Σ_k is the space of skew-symmetric polynomials in n variables of degree $\dfrac{n(n-1)}{2} + k$; such polynomials
can be represented in the form $\Delta \cdot P(N_1, N_2, \ldots)$, where Δ is
the product of all differences of variables, N_s is the sum
of the s-th powers of the variables, and P is a polynomial.
Suppose further $v_s: \Sigma_{k-s} \to \Sigma_k$ is the multiplication by N_s.

Divide the coefficients of the system $(7\mathrm{i})$ by λ and let
n and λ tend to ∞ so that the quotient $\rho = n/\lambda$ remains
constant. Then the matrix of the system $(7\mathrm{i})$ at the limit

will become the transposed matrix of the operator v_i (in the basis constituted by alternating monomials) multiplied by $\rho - (i + 1)$.   Note that, as a result of this limiting process, the systems (7_i) for $i \geqslant 3$ will no longer be linearly expressible in terms of (7_1) and (7_2).

The dimension of the solution space of the limit system equals, therefore, the codimension in Σ_k of the space

$$\sum_{i=1}^{k} \operatorname{Im}[\rho - (i + 1)] v_i. \tag{8}$$

If ρ does not equal $2, 3, \ldots, k + 1$, then the space (8) coincides with Σ_k; in the case $\rho = i + 1$ for some i, (8) is the space of polynomials of the form

$$\Delta (N_1 P_1 + \ldots \widehat{N_i P_i} \ldots + N_k P_k),$$

i.e., has codimension 1 in Σ_k when i divides k (in this case the polynomial $\Delta N_i^{k/i}$ does not appear in it) and coincides with Σ_k in the converse case.   Thus, the limit system has a unique (up to proportionality) nontrivial solution if $\rho = i + 1$, where i is a divisor of k, and has no solutions in the converse case.   Therefore, S_k can have no more than one asymptote of multiplicity one with slope $k_s + 1$ $(s = 1, \ldots, d)$ and cannot have any other asymptotes.   The proposition is proved.

3.   General position operators.   The problem of listing all (not necessarily skew-symmetric) invariant differential operators on the line consists in describing, for any $\lambda_1, \ldots,$ λ_n , the spaces

$$D (\lambda_1, \ldots, \lambda_n) = \bigoplus_{\lambda} \operatorname{Hom}_{W_1} (F_{\lambda_1} \otimes \ldots \otimes F_{\lambda_n}, F_\lambda).$$

It follows from our remarks in Subsection 1 that

$$D\ (\lambda_1,\ \ldots,\ \lambda_n)\ =\oplus_{k\geqslant 0}D^k\ (\lambda_1,\ldots,\ \lambda_n),$$

where $D^k\ (\lambda_1,\ \ldots,\ \lambda_n)\ =\ \mathrm{Hom}_{W_1}\ (F_{\lambda_1}\otimes\ \cdots\ \otimes F_{\lambda_n},\ F_{\lambda_1+\ldots+\lambda_n-k})$ is the space of order k operators. It follows from general considerations that $\dim D^k\ (\lambda_1,\ \ldots,\ \lambda_n)<\infty$ for each k (actually, we also have $\dim D\ (\lambda_1,\ \ldots,\ \lambda_n)<\infty$ and even $\dim D\ (\lambda_1,\ \ldots,\ \lambda_n)\leqslant (n+1)!$ see [57]). It follows from the same general considerations that there exists an open set (in the Zariski topology) $\Lambda\subset\mathbb{C}^n$, such that the dimensions $d_{n,k}\ =d_k$ of the space $D^k\ (\lambda_1,\ \ldots,\ \lambda_n)$ for $(\lambda_1,\ \ldots,\ \lambda_n)\in\Lambda$ do not depend on $\lambda_1,\ \ldots,$ λ_n. In other words, there exist d_k continuous families

$$\varphi_i(\lambda_1,\ldots,\ \lambda_n)\colon F_{\lambda_1}\otimes\ldots\otimes F_{\lambda_n}\to F_{\lambda_1+\ldots+\lambda_n-k}\quad (i=1,\ldots,\ d_k)$$

of invariant differential operators of order k, such that for $(\lambda_1,\ \ldots,\ \lambda_n)\in\Lambda$ the operators $\varphi_i\ (\lambda_1,\ \ldots,\ \lambda_n)$ constitute a basis in the space $D^k\ (\lambda_1,\ \ldots,\ \lambda_n)$.

Concerning the numbers d_k there is a rather plausible conjecture.*

Conjecture 3.3.3.

$$\sum_k d_{n,\,k}t^k=\prod_{i=1}^{n-1} (1+t+\ldots+t^i).\qquad(9)$$

Consider in more detail the case of small n.

The case $n=1$. According to the conjecture, $d_{1,0}=1$ and $d_{1,k}=0$ for $k>0$. Actually, for a generic λ the only invariant differential operator (up to proportionality) defined on F_λ is the identity operator $F_\lambda\to F_\lambda$; its order is 0.

---

*See Subsection 3 of the Appendix.

The case $n = 2$. According to the conjecture, $d_{2,0} = d_{2,1} = 1$ and $d_{2,k} = 0$ for $k > 1$. The binary operator of general position of the zeroth order $F_{\lambda_1} \otimes F_{\lambda_2} \to F_{\lambda_1 + \lambda_2}$ is the multiplication of tensors

$$(\varphi_1 \, dx^{-\lambda_1}) \otimes (\varphi_2 \, dx^{-\lambda_2}) \mapsto (\varphi_1 \, dx^{-\lambda_1})(\varphi_2 \, dx^{-\lambda_2}) = \varphi_1 \varphi_2 \, dx^{-\lambda_1 - \lambda_2}.$$

The binary general position operator, which is of the first order, is described just as simply:

$$(\varphi_1 \, dx^{-\lambda_1}) \otimes (\varphi_2 \, dx^{-\lambda_2}) \mapsto \{\varphi_1 \, dx^{-\lambda_1}, \varphi_2 \, dx^{-\lambda_2}\} = (\lambda_2 \varphi_1' - \lambda_1 \varphi_2') \, dx^{-\lambda_1 - \lambda_2 + 1}.$$

(In order to explain the origins of this operator, it is convenient to write y^λ instead of $dx^{-\lambda}$ and note that the operator is an ordinary Poisson bracket; this bracket is invariant not only with respect to the algebra W_1, but with respect to the Poisson algebra $P = \mathbb{C}[x, y]$ which contains it: the algebra W_1 is included in P as the subalgebra $y\mathbb{C}[x]$.)

The case $n = 3$. According to the conjecture, $d_{3,0} = 1$, $d_{3,1} = d_{3,2} = 2$, $d_{3,3} = 1$, and $d_{3,k} = 0$ for $k > 3$. The ternary general position operator, which is of zero order, is also a multiplication of tensors:

$$(\varphi_1 \, dx^{-\lambda_1}) \otimes (\varphi_2 \, dx^{-\lambda_2}) \otimes (\varphi_3 \, dx^{-\lambda_3}) \mapsto$$

$$(\varphi_1 \, dx^{-\lambda_1})(\varphi_2 \, dx^{-\lambda_2})(\varphi_3 \, dx^{-\lambda_3}) = \varphi_1 \varphi_2 \varphi_3 \, dx^{-\lambda_1 - \lambda_2 - \lambda_3}.$$

General position ternary operators, which are of first and second order, may be constituted by using the binary operators already constructed:

$$\xi_1 \otimes \xi_2 \otimes \xi_3 \mapsto \xi_1 \{\xi_2, \xi_3\}, \ \xi_2 \{\xi_3, \xi_1\}, \ \xi_3 \{\xi_1, \xi_2\};$$
$$\xi_1 \otimes \xi_2 \otimes \xi_3 \mapsto \{\xi_1, \{\xi_2, \xi_3\}\}, \{\xi_2, \{\xi_3, \xi_1\}\}, \{\xi_3, \{\xi_1, \xi_2\}\}.$$

These operators satisfy the relations

$$\xi_1 \{\xi_2, \xi_3\} + \xi_2 \{\xi_3, \xi_1\} + \xi_3 \{\xi_1, \xi_2\} = 0,$$
$$\{\xi_1, \{\xi_2, \xi_3\}\} + \{\xi_2, \{\xi_3, \xi_1\}\} + \{\xi_3, \{\xi_1, \xi_2\}\} = 0,$$

the first of which is similar to the Jacobi identity, while
the second is precisely the Jacobi identity. There are no
other relations, so that the previously mentioned operators
for generic λ_1, λ_2, λ_3 generate the spaces $D^1(\lambda_1, \lambda_2, \lambda_3)$,
$D^2(\lambda_1, \lambda_2, \lambda_3)$, which are two-dimensional in view of the above.
There also exists a ternary general position operator of the
third order $F_{\lambda_1} \otimes F_{\lambda_2} \otimes F_{\lambda_3} \to F_{\lambda_1+\lambda_2+\lambda_3-3}$, and clearly it cannot
be expressed in terms of the above-mentioned binary operators.
Here is the formula which determines it:

$$(\varphi_1 \, dx^{-\lambda_1}) \otimes (\varphi_2 \, dx^{-\lambda_2}) \otimes (\varphi_3 \, dx^{-\lambda_3})$$

$$\mapsto \Big\{ \sum_\tau \lambda_{\tau(2)} \lambda_{\tau(3)} (\lambda_{\tau(3)} - \lambda_{\tau(2)}) (\lambda_{\tau(3)} + \lambda_{\tau(2)} - 1) \; \varphi_{\tau(1)}''' \varphi_{\tau(2)} \varphi_{\tau(3)}$$

$$+ \sum_\sigma \operatorname{sgn} \sigma \, \lambda_{\sigma(3)} (\lambda_{\sigma(2)} + \lambda_{\sigma(3)} - 1) \, [(\lambda_{\sigma(3)} - \lambda_{\sigma(2)})$$

$$+ (3\lambda_{\sigma(2)} - 2)(\lambda_{\sigma(1)} + \lambda_{\sigma(3)} - 1)] \, \varphi_{\sigma(1)}'' \varphi_{\sigma(2)}' \varphi_{\sigma(3)}$$

$$- (\lambda_2 - \lambda_1)(\lambda_3 - \lambda_2)(\lambda_1 - \lambda_3) \, \varphi_1' \varphi_2' \varphi_3' \Big\} \, dx^{-\lambda_1 - \lambda_2 - \lambda_3 + 3},$$

where the first sum is taken over all cyclic permutations of
the set $\{1, 2, 3\}$, and the second over all the permutations
of this set. Note that for almost all triples $\lambda_1, \lambda_2, \lambda_3$ this
is the only, up to proportionality, invariant third-order
operator.

Note that although Conjecture 3.3.3 is not proved, the
validity of the statements concerning unary, binary, and ternary operators may be checked by a simple verification.

In conclusion, let us mention a generalization of conjecture 3.3.3.

Conjecture 3.3.4. For all $q, k, \lambda_1, \ldots, \lambda_n$ we have

$$\dim H_{(k)}^q (L_1; F_{\lambda_1} \otimes \ldots \otimes F_{\lambda_n}) \geqslant \sum_{i+j=k} d_i \dim H_{(j)}^q (L_2),$$

where the d_i are defined by formula (9) and, for fixed q and k, we have the equality on an open (in the sense of Zariski) set in $\mathbb{C}^n$ of finite sequences $(\lambda_1, \ldots, \lambda_n)$.

For $n = 1$ this statement follows from Theorem 2.3.5b.

The generalization of Conjectures 3.3.3 and 3.3.4 to the multidimensional case and some preliminary results on this topic are contained in Tabachnikov's note [87].

§4. COHOMOLOGY OF LIE ALGEBRAS AND COHOMOLOGY OF LIE GROUPS

The limited volume of this book allows us to touch only slightly on the cohomology of groups. This cohomology interests us only in its connection with the cohomology of Lie algebras, but it is actually an independent and important mathematical subject. The reader interested in details may refer to the specialized literature; he may familiarize himself with the algebraic side of the question by using any of the textbooks on homological algebra (for example [11]), while the topological aspects of the theory, which are closer to the contents of this section, are developed in [41].

1. **Main definitions**. In this subsection we concern ourselves with a given real Lie group G.

There exists a large number of cohomologies related (in one way or the other) with G. Even if we limit ourselves to coefficients in $\mathbb{R}$ we can say that there are at least five cohomology rings associated with G which might all be termed well known. Let us list them.

1°. The cohomology of the Lie algebra $\mathfrak{g}$ of the group G already known to us.

2°. The cohomology of the group G as a topological space $H^*(G; \mathbb{R})$. To distinguish it from the others, we shall write $H^*_{\text{top}}(G)$.

3°. The cohomology of the classifying space BG of the group G, $H^*(BG; \mathbb{R})$.

4°. The cohomology of the group G as an abstract group, which shall be denoted by $H^*_{\text{alg}}(G)$. They are defined by means of the multiplicative complex $C^{\cdot}_{\text{alg}}(G)$, which has the following two equivalent descriptions: (i) $C^q_{\text{alg}}(G)$ is the space of functions $c: \underbrace{G \times \ldots \times G}_{q+1} \to \mathbb{R}$, not assumed continuous, but invariant with respect to the left action of the group G,   i.e., such that $c(gg_0, \ldots, gg_q) = c(g_0, \ldots, g_q)$ for all $g, g_0, \ldots, g_q \in G$. The differential $d: C^q_{\text{alg}}(G) \to C^{q+1}_{\text{alg}}(G)$ is defined by the formula

$$dc(g_0, \ldots, g_{q+1}) = \sum_{i=1}^{q+1} (-1)^i c(g_0, \ldots \hat{g}_i \ldots, g_{q+1}). \tag{1}$$

The product $c_1 c_2$ of cochains $c_1 \in C^{q_1}_{\text{alg}}(G), c_2 \in C^{q_2}_{\text{alg}}(G)$ is defined by the formula

$$c_1 c_2 (g_0, \ldots, g_{q_1+q_2}) = c_1(g_0, \ldots, g_{q_1}) c_2(g_{q_1}, \ldots, g_{q_1+q_2}).$$

(ii) $C^q_{\text{alg}}(G)$ is the space of entirely arbitrary functions $\underbrace{G \times \ldots \times G}_{q} \to \mathbb{R}$, the differential $d: C^q_{\text{alg}}(G) \to C^{q+1}_{\text{alg}}(G)$ is defined by the formula

$$dc(g_1, \ldots, g_{q+1}) = c(g_2, \ldots, g_{q+1}) + \sum_{i=1}^{q} (-1)^i c(g_1, \ldots, g_i g_{i+1}, \ldots, g_{q+1})$$
$$+ (-1)^{q+1} c(g_1, \ldots, g_q), \tag{2}$$

and the multiplication of cochains by the formula

$$c_1 c_2 (g_1, \ldots, g_{q_1+q_2}) = c_1(g_1, \ldots, g_{q_1}) c_2(g_{q_1+1}, \ldots, g_{q_2}).$$

The isomorphism between the complexes (i) and (ii) is established by means of the correspondence

$$\{c: \underbrace{G \times \ldots \times G \to \mathbb{R}}_{q+1}\} \leftrightarrow \{c': \underbrace{G \times \ldots \times G \to \mathbb{R}}_{q}\},$$
$$c'(g_1, \ldots, g_q) = c(1, g_1, g_1 g_2, \ldots, g_1 g_2 \ldots g_n).$$

5°. If we modify the previous definition by including in the description of the complex $C_{alg}^{\cdot}(G)$ (in any of its two versions) the requirement that the cochains be functions of class $\mathscr{C}^\infty$, we come to the ring of continuous or Van Est cohomology (see [92]) $H_{alg, c}^{*}(G) = H_c^{*}(G)$. It can be shown that these cohomologies do not change if the requirement of belonging to the class $\mathscr{C}^\infty$ is weakened to that of continuity (but they will be completely different if, say, measurable cochains are allowed).

Definitions 2°-5° remain valid for any topological group G (in 5°, of course, the continuous version is understood). All five definitions are meaningful for some infinite-dimensional Lie groups such as the group of diffeomorphisms of a compact manifold or the current group (i.e., the group of smooth maps of a smooth manifold into a Lie group). It should be mentioned that for Lie groups of positive dimension, the cohomologies given by definition 4°, when they are not trivial, turn out to be huge as a rule. They are interesting not in themselves, but in connection with the other definitions; for example, an important invariant of the group G is the kernel of the homomorphism $H_{alg, c}^{*}(G) \to H_{alg}^{*}(G)$, induced by the inclusion of the complex 5° in the complex 4°.

Definitions 4° and 5° may be generalized by introducing coefficients: an arbitrary G-module A (i.e., an Abelian group in which G acts by group automorphisms). Each of the

definitions (i), (ii) generalizes to this case. In the generalized definition (i), the space of cochains $C_{alg}^q(G)$ is de-
fined as the set of G-maps $\underbrace{G \times \ldots \times G}_{q+1} \to A$, while the differential is defined by the same formula (1); in the generalized
definition (ii), the space of cochains is defined as the set
of arbitrary maps $\underbrace{G \times \ldots \times G}_{q} \to A$, while the differential is
defined by the formula

$$dc(g_1, \ldots, g_{q+1}) = g_1 c(g_2, \ldots, g_{q+1}) + \sum_{i=1}^{q} (-1)^i c(g_1, \ldots, g_i g_{i+1}, \ldots, g_{q+1})$$

$$+ (-1)^{q+1} c(g_1, \ldots, g_q). \tag{2}$$

The generalized definition 5° is obtained from the generalized
definition 4° by adding the smoothness or continuity requirement for cochains.

It is hardly worthwhile to list all the algebraic and
geometric interpretations of cohomology of groups. (Among
these interpretations there are some rather curious ones; for
example, elements of the group $H_c^*(G)$ correspond to characteristic classes of principal G-bundles with not necessarily
paracompact bases, ranging in the cohomology of the base with
the coefficients in the sheaf of germs of continuous functions — see [26, 81].) I will limit myself to the remark
that the group $H_c^2(G)$ is the set of classes of topological
central extensions of the group G by means of $\mathbb{R}$, i.e., group
structures in $G \times \mathbb{R}$, such that

$$(g_1, t_1)(g_2, t_2) = (g_1 g_2, t_1 + t_2 + \varphi(g_1, g_2)), \tag{3}$$

where φ is a smooth map $G \times G \to \mathbb{R}$. Any two such extensions are assumed equivalent if they are mapped into each
other by the map $G \times \mathbb{R} \to G \times \mathbb{R}$, acting according to the

formula $(g, t) \mapsto (g, t + \psi(g))$, where ψ is a smooth function on G. As shown by verification, the associativity condition for the multiplication (3) coincides with the requirement that the function φ be a cocycle in $C^2_c(G)$ [in the sense of definition 4° (ii)], while the equivalence of extensions means the co-cycles are cohomologous.

Let us add that it is sometimes necessary to consider central extensions not by means of $\mathbb{R}$, but by means of another Abelian group A (for example S^1). Such extensions are clas-sified by the cohomology $H^2_c(G; A)$ with coefficients in the group A, viewed as a trival G-module.

2. Relations between the main definitions. If the group G is discrete, then the definitions 3°, 4°, 5° deter-mine the same object [namely $H^*(K(G, 1); \mathbb{R})$], while defini-tions 1°, 2° are meaningless. In the general case, we have the natural homomorphisms $H^*(\mathfrak{g}) \to H^*_{\mathrm{top}}(G)$ (see 1.3.3 and 2.1.1) and $H^*_{\mathrm{alg,\,c}}(G) \to H^*_{\mathrm{alg}}(G)$ (see Subsection 1), already known to us, and the natural homomorphism $H^*_{\mathrm{alg,\,c}}(G) \to H^*(\mathfrak{g})$. The lat-ter can easily be described on the cochain level by using the first description of the complex $C_{\mathrm{alg,\,c}}(G)$; the cochain $c \in C^q_c(G)$ is mapped to the cochain $\bar{c} \in C^q(\mathfrak{g})$, defined by the formula

$$\bar{c}(\gamma_1, \ldots, \gamma_q) = \frac{d}{dt} c(1, \exp t\gamma_1, \ldots, \exp t\gamma_q)|_{t=0},$$

where $\{\exp t\gamma\}$ is the one-parameter subgroup of the group G, defined by the element γ of the algebra $\mathfrak{g}$; this map $c \mapsto \bar{c}$, commutes with the differential, as can be established by a direct verification. It is also easy to show that the same formula defines the homomorphism $H^*_c(G; A) \to H^*(\mathfrak{g}; A)$, where A is a linear G-module, simultaneously viewed as a $\mathfrak{g}$-module.

Note that our homomorphism is compatible with the interpretation of the cohomology $H^2(\mathfrak{g})$ and $H_c^2(G)$ as sets of central
extensions (see Subsections 1.4.6 and 1): if

$$1 \to \mathbb{R} \to \tilde{G} \to G \to 1$$

is a central extension of the group G and

$$0 \to \mathbb{R} \to \tilde{\mathfrak{g}} \to \mathfrak{g} \to 0$$

is the corresponding extension of the algebra $\mathfrak{g}$, then our
homomorphism $H_c^2(G) \to H^2(\mathfrak{g})$ sends the class corresponding
to the first extension into the class corresponding to the
second one. An analogous statement holds for the similar
homomorphism $H_c^2(G; S^1) \to H^2(\mathfrak{g})$.

A more serious relationship between the definitions of
Subsection 1 is established by the <u>Hochschild–Mostow spectral</u>
<u>sequence</u> [51], in which the cohomologies $H^*(\mathfrak{g})$, $H_{top}^*(G)$ and
$H_{alg,c}^*(G)$ participate. To construct this sequence, we shall
need two lemmas. In the statements and proofs of these lemmas, we use the second definition of the complex $C_{alg,c}^{\cdot}(G)$.

Lemma 1.

$$\dim H_c^q(G; \mathscr{C}^\infty(G)) = \begin{cases} 1 & \text{for } q = 0, \\ 0 & \text{for } q > 0. \end{cases}$$

The space $H_c^0(G; \mathscr{C}^\infty(G))$ consists of classes of cocycles $g \mapsto$
const.

Proof. Define the map $D: C_c^q(G; \mathscr{C}^\infty(G)) \to C_c^{q-1}(G; \mathscr{C}^\infty(G))$
by the formula $[Dc(g_1, \ldots, g_{q-1})](g) = [c(g_1, \ldots, g_{q-1}, g)](1)$.
As shown by direct verification, D is the homotopy joining
the identity map of the complex $C_c^{\cdot}(G; \mathscr{C}^\infty(G))$ with the map
which is trivial on chains of positive dimension and defined
on zero-dimensional ones by the formula $c \mapsto \{g \mapsto [c(1)](1)\}$.

Remark. Our lemma allows us to compute the cohomology
of the group G with coefficients in any G-module which is
a direct summand in $\mathscr{C}^\infty(G)$ as well: these cohomologies are
trivial if our summand does not contain constants and trivial
in positive dimensions if it does contain them. For example,
if the group G is compact, then the constants themselves constitute a direct summand in $\mathscr{C}^\infty(G)$. We see that if the group
G is compact, then $H_c^q(G) = 0$ for $q > 0$ and $\dim H_c^0(G) = 1$.
(We shall return to a discussion of this fact after the proof
of Theorem 3.4.1.)

Lemma 2.

$$H_c^q(G; \Omega^r(G)) = \begin{cases} 0, & \text{if} \quad q > 0, \\ C^r(\mathfrak{g}), & \text{if} \quad q = 0. \end{cases}$$

Indeed, the G-module $\Omega^r(G)$ can be represented in the
form $C^r(\mathfrak{g}) \otimes \mathscr{C}^\infty(G)$, where the G-module structure in the
first factor is trivial [$C^r(\mathfrak{g})$ can be included in $\Omega^r(G)$ as
the subspace of right-invariant forms — see 1.3.3].

Theorem 3.4.1. There exists a multiplicative spectral
sequence $\{E_r^{p,q}, \ d_r^{p,q}: \ E_r^{p,q} \to E_r^{p+r, q-r+1}\}$, such that

(i) $E_2^{p,q} = H_{\mathrm{alg},c}^p(G) \otimes H_{\mathrm{top}}^q(G)$;

(ii) the term E_∞ is associated (with respect to a certain filtration) to $H^*(\mathfrak{g})$;

(iii) the standard maps $E_\infty \to E_\infty^{0,*} \to E_2^{0,*}, \ E_2^{*,0} \to E_\infty^{*,0} \to E_\infty$
are induced by the maps described above: $H^*(\mathfrak{g}) \to H_{\mathrm{top}}^*(G)$,
$H_{\mathrm{alg},c}^*(G) \to H^*(\mathfrak{g})$.

This spectral sequence is constructed in accordance with
a method based on the multiplicative double complex (see [11],
Chapter 15, Subsection 6). We consider the double complex

$$C^{..} = \{C^{p,q} = C_c^p\,(G;\,\Omega^q\,(G))\}$$

and, as usual, construct two filtrations

$$F_r^{\mathrm{I}} = \bigoplus_{p \leqslant r} C^{p,q}, \qquad F_r^{\mathrm{II}} = \bigoplus_{q \leqslant r} C^{p,q},$$

to which correspond two spectral sequences

$$\{{}^{\mathrm{I}}E_r^{p,q},\ {}^{\mathrm{I}}d_r^{p,q}\}, \quad \{{}^{\mathrm{II}}E_r^{p,q},\ {}^{\mathrm{II}}d_r^{p,q}\}.$$

The second is quite simple: ${}^{\mathrm{II}}E_1^{p,q} = H_c^q\,(G;\,\Omega^p\,(G))$ and by Lemma 2,

$$
{}^{\mathrm{II}}E_1^{p,q} = \begin{cases} 0 & \text{for } q > 0, \\ C^p\,(\mathfrak{g}) & \text{for } q = 0. \end{cases}
$$

The differential ${}^{\mathrm{II}}d_1^{p,0}$ coincides, as can be easily seen, with $d\colon C^p\,(\mathfrak{g}) \to C^{p+1}\,(\mathfrak{g})$, so that

$$
{}^{\mathrm{II}}E_2^{p,q} = \begin{cases} 0 & \text{for } q > 0, \\ H^p\,(\mathfrak{g}) & \text{for } q = 0, \end{cases}
$$

${}^{\mathrm{II}}E_\infty = {}^{\mathrm{II}}E_2$, and the complete cohomology of the double complex coincides with the cohomology of the algebra $\mathfrak{g}$.

The spectral sequence $\{{}^{\mathrm{I}}E_r^{p,q},\ {}^{\mathrm{I}}d_r^{p,q}\}$ is the one in Theorem 3.4.1. Indeed, ${}^{\mathrm{I}}E_2^{p,q} = H_c^p\,(G; H_{\mathrm{top}}^q\,(G))$, which is $H_c^p\,(G) \otimes H_{\mathrm{top}}^q\,(G)$, since $H_{\mathrm{top}}^q\,(G)$ is a trivial G-module. The statement (ii) of our theorem is proved by the computation of the full cohomology of the double complex $C_c^{\cdot}\,(G;\ \Omega^{\cdot}\,(G))$, carried out above; the proof of statement (iii) is left to the reader as an exercise.

It follows from Theorem 3.4.1 that if the canonical map $H^*\,(\mathfrak{g}) \to H_{\mathrm{top}}^*\,(G)$ is an isomorphism, then $H_{\mathrm{alg},c}^*\,(G) = 0$. This is the case when the group G is compact (we already know that continuous cohomology of the compact group is trivial — see the remark after Lemma 1); such is also the case for certain noncompact groups, for example, for the group of currents G^{S^1} of a compact group G (see §2.5). In the general case, the intuitive meaning of Theorem 3.4.1 is that the continuous Lie group cohomology measures the difference between its topological cohomology and the cohomology of its Lie algebra.

Note also that in the case when the group G is contractible, Theorem 3.4.1 shows that the canonical map $H^*_{\mathrm{alg},c}(G) \to H^*(\mathfrak{g})$ is an isomorphism.

Now let us indicate an important generalization of Theorem 3.4.1.

Theorem 3.4.2. Suppose H is a compact subgroup of the group G. There exists a multiplicative spectral sequence $\{E_r^{p,q}, d_r^{p,q}: E_r^{p,q} \to E_r^{p+r,q-r+1}\}$, such that

(i) $E_2^{p,q} = H^p_{\mathrm{alg},c}(G) \otimes H^q(G/H; \mathbb{R})$,

(ii) the term E_∞ is adjoint to $H^*(\mathfrak{g}, H)$.

If the group H reduces to the unit, then the spectral sequence of Theorem 3.4.2 becomes the spectral sequence of Theorem 3.4.1.

The proof of Theorem 3.4.2 is similar to that of Theorem 3.4.1 and shall not be given here (it can be found in [11]). The reader who would like to recover it should be warned that certain difficulties will arise in proving the appropriate generalization of Lemma 2. To overcome these difficulties, it is necessary to take integrals over H in the appropriate place, and it is precisely there that the compactness of the group H is used.

Corollary. If the group G possesses a compact subgroup K such that the space G/K is acyclic, then $H^*_{alg,c}(G) = H^*(\mathfrak{g}, K)$.

Indeed, in this case the spectral sequence of Theorem 3.4.2 with $H = K$ satisfies

$$E_2^{p,q} = \begin{cases} 0 & \text{for } q > 0, \\ H_c^p(G) & \text{for } q = 0 \end{cases}$$

and $E_\infty = E_2$.

It is well known that such a subgroup exists for any finite-dimensional Lie group: it is its maximal compact subgroup. There are also such subgroups in infinite-dimensional Lie groups: such is, for example, the subgroup SO (2) of the group of orientation-preserving diffeomorphisms of the circle.

The previous corollary becomes especially attractive in the case when the group G possesses a "compact form," i.e., when the complexification $\mathbb{C}\mathfrak{g}$ of the algebra $\mathfrak{g}$ is at the same time the complexification of the Lie algebra $\bar{\mathfrak{g}}$ of some compact group $\bar{G}$ (such are all semisimple and even reductive Lie groups: see §1.1 and [94]). In this case the maximal compact subgroup K of the group G can naturally be included in $\bar{G}$, and we obtain the relation

$$H^*_{alg,c}(G) = H^*(\bar{G}/K; \mathbb{R}). \tag{4}$$

In conclusion, note that if the group G is connected, then in the situation of the previous corollary the spectral sequence of Theorem 3.4.1, beginning with the second term, is isomorphic to the Serre—Hochschild spectral sequence corresponding to the pair $\mathfrak{g}, \mathfrak{k}$, and if the group G has the compact form $\bar{G}$, then the spectral sequence of Theorem 3.4.1, beginning

with the second term, is isomorphic to the cohomology spectral
sequence of the canonical bundle $\bar{G} \to \bar{G}/K$. The proof is
left to the reader.

3. Computation of continuous cohomology. For the most
important finite-dimensional Lie groups, the problem of computing their continuous cohomology is satisfactorily solved
by formula (4). For example,

$$H^*_{\text{alg,c}} (\text{SL} (n, \mathbb{R})) = H^* (\text{SU} (n)/\text{SO} (n); \mathbb{R})$$

is, for $n = 2k$, the exterior algebra in k generators of dimensions $5, 9, \ldots, 4k - 3, 2k$ and for $n = 2k + 1$ is the exterior
algebra in k generators of dimensions $5, 9, \ldots, 4k + 1$. However, it is not always possible to indicate explicitly the
cocycle which represents some specific cohomology class. A
fairly simple construction of such a cocycle (due to Guichardet and Wigner [42]) exists for the two-dimensional cohomology class arising in the situation when the canonical homomorphism $H^1 (\mathfrak{g}) \to H^1_{\text{top}} (G)$ has a nontrivial cokernel.

Theorem 3.4.3. Suppose $\alpha \in H^1_{\text{top}} (G)$ is a class not contained in the image of the homomorphism $H^1 (\mathfrak{g}) \to H^1_{\text{top}} (G)$, and
let $f: G \to S^1$ be a continuous map such that $f^*: H^1 (S^1; \mathbb{R}) \to$
$H^1 (G; \mathbb{R})$ sends α into the canonical generator of the group
$H^1 (S^1; \mathbb{R})$. Then the multivalued function on $G \times G$

$$(g, h) \mapsto \text{Arg} (f (hg) - f (h) - f (g)) \tag{5}$$

has a single-valued branch which assumes the value 0 at the
point $(1, 1)$, and this single-valued branch is a two-dimensional cocycle of the group G [in the sense of the second description of the complex $C^{\cdot}_{\text{alg,c}} (G)$] representing the nonzero element of the group $H^2_{\text{alg,c}} (G)$. This element is the image of

the class α under transgression in the spectral sequence of Theorem 3.4.1.

The proof consists of an explicit calculation of transgression. On the cochain level, the latter is represented by the compound "map"

$$\Omega^1 (G) = C_c^0 (G; \; \Omega^1 (G)) \to C_c^1 (G; \Omega^1 (G))$$
$$\leftarrow C_c^1 (G; \Omega^0 (G)) \to C_c^2 (G; \Omega^0 (G)) \to C_c^2 (G),$$

where the first and third arrows are differentials of the cochain complex of the group G, the second arrow is induced by the exterior differential d: $\Omega^0 (G) \to \Omega^1 (G)$, and the last arrow is induced by the homomorphism $\Omega^0 (G) = \mathscr{C}^\infty (G) \to \mathbb{R}$, sending functions into their values at the unit of the group G. Denote by ω the standard volume form on the circle. The class α is represented by the form $f^*\omega$. The first arrow sends this form into the one-dimensional cochain $g \mapsto gf^*\omega - f^*\omega$. One of the inverse images of this cochain under the second homomorphism is the cochain $g \mapsto A_g$, where A_g is the function defined as the single-valued branch of the multivalued function

$$(g, h) \mapsto \mathrm{Arg} \, (f \, (gh) - f \, (h) - f \, (g)),$$

which assumes the value 0 at the point 1. The third arrow sends this one-dimensional cochain into the two-dimensional cochain $(g, h) \mapsto A_{g, h}$, where

$$A_{g, h} \, (k) = A_h \, (gk) - A_{gh} \, (k) + A_g \, (k).$$

Finally, if we put $k = 1$ on the right-hand side of the last relation, we obtain the cochain

$$(g, h) \mapsto A_h \, (g) - A_{gh} \, (1) + A_g \, (1) = A_h \, (g),$$

as required.

In certain cases, a satisfactory choice of the map f enables one to write formula (5) in a more attractive form. For example, if $G = \mathrm{SL}\,(2, \mathbb{R})$ and f acts in accordance with the formula

$$\left\| \begin{matrix} a & b \\ c & d \end{matrix} \right\| \mapsto \frac{ci+d}{|ci+d|}\,,$$

Then the cocycle (5) sends the elements g, h of the group G into the oriented area of the triangle of the Lobachevski plane whose vertices are an arbitrary point A and the points gA, hgA.

Now let us consider the case of infinite-dimensional groups.

Theorem 3.4.4. The ring $H_c^*\,(\mathrm{Diff}_+S^1)$ has two generators α, $\beta \in H_c^2\,(\mathrm{Diff}_+S^1)$, which satisfy the only relation $\beta^2 = 0$. The homomorphism $H_c^*\,(\mathrm{Diff}_+S^1) \to H^*\,(\mathrm{Vect}\,S^1)$ sends β into 0 and α into a nonzero element.

Proof. The cohomology of the algebra $\mathrm{Vect}\,S^1$ is known to us: the ring $H^*\,(\mathrm{Vect}\,S^1)$ is the free skew-symmetric algebra in two generators of dimension 2, 3 (see Theorem 2.4.2). In particular, $H^1\,(\mathrm{Vect}\,S^1) = 0$ and, therefore, in the spectral sequence of Theorem 3.4.1 we have $E_2^{1,0} = 0$ and the differential $d_2^{0,1} \colon E_2^{0,1} \to E_2^{2,0}$ is injective. Further, $E_2^{1,1} = E_2^{1,0} = 0$, and since $\dim H^2\,(\mathrm{Vect}\,S^1) = 1$, we have $\dim E_2^{2,0} = 2$. Choose the generator s, in $E_2^{0,1}$ put $\beta = d_2^{0,1}\,(s)$, and denote by α an arbitrary element of the space $E_3^{2,0}$, which constitutes, together with β, a basis in this space. The class α is sent into a nonzero element by the homomorphism $H_c^2\,(G) \to H^2\,(\mathfrak{g})$, and, therefore, the same is true for α^2, α^3, $\ldots$, while the class β is sent to zero and therefore the same is true of $\alpha\beta$, $\alpha^2\beta$, $\ldots$; In particular, α^n is not proportional to $\alpha^{n-1}\beta$.

Further we shall show that $d_2^{2,1}(\beta s) = 0$, i.e., that $\beta^2 = 0$. This relation yields a brief conclusion to the proof of the theorem: Since $\dim H^3(\text{Vect } S^1) = 1$, this implies that $d_2^{2,1}(\alpha s) \neq 0$ and $E_2^{3,0} = 0$. Therefore, $\alpha\beta \ [= d_2^{2,1}(\alpha s)] \neq 0$, so that in $E_2^{4,0}$ there are two linearly independent elements α^2 and $\alpha\beta$. Since, further, $\dim H^4(\text{Vect } S^4) = 1$, it follows that $\{\alpha^2, \alpha\beta\}$ is a basis in $E_2^{4,0}$ and $\{\alpha^2 s, \alpha\beta s\}$ is a basis in $E_2^{4,1}$; since $\dim H^5(\text{Vect } S^1) = 1$ and $d_2^{1,1}(\alpha\beta s) = \alpha\beta^2 = 0$, it follows that $d_2^{1,1}(\alpha^2 s) = \alpha^2\beta \neq 0$ and $E_2^{5,0} = 0$, etc. We see that $E_2^{2k+1,0} = 0$, while the space $E_2^{2k,0}$ is two-dimensional and is generated by α^k and $\alpha^{k-1}\beta$.

It remains to show that $d_2^{2,1}(\beta s) = 0$. To do this, consider the homomorphism $\text{SL}(2, \mathbb{R}) \to \text{Diff}_+ S^1$, defined by the standard action of $\text{SL}(2, \mathbb{R})$ in $\mathbb{R}P^1 = S^1$. This homomorphism is a homotopy equivalence, while the corresponding homomorphism $H^3(\text{Vect} S^1) \to H^3(\mathfrak{sl}(2, \mathbb{R}))$ is bijective. If $d_2^{2,1}(\beta s)$ were different from 0, then the space $E_\infty^{2,1}$ would be trivial and, therefore, the space $E_\infty^{3,0}$ would differ from 0. This would mean that $H_c^3(\text{Diff}_+ S^1) \to H^3(\text{Vect } S^1)$ is an epimorphism, and we would obtain an impossible commutative diagram

$$
\begin{array}{ccc}
H_c^3(\text{Diff}_+ \ S^1) & \overset{\text{epi}}{\longrightarrow} & H^3(\text{Vect } S^1) \\
\downarrow & & \downarrow \text{iso} \\
0 = H_c^3(\text{SL}(2, \mathbb{R})) & \longrightarrow & H^3(\mathfrak{sl}(2, \mathbb{R})) \neq 0.
\end{array}
$$

The theorem is proved.

The cocycle representing the class β is given by Theorem 3.4.3. Bott [7] found an effective formula for a cocycle representing α. Namely, for the diffeomorphism $f: S^1 \to S^1$ define the function μ_f on S^1 by the formula

$$
f^*\omega = \mu_f \omega, \tag{6}
$$

where ω is the standard volume form on S^1.

Theorem 3.4.5. The formula

$$(f_1, f_2) \mapsto \int_{S^1} \log \mu_{f_1} \, d \log \mu_{f_1 \circ f_2}$$

defines a cocycle in $C_c^2 (\text{Diff}_+ S^1)$, whose cohomology class does not vanish and has a nonzero image in $H^2 (\text{Vect}\, S^1)$.

The verification of this statement is not difficult and I leave it to the reader.

Bott also noticed that a similar cocycle may be constructed if we take an arbitrary manifold M instead of the circle: the formula

$$(f_1, \ldots, f_{n+1}) \mapsto \int_M \log \mu_{f_1} \, d \log \mu_{f_1 \circ f_2} \wedge \ldots \wedge d \log \mu_{f_2 \, \circ \ldots \circ f_n},$$

where $n = \dim M$, while μ is determined by formula (6) (in which ω now denotes an arbitrary volume form on M), gives us a cocycle from $C_c^{n+1} (\text{Diff}_+ M)$, whose cohomology class has a nonzero image in $H^{n+1} (\text{Vect}\, M)$.

Bott's paper contains also some other formulas for continuous cocycles of diffeomorphism groups. Nevertheless, for a majority of finite-dimensional and infinite-dimensional groups, the problem of explicit construction of cocycles representing classes of continuous cohomology remains open.

4. Kac–Moody groups. In this section, we develop some simple considerations relating to central extensions of group and current algebras, leading to unexpected results and interesting problems. These considerations arose in numerous conversations which involved A. Beilinson, I. Bernstein, B. Feigin, A. Vershik, and the author.

    As we already pointed out, the current algebra $\mathfrak{g}^{S^1}$ of
the simple Lie algebra $\mathfrak{g}$ possesses a central extension which
leads to a certain completion of the Kac–Moody algebra [to
the Kac–Moody algebra itself, if we extend the algebra of
polynomial currents $(\mathfrak{g}^{S^1})^{\text{pol}}$]. Does this extension correspond
to some extension of the current group G^{S^1} ? To answer this
question, the theory from Subsection 2 recommends looking at
the homomorphism $H^2_C(G^{S^1}) \rightarrow H^2(\mathfrak{g}^{S^1})$.   Comparing results of Subsections 2.5.3 and 2, we see that this homomorphism is often
trivial. For example, if the group G is compact, then the
group G^{S^1} has trivial continuous cohomology in general and
the homomorphism is trivial. So it is in some other cases
as well. One has the impression that for these groups G
there exists no "Kac–Moody group," whose Lie algebra would
be a Kac–Moody algebra, i.e., the nontrivial central exten-
sion of the algebra $\mathfrak{g}^{S^1}$ or $(\mathfrak{g}^{S^1})^{\text{pol}}$. This conclusion seemed
dubious from the outset, since it contradicts the intuition
from representation theory; luckily it turned out to be false.
The defect in the previous argument is that in it we limit ourselves to groups which topologically fall apart into a direct
product of the extended group and a one-dimensional group,
while we should actually consider locally trivial but nontrivial bundles. Note at once that no such topologically nontrivial central extensions exist in the finite-dimensional situation; hence the definition of central extension given in Subsection 1 suffices for the classical theory. But it is insuf-
ficient for our present goal. Neither are the five cohomology
theories listed in Subsection 1 satisfactory for us now; we
shall briefly describe one more, the sixth theory, devised
by Segal (see [80, 81]).

    This sixth theory assigns to every Lie group G and local-
ly contractible Hausdorff Abelian group A, in which G acts
by means of automorphisms, a sequence of cohomology groups
$H_S^q(G; A)$, possessing the usual properties of cohomology (in
particular, to the short exact sequence of coefficient groups
corresponds an infinite exact "coefficient sequence" of the
usual type). The actual definition of these groups is not
required for our aims, and we shall limit ourselves to indi-
cating that the functor $A \mapsto H_S^q(G; A)$ is the q-th derived
functor of the functor $A \mapsto A^G$ on the category of locally
contractible topological G-modules with respect to the class
of short exact sequences which are locally trivial bundles.

    It is much more important for us to note the following
properties of the cohomology H_S^*: (i) if the group G is discrete, then $H_S^*(G; A) = H_{alg}^*(G; A)$; (ii) if the module A is
contractible, then $H_S^*(G; A) = H_c^*(G; A)$; (iii) if the module
A is discrete and trivial, then $H_S^*(G; A) = H^*(BG; A)$; (iv)
if A is the Abelian group, considered as a trivial G-module,
then $H_S^2(G; A)$ is the group of classes of central extensions $1 \to$
$A \to \hat{G} \to G \to 1$, which are locally trivial bundles.

    Now suppose G is a compact simply connected Lie group.
Consider the segment

$$H_S^2(G^{S^1}; \mathbb{R}) \to H_S^2(G^{S^1}; S^1) \to H_S^3(G^{S^1}; \mathbb{Z}) \to H_S^3(G^{S^1}; \mathbb{R}) \qquad (7)$$

of the coefficient sequence of the Segal cohomology of the
group G^{S^1}, corresponding to the short exact sequence $0 \to \mathbb{Z} \to$
$\mathbb{R} \to S^1 \to 0$. According to the above, $H_S^2(G^{S^1}; \mathbb{R}) = H_c^2(G^{S^1};$
$\mathbb{R}) = 0$, similarly, $H_S^3(G^{S^1}; \mathbb{R}) = 0$ and $H_S^3(G^{S^1}; \mathbb{Z}) = H^3(B(G^{S^1});$
$\mathbb{Z})$. Thus, the sequence (7) acquires the form

$$0 \to H_S^2(G^{S^1}; S^1) \to H^3(B(G^{S^1}); \mathbb{Z}) \to 0.$$

Since the group G is simply connected, it is also 2-connect-
ed, and $H^3(G; \mathbb{Z}) \cong \mathbb{Z}$. Hence, $H^1(G^{S^1}; \mathbb{Z}) = 0$, $H^2(G^{S^1}; \mathbb{Z}) \cong$
$\mathbb{Z}$, and $H^3(B(G^{S^1}); \mathbb{Z}) \cong \mathbb{Z}$. Thus

$$H_S^2(G^{S^1}; S^1) \cong \mathbb{Z},$$

i.e., there exist infinitely many nonequivalent central extensions of the group G^{S^1} by means of S^1. Moreover, as
is known, integer-valued two-dimensioned cohomology of the
topological space X can be interpreted as classes of principal S^1-bundles with base X, and it is easy to verify that
our isomorphism

$$H_S^2(G^{S^1}; S^1) \to H^3(B(G^{S^1}); \mathbb{Z}) = H^2(G^{S^1}; \mathbb{Z})$$

sends the class of central extensions $1 \to S^1 \to (G^{S^1})^\sim \to G^{S^1} \to 1$
into the corresponding class of S^1-bundles $(G^{S^1})^\sim \to G^{S^1}$. We
come to the following conclusion.

Theorem 3.4.6. For any principal S^1-bundle $E \to G^{S^1}$
[classes of such bundles are in bijection with elements of
the group $H^2(G^{S^1}; \mathbb{Z}) \cong \mathbb{Z}$] there exists a unique (up to equivalence, i.e., up to an automorphism of the bundle which is
the identity on the base) topological group structure in E,
with respect to which the sequence $1 \to S^1 \to E \to G^{S^1} \to 1$ is a
central extension.

The group E may be viewed as an infinite-dimensional
Lie group and, if the bundle is nontrivial, its Lie algebra
is the Kac–Moody algebra (the Lie algebras of all these groups
are isomorphic, since they cover each other).

It is impossible, however, to say that this subject is
understood completely, since an explicit description of Kac–
Moody groups is absent. The uniqueness of the group structure
in the total space of the bundle $E \to G^{S^1}$ offers hope that

its description will be simple and natural. Nevertheless,
we have not succeeded in finding it.

§5.   COHOMOLOGY OPERATIONS IN COBORDISM THEORY

In this section we briefly describe the work of Bukhshtaber and Shokurov (see [10, 84]), who have found promising
relationships between the cohomology of infinite-dimensional
Lie algebras and the apparatus of contemporary algebraic topology. Namely, the universal enveloping algebra of the Lie
algebra L_1 (1) turned out to be isomorphic to the tensor product by $\mathbb{K}$ of the so-called Landweber–Novikov algebra, the
main ingredient in the ring of stable cohomology operations
in the theory of complex cobordisms. Thus, the cohomology
of the Landweber–Novikov algebra, extremely important in topology, turned out to be closely related to the cohomology
of the algebra L_1 (1).

The above-mentioned Bukhshtaber and Shokurov articles contain nontrivial topological results, but the relationship with
L_1 (1) which they have discovered is used in proofs to a limited extent. Hence it would be premature to speak about serious
applications of the cohomology of infinite-dimensional Lie
algebras to algebraic topology. Nevertheless, the interpretation of this cohomology found by Bukhshtaber and Shokurov is
doubtless of interest.

To understand this section, a certain preliminary knowledge of algebraic topology is required (although the definitions of the objects used most consistently are given here).
To acquaint oneself with cobordism theory, we recommend Stong's
book [85] (including the supplement to the Russian translation
written by Bukhshtaber). The proofs of the results from co-

bordism theory used here may be found in this book, in the
above-mentioned articles by Bukhshtaber and Shokurov, and in
the literature cited in them.

   1. Cohomology operations in cobordisms. A. Definition
of cobordisms. To the reader who is not familiar with the
general procedure of constructing extraordinary cohomology
theories, I prefer giving the geometric definition of cobordisms, due to Quillen. The theory of complex cobordisms maps
every finite cellular space X into the graded ring $U^*(X) = \bigoplus_{q \in \mathbb{Z}} U^q(X)$ and every continuous mapping $f: X \to Y$ into a
homogeneous ring homomorphism $f^*: U^*(Y) \to U^*(X)$, depending
only on the homotopy class of f. Since every finite cellular space is the retract of a closed smooth manifold, we can
limit ourselves to the description of the ring $U^*(X)$ in this
case. We say that a smooth map $h: Z \to Y$ of one smooth manifold into another possesses a complex orientation if we are
given a complex vector bundle ξ over Z, positive integers M
and N, and the equivalence $\tan g\, Z \oplus M\varepsilon \cong h^* \tan g\, Y \oplus N\varepsilon \oplus \mathbb{R}\xi$,
where $\mathbb{R}\xi$ is ξ viewed as a real bundle and ε is the standard
trivial real one-dimensional bundle over Z. [For example,
if h is the inclusion, then a choice of complex orientation
is equivalent to a choice of the complex structure in a stable
normal bundle of the image of $h(Z)$ in Y.] Smooth, complexly
oriented maps $h_0: Z_0 \to Y$, $h_1: Z_1 \to Y$ of closed manifolds of
the same dimension in Y are said to be cobordant if there
exists a compact manifold W satisfying $\partial W = Z_0 \bigsqcup Z_1$ and
a smooth map $H: W \to Y \times [0, 1]$, coinciding on Z_i with the
composition $Z_i \overset{h_i}{\to} Y = Y \times i \subset Y \times [0, 1]$ and transversally regular with respect to $Y \times (0 \bigcup 1)$; here we assume that the map
H also possesses a complex orientation and that this complex
orientation is compatible in the natural sense with the com-

plex orientations of the maps h_i. The set of classes of cobordant complexly oriented smooth maps $Z \to X$ of codimension
r (i.e., such that $\dim X - \dim Z = r$) is by definition
$U^r (X)$. For $\alpha_1, \alpha_2 \in U^r (X)$ the sum $\alpha_1 + \alpha_2$ is defined
as the class of the sum $h_1 \bigsqcup h_2 : Z_1 \bigsqcup Z_2 \to X$ of the maps h_1,
h_2, representing α_1, α_2. The product of elements α_1, α_2 of
the groups $U^{r_1} (X), U^{r_2} (X)$ is defined as the class of the map

$$Z = (h_1 \times h_2)^{-1} \operatorname{diag} X \xrightarrow{(h_1 \times h_2) \mid Z} \operatorname{diag} X = X,$$

where $h_1 : Z_1 \to X$, $h_2 : Z_2 \to X$ are representatives of the classes
α_1, α_2, such that the product $h_1 \times h_2$ is transversally regular with respect to the diagonal $\operatorname{diag} X$ of the product $X \times X$.
The identity map $X \to X$, supplied with the natural complex
orientation, represents the element of the group $U^0 (X)$,
which is the unit for the multiplication described above. The
map $f^* : U^* (X_2) \to U^* (X_1)$, where f is a smooth map of X_1 into
X_2, sends $\alpha \in U^* (X_2)$ into the class of the map

$$V = (f \times h)^{-1} \operatorname{diag} X_2 \xrightarrow{(f \times h) \mid V} \operatorname{diag} X_2 = X_2,$$

where h is a representative of the class α, such that the product $f \times h$ is transversally regular with respect to $\operatorname{diag} X_2$.

The map $(h : Z \to X) \mapsto \operatorname{Poi}^{-1} f_* [Z]$, where Poi is the Poincaré isomorphism, determines the homomorphism $U^* (X) \to H^* (X;$
$\mathbb{Z})$, compatible with multiplication and with pullback. We obtain, as one says, a homomorphism of the theory of cobordisms
into the theory of integer-valued cohomology.

By comparison with the theory of ordinary cohomology,
the theory of cobordisms is richer and more meaningful. For
example, already in the case when X is a point, the ring
$U^* (X)$ is sufficiently large: it is the polynomial ring in

generators of dimensions $-2, -4, -6, \ldots$ Note that in the
case of X consisting of one point, the definition of set $U^r(X)$
is noticeably simplified: this is simply the set of classes
of closed $(-r)$-dimensional manifolds with complex structure
in their stable normal bundle (i.e., in the normal bundle in
Euclidean space of sufficiently high dimension) which together bound compact manifolds with a similar structure.   The
ring U^* (point) is denoted also by $\Omega_U \ (=\oplus_q \Omega_U^q)$.

   B.   Definition of cohomology operations.   Cohomology
operations in cobordism theory are defined similarly to cohomology operations in ordinary cohomology.   We say that a
cohomology operation α of type (r_1, r_2) is given in the theory
of complex cobordisms if for any finite cellular space (or,
which is the same thing, for every closed smooth manifold) X
we have the map $\alpha_X \colon U^{r_1}(X) \to U^{r_2}(X)$, such that the diagram

$$
\begin{array}{ccc}
U^{r_1}(X) & \xrightarrow{\ \alpha_X\ } & U^{r_2}(X) \\[2pt]
\uparrow{\scriptstyle f^*} & & \uparrow{\scriptstyle f^*} \\[2pt]
U^{r_1}(Y) & \xrightarrow{\ \alpha_Y\ } & U^{r_2}(Y)
\end{array}
$$

is commutative for all Y and $f \colon X \to Y$.   We say that we are
given a stable cohomology operation of degree r if for each
$i \in Z$ the cohomology operation α_i of type $(i, r+i)$ is given
and these operations are in agreement in the sense that for
all X and i the diagram

$$
\begin{array}{ccc}
U^i(X) & \xrightarrow{\ (\alpha_i)_X\ } & U^{r+i}(X) \\[2pt]
\downarrow & & \downarrow \\[2pt]
U^{i+1}(X \times S^1) & \xrightarrow{\ (\alpha_{i+1})_{X \times S^1}\ } & U^{r+i+1}(X \times S^1),
\end{array}
$$

whose vertical arrows are defined by the map $(Z \xrightarrow{h} X) \to (Z \xrightarrow{h} X = X \times 1 \subset X \times S^1)$, is commutative.

Stable cohomology operations constitute a graded ring. Its completion with respect to this grading is denoted by A_U.

C. Fundamental examples of operations. Suppose V is a closed r-dimensional manifold of complex structure in a stable normal bundle. For all X and i let us define the homomorphism $U^i(X) \to U^{i-r}(X)$ by putting $(Z \xrightarrow{h} X) \mapsto (Z \times V \xrightarrow{pr} Z \xrightarrow{h} X)$. Obviously, such homomorphisms constitute a stable cohomology operation of degree $-r$, and this operation depends only on the element of the group Ω_U^r defined by V. Thus we have obtained the inclusion $\Omega_U \to A_U$.

More interesting stable cohomology operations may be constructed by means of the cobordism Chern classes. There is a construction similar to the classical construction of the ordinary Chern cohomology classes, which assigns to each complex vector bundle ξ with base X the sequence of classes $\sigma_i(\xi) \in U^{2i}(X)$ $[i = 0, 1, \ldots]$. These classes are determined by three axioms: (i) $\sigma_i(f^*\xi) = f^*(\sigma_i(\xi))$ for any continuous map f of an arbitrary finite cellular space into the base of the bundle ξ; (ii) $\sigma_i(\xi \oplus \eta) = \Sigma_{\alpha+\beta=i} \sigma_\alpha(\xi) \sigma_\beta(\eta)$; (iii) if ξ is the canonical linear bundle over $\mathbb{C}P^n$, then $\sigma_i(\xi) = 0$ for $i > 1$, $\sigma_0(\xi) = 1$ and $\sigma_1(\xi)$ (for $n \geqslant 1$) is represented by the canonical inclusion $\mathbb{C}P^{n-1} \to \mathbb{C}P^n$. It follows from these axioms that $\sigma_0(\xi) = 1$ for any ξ and that if the bundle ξ is trivial, then $\sigma_i(\xi) = 0$ for $i > 0$. In particular, the Chern classes σ_i are defined not only for complex, but also for stably complex bundles. [The relationship of Chern cobordism classes with the ordinary ones is obvious: the canonical homomorphism $U^*(X) \to H^*(X; \mathbb{Z})$ sends $\sigma_i(\xi)$ into $c_i(\xi)$.]

Now fix an arbitrary finite sequence $\omega = (\omega_1, \omega_2, \ldots)$ of nonnegative integers and put $|\omega| = \omega_1 + 2\omega_2 + \ldots$ Choose arbitrary X, q and $\alpha \in U^q(X)$ and a map $h: Z \to X$ representing α. According to Sub-subsection A, the manifold Z is the base of a complex bundle ξ with equivalence tang $Z \oplus M_\varepsilon \cong$ h^* tang $X \oplus N\varepsilon \oplus \mathbb{R}\xi$. Consider the monomial $\sigma_1(\xi)^{\omega_1} \sigma_2(\xi)^{\omega_2} \ldots$ in the cobordism Chern classes of the bundle ξ; this is an element of the group $U^{2|\omega|}(X)$, and it is represented by a certain map $g: V \to Z$, supplied with the appropriate structure. The composition $h \circ g: V \to X$ defines an element of the group $U^{q+2|\omega|}(X)$. This element depends only on α and ω, and shall be denoted by $s_\omega(\alpha)$. Obviously s_ω is a stable cohomology operation of degree $2|\omega|$.

D. The structure of the ring A_U. The following theorem is fundamental in the theory of complex cobordisms.

Theorem 3.5.1 (Landweber–Novikov). (i) The subgroup S of the group A_U, generated by the operations s_ω, is closed with respect to composition. (ii) The map $\Omega_U \otimes S \to A_U$, defined by the formula $\xi \otimes s \mapsto \xi \circ s$, is monomorphic and has a dense image.

The subring S of the ring A_U is called the Landweber–Novikov algebra.

2. Relationship between $L_1(1)$ and the Landweber–Novikov algebra. Denote by $\mathrm{Diff}_1(\mathbb{Z})$ the group of formal diffeomorphisms of the line of the form

$$t \mapsto t + x_1 t^2 + x_2 t^3 + \ldots, x_1, x_2, \ldots \in \mathbb{Z}.$$

The group $\mathrm{Diff}_1(\mathbb{Z})$ acts in the ring $P = \mathbb{Z}[x_1, x_2, \ldots]$ of functions on itself by means of left translations (this is a variant of the adjoint representation). Denote by $\mathscr{S}$ the algebra of all $\mathrm{Diff}_1(\mathbb{Z})$-invariant formal differential operators in P, i.e., of formal power series in $\frac{\partial}{\partial x_1}, \frac{\partial}{\partial x_2}, \ldots$ with coefficients in $\mathbb{Q}[x_1, x_2, \ldots]$, sending elements from P into P and commuting with the transformations effected by elements of the group $\text{Diff}_1(\mathbb{Z})$. For any finite sequence $\omega = (\omega_1, \omega_2, \ldots)$ of nonnegative integers, define the operator $D_\omega \in \mathcal{S}$ by means of the formula

$$P(x \circ y) = \sum_\omega D_\omega P(x) y^\omega,$$

in which $x = (x_1, x_2, \ldots)$, $y = (y_1, y_2, \ldots)$, and the operation $\circ$ corresponds to multiplication in the group $\text{Diff}_1(Z)$ [i.e., the sequence $x \circ y = (z_1, z_2, \ldots)$ is determined by the relation

$$t + \sum_i z_i t^{i+1} = (t + \sum_j y_j t^{j+1}) + \sum_k x_k (t + \sum_j y_j t^{j+1})^{k+1}]$$

and y^ω denotes $y_1^{\omega_1} y_2^{\omega_2} \ldots$. For example,

$$D_{(1,0,0,\ldots)} = \frac{\partial}{\partial x_1} + \sum_{k=2}^{\infty} k x_{k-1} \frac{\partial}{\partial x_k},$$

$$D_{(0,1,0,\ldots)} = \frac{\partial}{\partial x_2} + \sum_{k=3}^{\infty} (k-1) x_{k-2} \frac{\partial}{\partial x_k},$$

$$2D_{(2,0,0,\ldots)} = \frac{\partial^2}{\partial x_1^2} + 2 \sum_{k=2}^{\infty} k x_{k-1} \frac{\partial^2}{\partial x_1 \partial x_k}$$

$$+ \sum_{k=2}^{\infty} \sum_{l=2}^{\infty} k l x_{k-1} x_{l-1} \frac{\partial^2}{\partial x_k \partial x_l} + \sum_{k=3}^{\infty} (k-1)(k-2) x_{k-2} \frac{\partial}{\partial x_k}.$$

The main result of Bukhshtaber and Shokurov is the following:

Theorem 3.5.2 (see [10]). The map $s_\omega \leftrightarrow D_\omega$ determines an isomorphism between the rings S and $\mathcal{S}$.

The isomorphism $\mathcal{S} = S$ transforms P into an S-module and this module also has an interpretation in cobordisms. Namely, denote by $\Omega_U^r(\mathbb{Z})$ the part of the tensor product $\Omega_U^r \otimes \mathbb{Q} = U^r$ (point) $\otimes \mathbb{Q}$, in which the Chern numbers are integers. Thus, $\Omega_U^r \subset \Omega_U^r(\mathbb{Z}) \subset \Omega_U^r \otimes \mathbb{Q}$; it is well known that the quotient group $\Omega_U^r(\mathbb{Z})/\Omega_U^r$ is finite. The action of the operations from S can be carried over from Ω_U^* into $\Omega_U^*(\mathbb{Z})$.

Theorem 3.5.3 (see [10]). The isomorphism between S and $\mathcal{S}$ can be extended to an isomorphism between the S-module $\Omega_U(\mathbb{Z})$ and the $\mathcal{S}$-module P.

Actually in [10] the isomorphism between $\Omega_U(\mathbb{Z})$ and P is constructed first and then it is established that under this isomorphism the operations s_ω correspond to operators D_ω, which implies Theorem 3.6.2. An interesting description is then acquired by the part of P corresponding to $\Omega_U \subset \Omega_U(\mathbb{Z})$: this is the set of polynomials from P, which remain in P under right translation by the diffeomorphism $1 - \exp(-t) \in \mathrm{Diff}_1(\mathbb{Q})$.

Now note that after tensor multiplication by $\mathbb{K}$ the algebra $\mathcal{S}$ becomes the algebra of all left-invariant differential operators on the group $\mathrm{Diff}_1(\mathbb{K})$. This algebra coincides, as can be easily understood, with the universal enveloping algebra of the Lie algebra of the group $\mathrm{Diff}_1(\mathbb{K})$, i.e., with $U(L_1(1))$. We come to the following conclusion.

Theorem 3.5.4. There is an isomorphism of $\mathbb{K}$-algebras

$$S \otimes \mathbb{K} \cong U(L_1(1)).$$

In particular, the algebra S is included in $U(L_1(1))$. Under this inclusion, for example,

$$s_{(1,0,\dots)} \mapsto e_1, \quad s_{(0 \ 1,0,\dots)} \mapsto e_2, \quad s_{(2,0,\dots)} \mapsto \frac{1}{2}(e_1^2 - e_2).$$

The disagreement between the gradings in S and $U(L_1(1))$ should be noted: the operation s_ω is of degree $2|\omega|$, while the corresponding element of the algebra $U(L_1(1))$ is of degree $|\omega|$. Thus S^q is included into $U(L_1(1))^{(q/2)}$ and

$$S^q \otimes \mathbb{K} \cong U\,(L_1\,(1))^{(q/2)}.$$

3. **The Adams–Novikov spectral sequence and the Bukh-shtaber spectral sequence.** The main method for applying co-bordisms to the solution of the homotopy problems is the Adams–Novikov spectral sequence, which is defined for an ar-bitrary finite cellular space X ; it has the initial term $\mathrm{Ext}_{A_U}\,(U^*\,(X),\,\Omega_U)$ and converges to stable homotopy groups of the space $X \bigsqcup$ (point). In the case of a one-point space X we obtain the spectral sequence

$$\mathrm{Ext}_{A_U}\,(\Omega_U,\,\Omega_U) \Rightarrow \pi^S\,(S^0),$$

which converges to the stable homotopy groups of the sphere. The application of the spectral sequence is complicated by the fact that little is known about its initial term. Some progress in the computation of this initial term is achieved by means of the results presented above. Namely, it follows from part (ii) of Theorem 3.5.1 that

$$\mathrm{Ext}_{A_U}(\Omega_U,\,\Omega_U) = \mathrm{Ext}_S\,(\mathbb{Z},\,\Omega_U),$$

and Theorem 3.5.2 allows us to replace the ring S by the ring $\mathcal{S}$ in this equality. As can be easily shown,

$$\mathrm{Ext}_S\,(\mathbb{Z},\,\Omega_U\,(\mathbb{Z})) = \mathrm{Ext}_{\mathcal{S}}\,(\mathbb{Z},\,P) = 0$$

(this result resembles the theorem on the triviality of the homology of Lie algebras with coefficients in its universal enveloping algebra – see 1.3.4). For the computation of $\mathrm{Ext}_S\,(\mathbb{Z},\,\Omega_U)$ Bukhshtaber proposed the filtration

$$0 \subset N_0 \subset N_1 \subset \ldots \subset N_\infty = \Omega_U\,(\mathbb{Z}),$$

where $N_0 = \Omega_U$, $N_i^r = \{\sigma \in \Omega_U^r\,(\mathbb{Z}) \mid s_\omega \sigma \in N_{i-1}^r \text{ for } \mid \omega \mid > 0\}$. The quotient groups N_i^r/N_{i-1}^r are finite and trivial as S-

modules (at the present time N_1/N_0 and N_2/N_1 have been
calculated; see [84]). A trigraded spectral sequence is associated with the Bukhshtaber filtration

$$\{E_r^{s,\,t,\,q}, s \geqslant 0, q \geqslant 0, s + t \leqslant 0, d_r^{s,\,t,\,q}\colon E_r^{s,\,t,\,q} \to E_r^{s-r,\,t+r-1,\,q}\},$$

and in it $E_1^{0,\,*,\,*} = \mathrm{Ext}_S^{*,\,*}(\mathbb{Z}, \Omega_U), E_\infty = E_\infty^{0,0,0} = \mathbb{Z}$ while for $s > 0$

$$E_1^{s,\,t,\,\bullet} = \mathrm{Ext}_S^{-(s+t),\,\bullet}(\mathbb{Z}, N_s/N_{s-1}).$$

Since N_s/N_{s-1} is a trivial S-module, it follows from the
Kunneth formulas that

$$\mathrm{Ext}_S^{-(s+t),\,q}(\mathbb{Z}, N_s/N_{s-1}) = \bigoplus_{q_1+q_2=q} \{[\mathrm{Ext}_S^{-(s+t),\,q_1}(\mathbb{Z}, \mathbb{Z}) \otimes (N_s^{-q_2}/N_{s-1}^{-q_2}]$$

$$\oplus [\mathrm{Ext}_S^{-(s+t)+1,\,q_1}(\mathbb{Z}, \mathbb{Z}) * (N_s^{-q_2}/N_{s-1}^{-q_2})]\}.$$

Thus, to compute the initial term of the Bukhshtaber spectral
sequence, we must know the groups $\mathrm{Ext}_S(\mathbb{Z}, \mathbb{Z})$. Theorem
3.5.4 shows that

$$\mathrm{Ext}_S(\mathbb{Z}, \mathbb{Z}) \otimes \mathbb{K} = \mathrm{Ext}_{U(L_1(1))}(\mathbb{K}, \mathbb{K}) = H^*(L_1(1)),$$

or, in more detail, that

$$\mathrm{Ext}_S^{u,\,v}(\mathbb{Z}, \mathbb{Z}) \otimes \mathbb{K} = H_{(v/2)}^u(L_1(1)).$$

This in turn implies, by Theorem 2.3.1, that

$$\mathrm{rang}\ \mathrm{Ext}_S^{u,\,v}(\mathbb{Z}, \mathbb{Z}) = \begin{cases} 1, & \text{if}\quad v = 3u^2 \pm u, \\ 0, & \text{otherwise.} \end{cases}$$

We come to the following result.

Theorem 3.5.5. In the Bukhshtaber spectral sequence for
all s, u the group $E_1^{s,\,-(s+u),\,3u^2\pm u+q}$ contains a summand isomorphic to N_s^{-q}/N_{s-1}^{-q}.

To apply this theorem to the homotopy groups of spheres
it is necessary to determine, by using the Bukhshtaber spec-

tral sequence, what influence these groups have on $E_1^{0,*,*}$,
i.e., on the initial term of the Novikov—Adams spectral sequence and then, having computed the differentials of the latter, to process this information into information concerning
homotopy groups. At present this program has been carried
out only to a small extent (see [84]).

Note also that the problem of the complete calculation
of the groups $\mathrm{Ext}_S(\mathbb{Z}, \mathbb{Z})$ is the correct integer analog of
the problem of computing the cohomology of the algebra $L_1(1)$;
in this connection, the problem of carrying over the entire
cohomology theory of Lie algebras to the case of integer coefficients appears opportune at present.

References

1. D. V. Alekseevskii, D. A. Leitis, and I. M. Shchepoch-
   kina, "Examples of simple infinite-dimensional Lie superalgebras of vector fields," Dokl. Bolg. Akad. Nauk, 33,
   No. 9, 1187-1190 (1980).

2. D. Anderson, "A generalization of the Eilenberg–Moore
   spectral sequence," Bull. Am. Math. Soc., 78, 784-786
   (1972).

3. D. Baker, "On a class of foliations and the evaluation
   of their characteristic classes," Comment. Math. Helv.,
   53, No. 3, 334-363 (1978).

4. I. N. Bernstein and B. I. Rozenfeld, "On the characteristic classes of foliations," Funkts. Anal. Prilozhen.,
   6, No. 1, 68-69 (1972).

5. I. N. Bernstein and B. I. Rozenfeld, "Homogeneous spaces
   of infinite-dimensional Lie algebras and characteristic
   classes of foliations," Usp. Mat. Nauk, 28, No. 5, 103-
   138 (1973).

6. A. Borel, "Compact Clifford–Klein forms of symmetric
   spaces," Topology, 2, 111 (1963).

7. R. Bott, "On the characteristic classes of groups of diffeomorphisms," Enseign. Math., 23, No. 3-4, 209-220 (1977).

8. R. Bott and A. Haefliger, "On characteristic classes of
    Γ-foliations," Bull. Am. Math. Soc., 78, No. 6, 1039-
    1044 (1972).

9. R. Bott and G. Segal, "The cohomology of the vector
    fields on a manifold," Topology, 16, 285-298 (1977).

10. V. M. Bukhshtaber and A. V. Shokurov, "The Landweber—
    Novikov algebra and formal vector fields on the line,"
    Funkts. Anal. Prilozhen., 12, No. 3, 1-11 (1978).

11. H. Cartan and S. Eilenberg, Homological Algebra, Prince-
    ton Univ. Press, Princeton (1956).

12. H. Cartan, Calcul Différentiel, Hermann, Paris (1968);
    Formes Différentielles, Hermann, Paris (1968).

13. J. Dyson, "Missed opportunities," Bull. Am. Math. Soc.,
    78, 635-652 (1972).

14. B. L. Feigin, "Cohomology of groups and current algebras,"
    Usp. Mat. Nauk, 35, No. 2, 225-226 (1980).

15. B. L. Feigin, "Cohomology of contact Lie algebras,"
    Dokl. Bolg. Akad. Nauk, 35, No. 10, 1294-1296 (1982).

16. B. L. Feigin and D. B. Fuks, "On invariant differential
    operators on the line," Funkts. Anal. Prilozhen., 13,
    No. 4, 91-92 (1979).

17. B. L. Feigin and D. B. Fuks, "Homology of Lie algebras
    on vector fields on the line," Funkts. Anal. Prilozhen.,
    14, No. 3, 45-60 (1980).

18. B. L. Feigin and D. B. Fuks, "Skew-symmetric invariant
    differential operators on the line and Vermat modules
    over the Virasoro algebra," Funkts. Anal. Prilozhen.,
    16, No. 2, 47-63 (1982).

19. D. B. Fuks, "Characteristic classes of foliations," Usp.
    Mat. Nauk, 28, No. 2, 3-17 (1973).

20. D. B. Fuks, "Finite-dimensional Lie algebras of formal

vector fields and characteristic classes of homogeneous foliations," Izv. Akad. Nauk SSSR, Ser. Mat., $\underline{40}$, No. 1, 57-64 (1976).

21. D. B. Fuks, "Cohomology of infinite-dimensional Lie algebras and characteristic classes of foliations," in: Contemporary Problems in Mathematics, Vol. 10 [in Russian], VINITI, Moscow (1978), pp. 179-285.

22. D. B. Fuks, "Foliations," in: Algebra. Topology. Geometry, Vol. 18, from the series Progress in Science and Technology [in Russian], VINITI, Moscow (1981), pp. 151-213.

23. D. B. Fuks, "On the removal of parentheses, on Euler, Gauss, Macdonald, and missed opportunities," Kvant, No. 8, 12-20 (1981).

24. H. Garland and J. Lepowsky, "Lie algebra homology, and the Macdonald—Kac formulas," Invent. Math., $\underline{34}$, 37-76 (1976).

25. I. M. Gelfand and G. E. Shilov, Generalized Functions and Operations on Them [in Russian], Fizmatgiz, Moscow (1958).

26. I. M. Gelfand and D. B. Fuks, "The topology of noncompact Lie groups," Funkts. Anal. Prilozhen., $\underline{1}$, No. 4, 33-45 (1967).

27. I. M. Gelfand and D. B. Fuks, "Cohomology of Lie algebras of vector fields on the circle," Funkts. Anal. Prilozhen., $\underline{2}$, No. 4, 92-93 (1968).

28. I. M. Gelfand and D. B. Fuks, "Cohomology of Lie algebras of tangent vector fields of a smooth manifold," Funkts. Anal. Prilozhen., $\underline{3}$, No. 3, 32-52 (1969).

29. I. M. Gelfand and D. B. Fuks, "Cohomology of Lie algebras of formal vector fields," Izv. Akad. Nauk SSSR, Ser. Mat., $\underline{34}$, No. 2, 322-337 (1970).

30. I. M. Gelfand and D. B. Fuks, "Cohomology of Lie algebras of vector fields with nontrivial coefficients," Funkts. Anal. Prilozhen., 4, No. 3, 10-25 (1970).

31. I. M. Gelfand and D. B. Fuks, "On cycles representing cohomology classes of Lie algebras of formal vector fields," Usp. Mat. Nauk, 25, No. 5, 239-240 (1970).

32. I. M. Gelfand and D.B. Fuks, "Estimates from above for the cohomology of infinite-dimensional Lie algebras," Funkts. Anal. Prilozhen., 4, No. 4, 70-71 (1970).

33. I. M. Gelfand, D. I. Kalinin, and D. B. Fuks, "On the cohomology of Lie algebras of Hamiltonian formal vector fields," Funkts. Anal. Prilozhen., 6, No. 3, 25-29 (1972).

34. I. M. Gelfand, B. L. Feigin, and D. B. Fuks, "Cohomology of Lie algebras of formal vector fields with coefficients in adjoint space and variations of characteristic classes of foliations," Funkts. Anal. Prilozhen., 8, No. 2, 13-29 (1974).

35. I. M. Gelfand, B. L. Feigin, and D. B. Fuks, "Cohomology of infinite-dimensional Lie algebras and Laplace operators," Funkts. Anal. Prilozhen., 12, No. 4, 1-5 (1978).

36. C. Godbillon and J. Vey, "Un invariant des feuilletages de codimension 1," C. R. Acad. Sci., 273, No. 2, 92-95 (1971).

37. C. Godbillon, "Chohomologies d'algèbres de Lie de champs de vecteurs formels," Lect. Notes Math., 383, 69-87 (1974).

38. L. V. Goncharova, "Cohomology of Lie algebras of formal vector fields on the line," Usp. Mat. Nauk, 27, No. 5, 231-232 (1972).

39. L. V. Goncharova, "Cohomology of Lie algebras of formal vector fields on the line," Funkts. Anal. Prilozhen., 7, No. 2, 6-14 (1973).

40. P. Ya. Grozman, "Classification of bilinear invariant
    operators over tensor fields," Funkts. Anal. Prilozhen.,
    14, No. 2, 58-59 (1980).

41. A. Guichardet, Cohomologie des Groupes Topologiques et
    des Algèbres de Lie, Nathan (1980).

42. A. Guichardet and D. Wigner, "Sur la cohomologie réelle
    des groupes de Lie simples réels," Ann. Sci. Ec. Norm.
    Sup., 11, 277-292 (1978).

43. V. W. Guillemin, "Cohomology of vector fields on a mani-
    fold," Adv. Math., 10, 192-220 (1973).

44. V. W. Guillemin, Notes on Gelfand—Fuks Cohomology, MIT
    (1971).

45. V. W. Guillemin and S. D. Shnider, "Some stable results
    on the cohomology of classical infinite-dimensional Lie
    algebras," Trans. Am. Math. Soc., 179, 275-280 (1973).

46. A. Haefliger, "Sur les feuilletages analytiques," C. R.
    Acad. Sci., 242, No. 5, 2908-2910 (1956).

47. A. Haefliger, "Sur la cohomologie de Gelfand—Fuchs,"
    Lect. Notes Math., 484, 121-152 (1975).

48. A. Haefliger, "Sur la cohomologie de l'algèbre de Lie
    des champs de vecteurs," Ann. Sci. Ec. Norm. Sup., 9,
    503-532 (1976).

49. M. Hall, Jr., Combinatorial Theory, Blaisdell Publ.,
    Waltham—Toronto—London (1967).

50. J. Heitsch, "Deformations of secondary characteristic
    classes," Topology, 12, No. 4, 381-388 (1973).

51. G. P. Hochschild and G. D. Mostow, "Cohomology of Lie
    groups," Illinois J. Math., 6, 367-401 (1962).

52. G. P. Hochschild and J.-P. Serre, "Cohomology of Lie al-
    gebras," Ann. Math., 57, No. 3, 591-603 (1953).

53. V. G. Kac, "Simple irreducible graded Lie algebras of

finite growth," Izv. Akad. Nauk SSSR, Ser. Mat., $\underline{32}$,
1323-1367 (1968).

54. V. G. Kac, "Contravariant form for infinite dimensional
    Lie algebras and superalgebras," Lect. Notes Phys., $\underline{94}$,
    441-445 (1979).

55. F. W. Kamber and P. Tondeur, "Nontrivial characteristic
    invariants of foliated bundles," Ann. Sci. Ec. Norm.
    Sup., $\underline{8}$, 433-486 (1975).

56. F. W. Kamber and P. Tondeur, "Foliated bundles and char-
    acteristic classes," Lect. Notes Math., $\underline{493}$ (1975).

57. A. A. Kirillov, "Invariant operators over geometric
    quantities," Contemporary Problems in Mathematics, Vol. 16
    [in Russian], VINITI, Moscow (1980), pp. 3-29.

58. H. B. Lawson, "Codimension one foliations on spheres,"
    Ann. Math., $\underline{94}$, No. 3, 494-503 (1971).

59. H. B. Lawson, "Foliations," Bull. Am. Math. Soc., $\underline{80}$,
    No. 3, 369-418 (1974).

60. H. B. Lawson, The Quantitative Theory of Foliations, Re-
    gional Conf. Ser. in Math., Vol. 27, Am. Math. Soc.,
    Providence (1977).

61. D. A. Leites and D. B. Fuks, "Cohomology of Lie super-
    algebras," Dokl. Bolg. Akad. Nauk, $\underline{37}$, No. 10, 1294-
    1296 (1984).

62. J. Lepowsky, "Generalized Vermat modules, loop space co-
    homology, and Macdonald-type identities," Ann. Sci. Ec.
    Norm. Sup., $\underline{12}$, 169-234 (1979).

63. M. V. Losik, "On the cohomology of infinite-dimensional
    Lie algebras of vector fields," Funkts. Anal. Prilozhen.,
    $\underline{4}$, No. 2, 45-53 (1970).

64. M. V. Losik, "On the cohomology of Lie algebras of vector
    fields with coefficients in the trivial unit representa-

tion," Funkts. Anal. Prilozhen., 6, No. 1, 24-36 (1972).

65. M. V. Losik, "Topological interpretation of the diagonal complex homology of Lie algebras of vector fields with coefficients in the trivial unit representation," Funkts. Anal. Prilozhen., 6, No. 3, 79-80 (1972).

66. M. V. Losik, "On the cohomology of Lie algebras of vector fields with nontrivial coefficients," Funkts. Anal. Prilozhen., 6, No. 4, 44-46 (1972).

67. I. G. Macdonald, "Affine root systems and Dedekind's function," Invent. Math., 15, 91-143 (1972).

68. I. G. Macdonald, Symmetric Functions and Hall Polynomial, Oxford Math. Monogr., Clarendon Press, Oxford (1979).

69. R. V. Moody, "A new class of Lie algebras," J. Algebra, 10, 211-230 (1968).

70. S. P. Novikov, "Topology of foliations," Tr. Mosk. Mat. Ova., 14, 248-278 (1965).

71. J. Perchik, "Cohomology of Hamiltonian and related formal vector fields with Lie algebras," Topology, 15, No. 4, 395-404 (1976).

72. H. V. Pittie, "Characteristic classes of foliations," Research Notes in Mathematics, Pitman, London (1976).

73. H. V. Pittie, "The secondary characteristic classes of parabolic foliations," Comment. Math. Helv., 54, 601-614 (1979).

74. G. Polya, Mathematics and Plausible Reasoning, Princeton Univ. Press, Princeton (1954).

75. G. Reeb, "Sur certaines propriétés topologiques des variétés feuilletées," Actual. Sci. Ind., No. 1183, Hermann, Paris (1952).

76. B. L. Reinhart and J. W. Wood, "A metric formula for the Godbillon—Vey invariant for foliations," Proc. Am. Math. Soc., 38, No. 2, 427-430 (1973).

77. V. S. Retakh and B. L. Feigin, "On the cohomology of certain Lie algebras and superalgebras of vector fields," Usp. Mat. Nauk, 37, No. 2, 233-234 (1982).

78. B. I. Rozenfeld, "Cohomology of certain infinite-dimensional Lie algebras," Funkts. Anal. Prilozhen., 5, No. 4, 84-85 (1971).

79. A. N. Rudakov, "Irreducible representations of infinite-dimensional Lie algebras of the Cartan type," Izv. Akad. Nauk SSSR, Ser. Mat., 38, No. 4, 835-866 (1974).

80. G. Segal, The Cohomology of Topological Groups. Symposia Math., Vol. 4, London—New York (1973), pp. 377-387.

81. G. B. Segal, "The classifying space of a topological group in the sense of Gelfand—Fuks," Funkts. Anal. Prilozhen., 9, No. 2, 48-50 (1975).

82. F. Sergeraert, "Feuilletages et difféomorphismes infinitement tangents à l'identité," Invent. Math., 17, No. 4, 367-382 (1978).

83. S. D. Shnider, "Invariant theory and the cohomology of infinite Lie algebras," Thesis, Harvard (1972).

84. A. V. Shokurov, "On relations between Chern numbers of quasicomplex manifolds," Mat. Zametki, 26, No. 1, 137-148 (1979).

85. R. E. Stong, Notes on Cobordism Theory, Princeton Univ. Press, Princeton, New Jersey (1968).

86. D. Sullivan, "Differential forms and topology of manifolds," Proc. Int. Conf. Topology and Related Topics, Univ. of Tokyo (1973), pp. 172-188.

87. S. Tabachnikov, "On invariant differential operators in general position," Funkts. Anal. Prilozhen., 16, No. 2, 73-74 (1982).

88. I. Tamura, The Topology of Foliations, Iwanami Shoten Publ., Tokyo (1976).

89. W. Thurston, "Noncobordant foliations of S^3," Bull. Am. Math. Soc., 78, No. 4, 511-514 (1972).

90. W. Thurston, "Existence of codimension-one foliations," Ann. Math., 104, No. 2, 249-268 (1976).

91. T. Tsujishita, "On the continuous cohomology of the Lie algebra of vector fields," Proc. Jpn. Acad., A53, No. 4, 134-138 (1977).

92. W. T. Van Est, "Group cohomology and Lie algebra cohomology in Lie groups," Indagationes Math., 15, 484-504 (1953).

93. H. Weyl, The Classical Groups, Their Invariants and Representations, Princeton Univ. Press, Princeton, New Jersey (1939).

94. Théorie des algèbres de Lie. Topologie des groupes de Lie. Séminaire "Sophus Lie." Sécrétariat Mathématique, Paris (1955).

95. F. V. Vainshtein, "Filtering bases, cohomology of infinite-dimensional Lie algebras, and Laplace operators," Funkts. Anal. Ego Prilozhen., 19, No. 4, pp. 11-12 (1985).

Appendix

When the basic text of the work was already printed, the possibility of supplementing the book with some of the most recent results on the cohomology of infinite-dimensional Lie algebras occurred to the author. The separate sections of the Appendix are not connected with one another.

1. Cohomology of Lie Algebras of Generalized Jacobian Matrices

A. Introduction. We recall (see §2.1) that $H^*(\mathfrak{gl}(n, \mathbb{K}))$ is an exterior algebra on generators of dimensions 1, 3, ..., $2n - 1$;   in particular, $H^2(\mathfrak{gl}(n, \mathbb{K})) = 0$, i.e., $\mathfrak{gl}(n, \mathbb{K})$ has no nontrivial central extensions. These results withstand passage to the limit as $n \to \infty$, if one assumes that $\mathfrak{gl}(\infty, \mathbb{K})$ is the Lie algebra of infinite finitary matrices, i.e., $\bigcup_n$ $\mathfrak{gl}(n, \mathbb{K})$.   Namely, $H^*(\mathfrak{gl}(\infty, \mathbb{K}))$ is an exterior algebra on generators of dimensions 1, 3, 5, ..., and, in particular,

$$H^2(\mathfrak{gl}(\infty, \mathbb{K})) = 0.$$

It turns out that a reasonable relaxation of the condition of being finitary leads to a radical change in the cohomology of the Lie algebra of infinite matrices and, in par-

329

ticular, to the appearance of nontrivial central extensions of them. This phenomenon was discovered by a group of Japanese authors [1, 2],* but the definitive results are due to Feigin and Tsygan [3].

The bilaterally infinite matrix $\| a_{ij} \|, {}_{i,j \in \mathbf{Z}}$ is called gen-eralized Jacobian if it has only a finite number of nonzero diagonals, i.e., if there exists an N such that $a_{ij} = 0$ for $|j - i| < N$. Obviously, the generalized Jacobian matrices constitute a Lie algebra with respect to the usual commuta-tion; we denote this algebra by $\mathfrak{gl}_J (\mathbb{K})$. It is also obvious that $\mathfrak{gl}_J (\mathbb{K}) \supset \mathfrak{gl} (\infty, \mathbb{K})$.

B. Central extension. For $\alpha = \| a_{ij} \| \in \mathfrak{gl}_J (\mathbb{K})$, we denote by $\pi (\alpha)$ the matrix $\| b_{ij} \| \in \mathfrak{gl}_J (\mathbb{K})$ with $b_{ij} = a_{ij}$ for $i \leqslant 0$, $j \leqslant 0$ and $b_{ij} = 0$ otherwise. Although the operator π: $\mathfrak{gl}_J (\mathbb{K}) \to \mathfrak{gl}_J (\mathbb{K})$ is not an endomorphism of Lie algebras, never-theless for any $\alpha, \beta \in \mathfrak{gl}_J (\mathbb{K})$

$$\rho (\alpha, \beta) = \pi ([\alpha, \beta]) - [\pi (\alpha), \pi (\beta)] \in \mathfrak{gl} (\infty, \mathbb{K}).$$

We set

$$\tau (\alpha, \beta) = \operatorname{Tr} \rho (\alpha, \beta).$$

Theorem A.1 [2, 3]. The cochain $\tau \in C^2 (\mathfrak{gl}_J (\mathbb{K}))$ is a co-cycle, which is not cohomologous to zero. The space $H^2 (\mathfrak{gl}_J (\mathbb{K}))$ is one-dimensional and is generated by the class of this cocycle.

[That τ is a cocycle can be verified directly. That this cocycle is not cohomologous to zero can be proved as follows: let $\sigma = \| \delta_{i,j-1} \| \in \mathfrak{gl}_J (\mathbb{K})$; then $\sigma^{-1} = \| \delta_{i-1, j} \|$, and an obvious calculation shows that $\tau (\sigma, \sigma^{-1}) = 1$; at the same time, $\delta \omega (\sigma,$

---

*See the references cited on pp. 338 and 339.

$\sigma^{-1}) = \omega([\sigma, \ \sigma^{-1}]) = \omega(0) = 0$ for any cochain $\omega \in C^1(\mathfrak{gl}_J(\mathbb{K}))$. The proof that the space $H^2(\mathfrak{gl}_J(\mathbb{K}))$ is one-dimensional is left to the reader.]

According to 1.4.6, to the cohomology class of the co-cycle τ corresponds a nontrivial one-dimensional central ex-tension of the Lie algebra $\mathfrak{gl}_J(\mathbb{K})$; we denote the algebra so extended by $\mathfrak{gl}_J^\wedge(\mathbb{K})$.

Remark. It follows from what was said above that the restriction of the cocycle τ to $C^2(\mathfrak{gl}(\infty, \mathbb{K}))$ is cohomologi-cally zero. In fact, it is the coboundary of the cochain $\alpha \mapsto \mathrm{Tr}\ \pi(\alpha)$.

C. Imbedding in $\mathfrak{gl}_J^\wedge(\mathbb{K})$ of Kac–Moody and Virasoro algebras. Two facts point to the importance of the algebra $\mathfrak{gl}_J^\wedge(\mathbb{K})$ in representation theory. First, many of the clas-sical constructions of the theory of representations of the algebra $\mathfrak{gl}$ are applicable to this algebra, which creates a sizable supply of $\mathfrak{gl}_J^\wedge(\mathbb{K})$-modules (see, e.g., [4]). Second, many important infinite-dimensional Lie algebras (including those we considered above) can be imbedded in $\mathfrak{gl}_J^\wedge(\mathbb{K})$, so that the representations of the algebra $\mathfrak{gl}_J^\wedge(\mathbb{K})$ cited become representations of these algebras.

Some of these imbeddings are also of interest from the cohomological point of view. For example, the subalgebra of the algebra $\mathfrak{gl}_J(\mathbb{K})$, composed of "$n$-periodic" matrices, i.e., matrices $\| a_{ij} \|$ with $a_{i+n, \ j+n} = a_{ij}$, is isomorphic to the al-gebra of currents $(\mathfrak{gl}(n, \ \mathbb{K})\ S^1)^{\mathrm{pol}}$: the isomorphism is defined by the formula

$$\| a_{ij} \|_{i, \ j=-\infty}^{\infty} \mapsto \sum_{k=-\infty}^{\infty} \| a_{kn+i, \ j} \|_{i, \ j=1}^{n} \varphi^k, \qquad (1)$$

where φ is an angular parameter on the circle. We note that under this imbedding the cocycle τ goes into a cocycle of the algebra of currents which is not cohomologous to zero [this follows from what was said above, since the matrix $\sigma = \| \delta_{i,j-1} \|$ lies in the image of the homomorphism (1)]. Thus, a nontrivial central extension of the algebra of currents, the Kac–Moody algebra (see 1.4.6 and 2.5.3A), is imbedded in $\mathfrak{gl}_j^{\wedge}(\mathbb{K})$.

Another important Lie algebra is also imbedded in $\mathfrak{gl}_j(\mathbb{K})$ (with $\mathbb{K} = \mathbb{C}$): the algebra $\mathcal{L} = \mathbb{C}(\text{Vect } S^1)^{\text{pol}}$ of (complex) polynomial vector fields on the circle, where this imbedding can be realized in many different ways. We recall that $\mathcal{L}$ has a basis $\{e_i \mid i \in \mathbb{Z}\}$ in which the commutator can be described by the formula $[e_i, e_j] = (j - i) e_{i+j}$ (see 1.1.2). The cohomology of the algebra $\mathcal{L}$ is known (see 2.4.1D; although the Lie algebra considered there differs somewhat from $\mathcal{L}$, the result, as is easy to see, remains valid in the case of interest to us); in particular, $H^2(\mathcal{L}) = \mathbb{C}$. The nontrivial central extension of the algebra $\mathcal{L}$ supplies an infinite-dimensional Lie algebra, which is denoted by $\widehat{\mathcal{L}}$, and is called the Virasoro algebra. In order to construct an imbedding of the algebra $\mathcal{L}$ in $\mathfrak{gl}_j(\mathbb{C})$, we consider the family $\{\mathcal{F}_{\lambda,\mu}\}$ of $\mathcal{L}$-modules with $\mathbb{C}$-basis $\{f_j \mid j \in \mathbb{Z}\}$ and the action of the algebra $\mathcal{L}$, described by the formula

$$e_i f_j = (\mu + j - (i + 1)\lambda) f_{i+j} \qquad (2)$$

[these modules were considered in 2.3.3, but only as modules over the algebra $(W_1)^{\text{pol}} \subset \mathcal{L}$]. The imbedding $i_{\lambda,\mu}: \mathcal{L} \to \mathfrak{gl}_j(\mathbb{C})$ carries e_i into the matrix of the operator (2): $i_{\lambda,\mu}(e_i) = \| c_{jk} \|$, where

$$c_{jk} = (\mu + j - (i + 1)\lambda) \delta_{i+j,k}.$$

Routine calculation shows that if $6\lambda^2 + 6\lambda + 1 \neq 0$, then the cocycle $i^*_{\lambda,\mu}(\tau) \in C^2(\mathscr{L})$ is not cohomologous to zero; thus, under this condition, $i_{\lambda,\mu}$ defines an imbedding of the Virasoro algebra in $\mathfrak{gl}_J^{\wedge}(\mathbb{C})$.

D. Higher cohomology of the Lie algebra $\mathfrak{gl}_J(\mathbb{K})$

Theorem A.2 [3]. $H^*(\mathfrak{gl}_J(\mathbb{K}))$ is a polynomial algebra on generators of dimensions 2, 4, 6,..., one in each dimension. A generator of dimension $2k$ is represented by the cocycle

$$\alpha_1, \ldots \cdot \alpha_{2n} \mapsto \sum_{\sigma \in \mathrm{Symm}\,(2n)} \mathrm{sgn}\,(\sigma)\, P_n\,(\rho\,(\alpha_{\sigma(1)},\, \alpha_{\sigma(2)}),\, \ldots,\, \rho\,(\alpha_{\sigma(2n-1)},\, \alpha_{\sigma(2n)})),$$

where P_n denotes the n-th basic invariant symmetric function on $\mathfrak{gl}\,(\infty,\, \mathbb{K})$ (the polarization of the n-th elementary symmetric polynomial in the eigenvalues), and ρ has the same meaning as above (see B).

We turn our attention to the striking difference between $H^*(\mathfrak{gl}_J(\mathbb{K}))$ and $H^*(\mathfrak{gl}\,(\infty,\, \mathbb{K}))$.

2. Stable Cohomology of the Algebra W_n with Tensor Coefficients and Generators and Relations for the Algebra $L_1(n)$

A. Notation. As usual, $L_1(n)$ denotes the subalgebra of the algebra W_n, composed of formal vector fields with trivial 1-jet, and T_q^p denotes the $\mathfrak{gl}\,(n,\, \mathbb{K})$-module $\underbrace{V \otimes \ldots \otimes V}_{p} \otimes \underbrace{V' \otimes \ldots \otimes V'}_{q}$, where V is the standard n-dimensional $\mathfrak{gl}\,(n,\, \mathbb{K})$-module. We denote by A some submodule of the module T_q^p and by $\mathcal{A}$ the corresponding space of formal tensor fields, i.e., $\mathrm{Coind}_{W_n} A$. For $A = T_q^p$ we set $\mathcal{A} = \mathscr{T}_q^p$; the notation $\mathscr{T}_q^0$ is shortened to $\mathscr{T}_q$. Further, for $A = S^q V'$ we set $\mathcal{A} = \mathscr{S}_q$.

The sums $\mathcal{J}_* = \oplus_q \mathcal{J}_q$, $\mathcal{S}_* = \oplus_q \mathcal{S}_q$ are graded W_n-algebras (i.e., algebras in which the transformations effected by the elements W_n, are differentiations).

    B. **Theorems on stable cohomology.** The recent papers [5, 6] contain the following results, which generalize and supplement Theorem 2.2.9 (compare with the discussion of this theorem in 2.2.4).

    <u>Theorem A.3</u>. If $n > 2r$, then $H^r_{(m)}(L_1(n)) = 0$ for $r \neq m$.

    <u>Theorem A.4</u>. If $n > 2r$, then

$$H^r(W_n, \mathfrak{gl}(n, \mathbb{K}); \mathcal{A}) = 0 \qquad \text{for} \qquad r \neq q - p.$$

    <u>Theorem A.5</u>. If $n > \min(2r, \, \imath + q)$, then

$$H^r(W_n; \mathcal{J}_q) \cong H^{r-q}(\mathfrak{gl}(n, \mathring{\mathbb{K}})) \otimes \mathrm{Inv}_{\mathfrak{gl}(n, \mathbb{K})}(\otimes^q \mathfrak{gl}(n, \mathbb{K})),$$
$$H^r(W_n; \mathcal{S}_q) \cong H^{r-q}(\mathfrak{gl}(n, \mathring{\mathbb{K}})) \otimes \mathrm{Inv}_{\mathfrak{gl}(n, \mathbb{K})}(\Lambda^q \mathfrak{gl}(n, \mathbb{K}))$$

(we recall that $\mathrm{Inv}\,\Lambda^q \mathfrak{gl}(n, \mathring{\mathbb{K}}) = H^q(\mathfrak{gl}(n, \mathbb{K}))$; both isomorphisms are multiplicative).

    <u>Theorem A.6</u>. If $n > 2r$, then

$$\dim H^r(W_n, \mathfrak{gl}(n, \mathbb{K}); \mathcal{J}_q^p) = \begin{cases} \sum \dfrac{q!}{q_1 \cdots q_p} & \text{for } r = q - p, \\ 0 & \text{for } r \neq q - p, \end{cases}$$

where the summation is over all collections $q_1, \ldots, q_p$ of positive integers with $q_1 + \ldots + q_p \leqslant q$ (collections which differ in order of the summands are considered different).

    Now let A be an irreducible component of the module T_q^0 (with $q \leqslant n$), corresponding to some Young diagram, and let $\bar{A}$ be the component corresponding to the symmetric Young diagram. To the components A, $\bar{A}$ correspond functors $a, \bar{a}$

from the category of linear spaces and linear isomorphisms
to the same category, such that $a(V') = A$, $\bar{a}(\bar{V}') = \bar{A}$.

Conjecture A.7.   If $n > \min(2r, \ r + q)$, then

$$H^r(W_n; \ \mathcal{A}) \cong H^{r-q}(\mathfrak{gl}(n, \mathbb{K})) \otimes \operatorname{Inv}_{\mathfrak{gl}(n, K)}\bar{a}(\mathfrak{gl}(n, \mathbb{K})).$$

Conjecture A.8.   In all assertions of this point, the
conditions on n can be replaced by the inequality $n > r$.

We note that Theorem A.4 is a direct consequence of
Theorems A.3 and 2.2.8, and Theorem A.5 differs from Theorem
2.2.9 and its supplement only in the conditions on n.

C.   Cohomological interpretation of generators and relations of graded nilpotent Lie algebras.

Proposition A.9.   Let $\mathfrak{g} = \oplus_{m>0} \mathfrak{g}_{(m)}$ be a graded Lie algebra and let $\{a_1^{(m)}, \ \ldots, \ a_{p_m}^{(m)}\}$, $\{b_1^{(m)}, \ \ldots, \ b_{q_m}^{(m)}\}$ be bases in
$H_1^{(m)}(\mathfrak{g})$, $H_2^{(m)}(\mathfrak{g})$.   In addition, let $\alpha_i^{(m)} \in \mathfrak{g}_{(m)}$ be a cycle, representing $a_i^{(m)}$, and $\sum_j \lambda_{ij}^{(m)} \beta_{ij}'^{(m)} \wedge \beta_{ij}''^{(m)} \in (\mathfrak{g} \wedge \mathfrak{g})_{(m)}$ be a cycle, representing $b_i^{(m)}$. Then

$$\{\alpha_i^{(m)} \mid m = 1, 2, \ldots; \ i = 1, \ldots, p_m\} \tag{3}$$

is a minimal system of generators of the algebra $\mathfrak{g}$, and

$$\{\sum_j \lambda_{ij}^{(m)} [\beta_{ij}'^{(m)}(\alpha), \ \beta_{ij}''^{(m)}(\alpha)] = 0 \mid m = 1, 2, \ldots; \ i = 1, \ldots, q_m\}, \tag{4}$$

where $\beta_{ij}'^{(m)}(\alpha)$, $\beta_{ij}''^{(m)}(\alpha)$ are expressions for the elements $\beta_{ij}'^{(m)}$,
$\beta_{ij}''^{(m)}$ of the algebra $\mathfrak{g}$ in terms of the generators (3), is the
corresponding minimal system of defining relations.

The proof, whose details I leave to the reader, consists
of the inductive construction of a minimal system of homogeneous generators and the corresponding minimal system of homogeneous defining relations and the comparison of these systems with the systems (3) and (4).

D. Generators and relations in the algebra $L_1(n)^{pol}$.

Since $H_q^{(m)}(L_1(n)^{pol}) = [H_{(m)}^q(L_1(n))]'$, Proposition A.9 shows that
minimal systems of generators and relations for the Lie al-
gebra $L_1(n)^{pol}$ contain, respectively, dim $H_{(m)}^1(L_1(n))$ generators of degree m and dim $H_{(m}^2(L_1(n))$ relations of degree m.
Thus, if we set $r = 1$ in Theorem A.3, we get that for $n > 2$
the algebra $L_1(n)^{pol}$ is generated by elements of degree 1.
However, this is completely obvious and direct; moreover,
this is also true for $n = 2$ (see Conjecture A.8). Now
the algebra $L_1(1)^{pol}$ is obviously generated by two generators
of degrees 1 and 2. Further, if we let $r = 2$ in Theorem A.3,
then we get that for $n > 4$ these generators are connected
only by relations of degree 2. In fact, this is also true
for $n = 3, 4$ (see again Conjecture A.8). For $n = 1$ relations two and one have degrees 5 and 7, and for $n = 2$ the
generators (six of them) are connected by twenty-five relations,
seven of which have degree 2, and the rest, degree 3.

The relations of degree 2 between generators of degree
1 are linear relations between commutators. We arrive at the
following assertion.

Theorem A.10. For $n \geqslant 2$ the algebra $L_1(n)^{pol}$ is generated by the elements of any basis of the space $L_1(n)_{(1)}$. For
$n \geqslant 3$ a defining system of relations is made up of the elements of any basis of the kernel of the operator $[,]$:
$\Lambda^2 L_1(n)_{(1)} \to L_1(n)_{(2)}$.

3. Invariant Differential Operators in
General Position on the Line (see 3.3.3)

The result of Tabachnikov [7], disproving Conjecture
3.3.3 (and with it Conjecture 3.3.4), can be considered unexpected. We recall that we are concerned here with W_1-homo-

morphisms $F_{\lambda_1} \otimes \ldots \otimes F_{\lambda_n} \rightarrow F_{\lambda_1 + \ldots + \lambda_n - k}$ (see 1.2.1 for the notation). The dimension of the space of such homomorphisms de-
pends on $\lambda_1, \ldots, \lambda_n$, but is constant in a Zariski open subset
of the space $\mathbb{C}^n$ (it jumps outside this subset); as in 3.3.3,
we denote this constant dimension by $d_{n,k}$. Conjecture 3.3.3
was that

$$\sum d_{n,k} t^k = (1 + t)(1 + t + t^2)\ldots(1 + t + \ldots + t^{n-1}).$$

We denote the left-hand side of the last equation by $D_n(t)$,
the right-hand side by $C_n(t)$, the coefficient with which t^k
occurs in $C_n(t)$ by $c_{n,k}$. Tabachnikov's results are the following.

Theorem A.11. If $k \leqslant n$, then $d_{n,k} = c_{n,k}$.

Theorem A.12. If $n < k < n(n-1)/2$, then $d_{n,k} < c_{n,k}$.
Moreover, the coefficients of the difference $C_n(t) - D_n(t)$
are not less than the corresponding coefficients of the poly-
nomial $(t^{n+1} + t^{n+2} + \ldots + t^{2n-3}) D_{n-2}(t)$; in particular, $d_{n, \frac{n(n-1)}{2}} = 0$, and if $n \neq 4$, then $d_{n, \frac{n(n-1)}{2}} \leqslant 1$ (according to Conjecture
3.3.3, $d_{n, \frac{n(n-1)}{2}} = 1$, $d_{n, \frac{n(n-1)}{2} - 1} = n - 1$).

In Tabachnikov's paper the results of machine calculation of the numbers $d_{n,k}$ for $n = 4, 5$ are given. These results are given below in tables which also contain the values
of $c_{n,k}$ and estimates for $d_{n,k}$, supplied by Theorem A.12 (these
estimates are only given for $n = 5$: for $n = 4$ they do not
differ from $d_{n,k}$).

It is worth saying that the values of $d_{n,k}$ given in the
tables are calculated by the machine for several randomly
chosen collections $\lambda_1, \ldots, \lambda_n$. In principle, it could happen

$n = 4$:

k	0	1	2	3	4	5	6	>6
$d_{n,k}$	1	3	5	6	5	2	0	0
$c_{n,k}$	1	3	5	6	5	3	1	0

$n = 5$:

k	0	1	2	3	4	5	6	7	8	9	10	>10
$d_{n,k}$	1	4	9	15	20	22	19	10	0	0	0	0
(A.12)	1	4	9	15	20	22	19	12	5	1	0	0
$c_{n,k}$	1	4	9	15	20	22	20	15	9	4	1	0

that all these collections lie in the complement of the Zariski open set mentioned above; hence, strictly speaking, these "values" are also, really, only upper bounds for $d_{n,k}$, although one need not doubt the accuracy of these estimates. The second of the tables also shows that the estimate for the numbers $d_{n,k}$, supplied by Theorem A.12, is also strongly overstated.

REFERENCES

1. E. Date, M. Jimbo, T. Miwa, and M. Kashiwara, "Transformation groups for soliton equations," RIMS, 394, Kyoto, February (1982).

2. J. L. Verdier, "Les representations des algèbres de Lie affines: applications à quelques problèmes de physique (d'après E. Date, M. Jimbo, M. Kashiwara, T. Mawa)," Sem. Bourbaki, Exp. 596, Vol. 1981/82.

3. B. L. Feigin and B. L. Tsygan, "Cohomology of Lie algebras of generalized-Jacobian matrices," Funkts. Anal. Prilozhen., 17, No. 2, 86-87 (1983).

4. A. M. Vershik, "Metagonal and metaplectic infinite-dimensional groups. I. General concepts and the metagonal group," in: Differential Geometry, Lie Groups, and Me-

chanics, V. Zap. Nauchn. Sem. Leningr. Otd. Mat. Inst., 123, 3-35 (1983).

5. D. B. Fuks, "Stable cohomology of the Lie algebra of formal vector fields with tensor coefficients," Funkts. Anal. Prilozhen., 17, No. 4, 62-69 (1983).

6. B. L. Feigin and D. B. Fuks, "Stable cohomology of the algebra W_n and relations in the algebra L_1," Funkts. Anal. Prilozhen., 18, No. 3, 94-95 (1984).

7. S. L. Tabachnikov, "Homology of general position of the Lie algebra of vector fields on the line," Dokl. Akad. Nauk SSSR, 275, No. 2, 310-313 (1984).